T0257793

Encyclopedia of Soybean: Integrated Study

Volume III

Encyclopedia of Soybean: Integrated Study
Volume III

Edited by **Albert Marinelli and Kiara Woods**

New York

Published by Callisto Reference,
106 Park Avenue, Suite 200,
New York, NY 10016, USA
www.callistoreference.com

Encyclopedia of Soybean: Integrated Study
Volume III
Edited by Albert Marinelli and Kiara Woods

International Standard Book Number: 978-1-63239-298-5 (Hardback)

This book contains information obtained from authentic and highly regarded sources. Copyright for all individual chapters remain with the respective authors as indicated. A wide variety of references are listed. Permission and sources are indicated; for detailed attributions, please refer to the permissions page. Reasonable efforts have been made to publish reliable data and information, but the authors, editors and publisher cannot assume any responsibility for the validity of all materials or the consequences of their use.

The publisher's policy is to use permanent paper from mills that operate a sustainable forestry policy. Furthermore, the publisher ensures that the text paper and cover boards used have met acceptable environmental accreditation standards.

Trademark Notice: Registered trademark of products or corporate names are used only for explanation and identification without intent to infringe.

Printed in the United States of America.

Contents

Preface

This book provides a comprehensive description and analysis of Soybean as an essential diet requirement and its therapeutic qualities. Legumes are an essential part of a healthy diet. They are an excellent source of carbohydrates, protein, vitamins and minerals. Soybean has its significance in the overall agriculture and trade as well as in its contribution to food supply. Soybean is the richest in terms of protein and has no cholesterol in comparison with animal food sources and conventional legume. In addition to this, soybean is a cheap source of food and has medicinal qualities because of its photochemical, genistein, isoflavones content. It has been found that Soybean is very helpful in fighting diabetes, heart disease and cancer. Presently, Soybean protein and calories are being employed for the prevention of body wasting which is mostly because of HIV. Soybean's nutritional value is of extreme importance in places where medication facilities are not available. Therefore, this book covers some important topics such as techniques in Soybean biodiesel, Soybean & prostate cancer, chemical feedstocks, etc.

The researches compiled throughout the book are authentic and of high quality, combining several disciplines and from very diverse regions from around the world. Drawing on the contributions of many researchers from diverse countries, the book's objective is to provide the readers with the latest achievements in the area of research. This book will surely be a source of knowledge to all interested and researching the field.

In the end, I would like to express my deep sense of gratitude to all the authors for meeting the set deadlines in completing and submitting their research chapters. I would also like to thank the publisher for the support offered to us throughout the course of the book. Finally, I extend my sincere thanks to my family for being a constant source of inspiration and encouragement.

Editor

Advanced Techniques in Soybean Biodiesel

Mauricio G. Fonseca, Luciano N. Batista,
Viviane F. Silva, Erica C. G. Pissurno, Thais C. Soares,
Monique R. Jesus and Georgiana F. Cruz

Additional information is available at the end of the chapter

1. Introduction

The planet where we inhabit, live, has experienced a great transformation period in the most different fields. The evolution followed by a great technological development, on the other side has caused an imbalance both in society itself, as in the material environment in which we live. The planet earth has visibly demonstrated how has been affected by this imbalance and how it has naturally reacted. In recent years, all this has being reported through the literature, studied by different scientific research groups, as also observed by what is reported in the media in general, even in the form of documentaries and films, as the documentary performed by former USA vice president Al Gore, an inconvenient truth (an inconvenient truth, 2006). All learning takes a certain time to begin to be assimilated and been put in practice effectively, so humanity has learned, been advised by the latest natural disasters of this century, as in the case of Japan's earthquake, tsunami, the strong hurricanes that plague the northern hemisphere summer, as in fact glaciers melting that were called eternal, the poles of this planet, strong climate change experienced over the past years and the major pollution in large urban cities where population are forced to live in many different fields, has signaling how much real acts, changes are necessary to continue to be possible living an inhabited planet.In this century, XXI, the world main problems, which it has experienced, are related to the scarcity of natural resources such as water, which had been mismanaged, contaminated by urban and industrial solid waste disposal, and in relation to generation and use of energy the most diverse shapes.

These energy sources can be broadly classified into three categories: fossil fuels (coal, oil and natural gas), renewable (hydroelectric, wind, solar and biomass) and nuclear sources. Among those can be highlighted Biomass, where all organic matter that is produced by this process is called biomass. This has a great advantage over fossil fuels, it's less pollut-ing, because its processes do not add carbon dioxide to the atmosphere, the environment. The biomass process reduces the carbon dioxide amount in atmosphere through the pho-tosynthesis, performed by increasing the planted green areas, to cultivate the seeds crops. Research and development departments have been engaged in fuels discovery that do not cause much environment damage and that can replace fossil fuels, reducing the toxic emissions level, replacing the rare fossil fuel used to date. In the midst of these researches has been observed that the use of vegetable oils has shown great ability to make this one a possible alternative renewable energy (Agarwal & Das, 2000).A related problem in the replacement of diesel for oil plant was related to physical and chemical factors such as high viscosity, low volatility which results in incomplete combustion, leading to forma-tion of carbon deposits in the engine and a high unsaturations degree (Meher et al, 2008), factor that reduces the power of the fuel at the lowest level of cetane,but also favors oxi-dation. Studies have shown that vegetable oils characteristics can be modified through four ways (Shrivastava & Presad, 2002): By pyrolysis, microemulsification, dilution and transesterification process. The latter originates the alkyl esters that constitute what is called biodiesel.

2. Biodiesel

The use of vegetable oils as an alternative fuel for diesel engine was discovered more than100 years ago, in the Paris world exhibition in 1900, when Rudolph Diesel used pea-nut oil in an engine ignition (Shay, 1993).This predicted saying, "The use of vegetable oils as fuel engine may be negligible in the present moment, but in the future may become so important as oil and coal as energy sources. The biodiesel term is a subject still under dis-cussion. Some definitions consider biodiesel as a mixture of any vegetable oils with fuel, diesel and fossil derivative others consider the alkyl esters mixture from vegetable oils or animal fats with fossil fuels. Under the chemical aspect biodiesel, an alternative fuel can be defined as alkyl esters derived from fatty acids obtained from oils, vegetables or ani-mal fats, which suffering a chemical reaction, transesterification with short chain alcohols such as methanol and ethanol (Pinto et al, 2005). Transesterification: Chemical reaction be-tween an ester (RCOOR ') and an alcohol (R"COH) resulting in a new ester (R"COOR') and an alcohol (RCOH).

This reaction type, used in biodiesel production is the reaction between the triglycerides, main components of vegetable oils and fats that react with short chain alcohols, methanol and ethanol, resulting in two products, methyl esters derived from fatty acids, and the second product glycerol formation. Transesterification reaction rates can be affected by some aspects: The catalyst type (acid or alkaline), purity of reactants (mainly water con-tent), free fatty acid content and alcohol/vegetable oil molar ratio (Helwani et al, 2009). The biodiesel reaction can be optimized specially by three factors: The first is an increase in the temperature. An increase in temperature increases reaction rate in exponential, al-lowing the reactants to be more miscible, obtaining a higher reaction rate to take place.

This parameter is limited by the solvent, reactant boiling point. The second factor to improve reaction yield, vigorous mixing, possibilities a higher collision rate between the reactants, been obtained a reaction mixture plus homogenized, yielding a higher rate of methyl esters obtained. In general alcohols and triglyceride sources are immiscible, vigorous mixing possibilities the obtaining of alcohol dispersed as fine droplets, increasing the contact surface between the two immiscible reactants (Stamenkovic et al, 2008). The use of a secondary solvent, a co-solvent as THF, possibilities a higher miscibility of the alcohol in the triglyceride phase, obtaining a better mixing of the two phases and hence a more reactions to take place, improving the biodiesel yield. The following Figure 1 illustrates a biodiesel type of reaction.

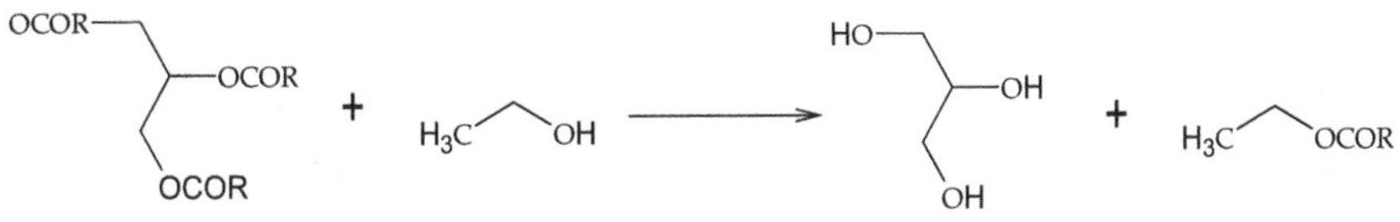

Figure 1. General transesterification reaction to produce biodiesel

In general terms, these reactions take place under homogeneous catalysts, acid or base catalysts, enzyme or through the use of heterogeneous catalysts. The selection of appropriate catalyst depends on the amount of free fatty acids in the oil. Heterogeneous catalyst provides high activity; high selectivity, high water tolerance properties and these properties depend on the amount and strengths of active acid or basic sites. Basic catalyst can be subdivided based on the type of metal oxides and their derivatives. Similarly, acidic catalyst can be subdivided depending upon their active acidic sites (Singh & Sarma, 2011). Generally, a basic catalyst gives better yields than the acids catalysts in both homogeneous and heterogeneous catalysts. The better results of homogeneous catalysts are related to the fact that base catalysts are kinetically much faster than heterogeneously catalyzed transesterification and are economically viable. There are many factors which govern the path of transesterification reactions, between these can be stand out the following parameters: the nature o raw material, the optimum experimental conditions, as the ratio oil/methanol, the temperature and the catalyst concentration, for example. Comparing heterogeneous catalyst with homogeneous catalysts can be observed that the use of solid heterogeneous use more extreme reaction conditions, higher pressure and temperature due the fact of the difficulty in the limited mass transfer between the three phase system solid-liquid-liquid immiscible (catalyst, oil, methanol). The main advantages in the use of solid catalysts are related to the easy work up when compared with homogeneous catalysts. Solid catalysts are separated just by filtration and centrifugation and are environmentally friendly, because they are reusable and reduce the amount of wasted, treated water used. Among the heterogeneous catalysts, we can highlight the use of zeolites (Suppes et al, 2004), clays, ion exchange resins and oxides.

3. Catalysts

3.1. Heterogeneous catalysts

The useof heterogeneous catalysts (Wang & Yang, 2007 and Leclercq et al, 2001) has as major advantage the reaction work-up, i.e., post-treatment reaction, separation and purification steps, since these can be easily removed and can be reused. Another interesting factor is the fact that this type of catalysis, there is no formation of by products, such as saponification (Suppes et al, 2001; Tomasevic et al, 2003 and Gryglewicz, 1999). The greatest difficulty encountered in using this reaction type is directly related to problems in relation between the diffusion systems, oil /catalyst /methanol.

3.2. Homogeneous catalysts

3.2.1. Basic catalysts

Basic Catalysis (Zhou et al, 2003) are procedures that use in general alkoxides of sodium and potassium, carbonates and hydroxides of these elements. Among these three groups it is found that alkoxides catalysts are financially unfavorable because they are more expensive but also difficult to handle because they are hygroscopic, and facilitate the achievement of side products such as derivatives of saponification, but have the advantage of carrying out the reactions in milder temperatures, produces high levels of esters derived from fatty acids and do not have corrosive properties as acid catalysts. A solution used to minimize the soap formation when biodiesel has a high free fatty acid content or water is the use of 2 or 3% mol of K_2CO_3 that will form the corresponding bicarbonate salt instead of water.

In the following, table 1, it's possible to find diverse types of heterogeneous catalysts used to obtain biodiesel of soybean, cotton seed, *Jatropha curcas*, palm, rape oil and sunflower.

Among the studies using soybean oil to obtain biodiesel, can be stand out the work developed by Wang et al, using CaO, SrO as a solid catalyst used in a heterogeneous process to obtain biodiesel. Cao, is a typical basic solid catalyst used in the most different ways. This compound has many advantages as a reusable due to its long catalyst lifetime, higher activity and requirement of only mild reaction conditions. At the example of table 1, is observed that in the best conditions to obtain biodiesel in yield of 95% is necessary a temperature of 65ºC, a molar ratio of MeOH/Oil of 5 and even a little reaction time from 0.5 to 3 hours. Even with all these specific positive factors, solid acid catalysts have been very useful at many industrial processes. Acid catalysts contain a large variety of acid sites with different strength of Bronsted, Lewis acidity, which is considered a good advantage at the transesterification process. These catalysts are even very useful, when is necessary to obtain biodiesel from oils rich in FFA, free fatty acids, because they convert the FFA into FAME prior to the biodiesel production, avoiding by this way the problem encountered at base catalysts, the soap formation.

Vegetable oil	Catalysts	Ratio MeOH/Oil	Reaction time (h)	Temperature (°C)	Conversion (%)	References
Soybean	Calcined LDH (Li-Al)	15	1-6	65	71.9	Li
Soybean	La/zeolite beta	14.5	4	160	48.9	Furata
Soybean	MgOMgAl$_2$O$_4$	3	10	65	57	Schumaker
Soybean	MgO, ZnO, Al$_2$O$_3$	55	7	70-130	82	Trakarnpruk
Soybean	Cu and Co	5	3	70		Shu
Soybean	CaO, SrO	12	0.5-3	65	95	Wang
Soybean	ETS-10	6	24	120	94.6	Arzamendi
Cotton seed	Mg-Al-CO$_3$ HT	6	12	180-210	87	Wang
Jatropha Curcas	CaO	9	2.5	70	93	Albuquerque
Palm	Mg-Al-CO$_3$ (hydrotalcite)	30	6	100	86.6	Huaping
Rape	Mg-Al HT	6	4	65	90.5	Zeng
Sunflower	NaOH/Alumina	6-48	1	50	99	Liu
Sunflower	CaO/SBA-14	12	5	160	95	Suppes
Blended vegetable	Mesoporous silica loaded with MgO	8	5	220	96	Barakos

Table 1. Different heterogeneous catalysts used for transesterification of vegetable oils.

In the following figure 2, is exemplified the mechanism of base catalyzed transesterification. The mechanism can be resumed in the following way. In the first step the methoxide anion attaches to the carbonyl carbon atom of the triglyceride. In the second step, the oxygen picks up an acid H$^+$ from the alcohol. In the last step a rearrangement of the tetrahedral intermediate results in the formation of biodiesel and glycerol.

3.2.2. Acid catalysts

Sulfur and chlorides compounds are the most commonly used acid catalysts. This type of catalysis[(Mohamad & Ali, 2002) has as main advantages the absence of products derived from saponification reactions, higher yields but has some disadvantages such as the fact that the reactions are performed in a highly corrosive and reactive post-treatment, where the medium, the rinse water should be neutralized.

4. Enzymes

Enzimes are a fourth class of compounds used to produce biodiesel (Fukuda et al, 2001). In general its use is complicated by the fact that the enzyme generally are a specific material,

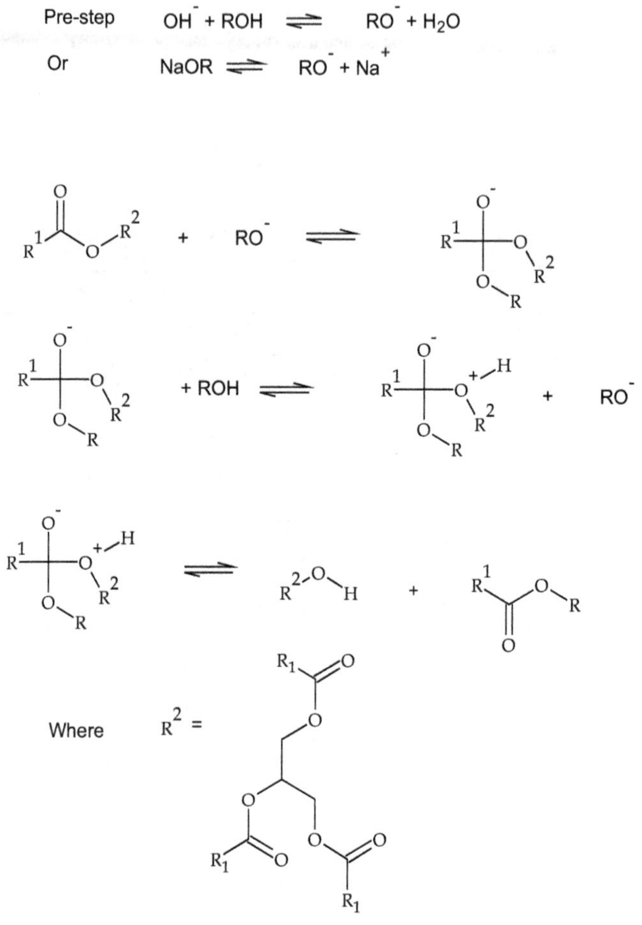

Pre-step $OH^- + ROH \rightleftharpoons RO^- + H_2O$

Or $NaOR \rightleftharpoons RO^- + Na^+$

Where $R^2 =$

$R_1 =$ Carbon chain of fatty acid

$R =$ alkyl group pf alcohol

Figure 2. Mechanism of base catalyzed transesterification.

and extremely expensive in relation to this type of reaction, are sensitive to the presence of methanol and ethanol, which causes deactivation of the same (Salis et al, 2008). Literature (Modi et al, 2007) shows that this problem can be circumvented by the water (Kaieda et al, 2001 & Kaieda et al, 1998) use and organic (Raganathan, 2008 & Harding et al, 2008) solvents such as dioxanes and petroleum ether, for example.

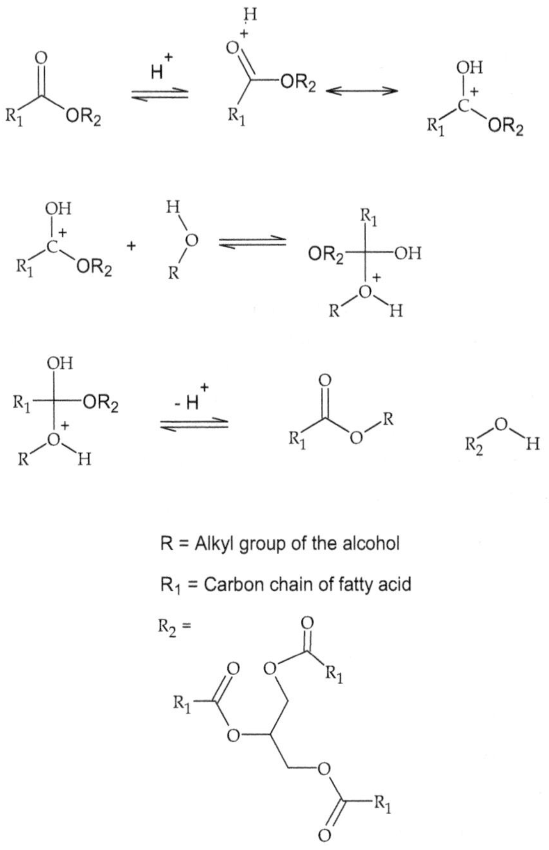

Figure 3. Mechanism of acid catalyzed transesterification

5. Non catalytic fatty acid alkyl ester production

The use of supercritical methanol process (Marulanda, 2012) to obtain biodiesel has been a useful method when the feed stock oil contains high amount of free fatty acids. This methodology has solved the problem encountered at the use of solid catalysts, low reaction rates due to low level of mass transfer, limitation between liquid and solid phase of catalysts and reactants. By this process, the dielectric constant of liquid methanol which tends to decrease in the supercritical state, increase the oil in to methanol solubility, resulting in a single phase oil/methanol system (Lee et al, 2012). In the next page, is exemplified the process of obtaining biodiesel by supercritical method, figure 4.

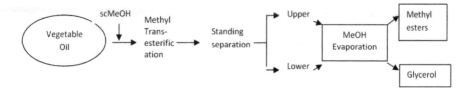

Figure 4. Schematic process of biodiesel production by supercritical method.

The methodology using supercritical methanol to obtain biodiesel by transesterification has reached the possibility to realize the reactions under mild, relatively moderate reactions to avoid the thermal degradation of fatty acid methyl esters (FAME). The reactions has been investigated in a wide range of reaction conditions (T = 200 – 425 °C, time = 2 – 40 min. and P = 9.6 – 43 MPa). Thermal stability studies of methyl esters has showed that the best reaction conditions by supercritical methanol methodology to obtain biodiesel, consists in the temperature of 270°C and reaction pressure of 8.09 MPa. The use of a co-solvent, as propane, CO_2 and heptane, diminishes, decrease the reaction temperature and the pressure needed to achieve a high yield of biodiesel obtained. The following table 2 exemplifies the investigated reactions conditions to obtain biodiesel.

Oil (co-solvent)	T (°C)	P (MPa)	MeOH/Oil ratio	Time (min)	B/C	Yield (%)	Refs.
Soybean	100 -320	32	40	25	C	96	He H. et al
Soybean (CH_3H_8/MeOH = 0.05)	280	12.8	24	10	B	98	Cao W. et al
Soybean (CH_3H_8/MeOH = 0.05)	288	9.6	65.8	10	B	99	Hegel P. et al
Soybean (CO_2/MeOH = 0.1)	280	14.3	24	10	B	98	Han H. et al
Soybean (CO_2/MeOH = 0.1)	350-425	10 - 25	3-6	2 - 3	C	100	Anitescu G. et al

Table 2. Examples of biodiesel production, experiment data using supercritical methanol.

6. Sources to obtain biodiesel

The sources for biodiesel production are chosen according to availability of the same in each country, region, taking into account the relative low cost of production and favorable economies of scale. For example, the use of refined oil would not be favorable due to high production costs and low production scale, on the other hand the use of seeds, algae and fat have a low production cost and greater availability than refined oils or recycled, which is a favorable factor for the production of biodiesel from these elements. When choosing a source of biodiesel production plants, a relationship is taken into account is how much they produce and the yield of oil per hectare. Following some examples of studied seeds: soybean, Babassu(*Orbiginiasp.*), castor oil, fish oil, microalgae(*Chorellavulgaris*) (Miao & Wu, 2006), tobacco

(Usta, 2005), *JatropaCurcas* (Berchmans & Hirata, 2008), Karanja (*Pongamiaglabra*) (Meher et al, 2006), salmon (El-Mashad et al, 2008), cooking oils, among others. All biodiesel sources are chosen according to the chemical composition of their fatty acids in relation to the size of their chains, unsaturation degree and the presence of other chemical functions, as these factors influence the biodiesel quality.

7. New advanced techniques to obtain biodiesel

In recent years soybean biodiesel has achieved a high level of advanced techniques to improve its production. Has been developed a methodology using microwave assistance to improve the esters conversion rates,, using heterogeneous catalysts, nano Cao, for example which facilitates the interaction between the molecules (Hsiao et al, 2011). Dr Hsiao and his research group has proved that by this methodology, is possible to obtain a higher biodiesel yield in less time. There are two factors that influence this reaction type. The use of nano-compounds facilitates the interaction between the molecules, nanocompounds possibility a high contact surface between the molecules. The microwave methodology reduces even the reaction time due to changing the electrical field activates the smallest variance degree of ions and molecules leading to molecular friction, enabling the initiation of molecule, chemical reaction. This methodology also provides an easier access to susceptible bonds, so, increases the chemical interaction, been obtained products in less time and higher yields. Microwave methodologies has proved to be part of desirable green chemistry, cause it is a safe, comfortable and clean way of working with chemical reactions. Microwave flow system assistance through homogeneous catalysis is another example which has improved the biodiesel production in less time, depending of some factors as reaction residence time, catalyst amount and temperature at the exit point (Encinar et al, 2012). In attempt to improve the microwave assisted methodology to obtain biodiesel another technique was added, used, the ultrasound. Microwave and ultrasound developed methodology has proved to be very efficient when used together. In this process, the first step used is ultrasound, cause ultrasonic field induced an effective emulsification and mass transfer that increases the rate of ester formation due to ultrasonic mixing causes cavitations of bubble near the phase boundary between the methanol and seeds oil, facilitating the thoroughly mixing, interaction between the oil and the reactant, methanol (Hsiao et al, 2010).Another technique has been also developed using ultrasonic irradiation with vibration ultrasonic. Instead of using heterogeneous or homogeneous transesterification catalyst, was used the enzyme methodology, through the application of Novozym 435. This methodology has proved to be efficient in enzymatic reaction to obtain biodiesel. Was observed that the use of ultrasonic added to vibration is a further factor to obtain higher values of biodiesel, cause the movement increase might facilitate the interaction between the substrate and the active site (Yu et al, 2010).Transesterification of soybean oil was achieved using ultrasonic water bath and two different commercial lipases in organic solvents (*n*-hexane) for example (Batistella et al, 2012). Dimethyl carbonate is a useful alternative to obtain biodiesel, this one is nontoxic, cheapness product and the reaction obtained product, glycerol carbonate is a value added substance with various useful applications. This can be obtained through enzymatic transes-terification of soybean oil in organic solvents in mild conditions (Seong et al, 2011).The products, biodiesel are obtained in more time, but by other side this methodology has many advantages: It is an easy to use methodology, the enzyme can be reusable and the reaction work up is chemically friendly, cause it's not necessary the treatment of water used to purify biodiesel by the traditional, usual transesterification by homogeneous basic catalyst and is obtained an added value product, the glycerol carbonates. Heterogeneous catalysis using subcritical methanol is an advance in the soybean biodiesel obtaining methodology. By this technique is possible to use less amount of catalyst and have as main advantages the catalyst reusable and the separation, obtaining from reaction medium through centrifugation. The use of small amount of a catalyst, K_3PO_4, 0.1%wt, insoluble in methanol has transformed the reaction in subcritical methanol more available, cause has reduced the temperatures from 350ºC to 160ºC and the methanol molar ratio from 42 to 24, for example. The catalyst can be reusable at least three times. (Yin et al, 2012). KF Modified calcium magnesium oxide cata-lyst is an example of heterogeneous catalysis to obtain soybean biodiesel and even to recycle the catalysts due to be easily removed from the reaction through centrifugation and the use of a reaction under atmosphere pressure and 65ºC of temperature. This new catalyst has even improved the ester methyl yield from 63.6% (CaO-MgO catalyst) to 97.9% (KF-MgO-CaO) (Fan et al, 2012). Response surface methodology is an applicable technique to improve the results in obtaining soybean biodiesel. This methodology verifies the main parameters to optimize the biodiesel production process (Silva et al, 2011).The use of a process entirely in-dependent from petroleum has been reached by the use of ethanol, obtained from a renewa-ble source, sugar cane and seed oil. Gomes and his research group has developed a methodology to obtain ethyl biodiesel and even has developed a methodology to simplify the work-up process. In order to optimize the separation step of glycerol from biodiesel, many techniques have been studied. The microfiltration through ceramic membrane has demonstrated to be a useful technique to obtain biodiesel. This methodology is environmen-tally friendly cause reduces the amount of used water to purify the biodiesel. This technique simplify the entire purification process, the biodiesel is obtained by transesterification, after the end of reaction is added acidified water, this process facilitates the separation in two phases, the organic one, rich in oil and the aqueous, which posses the soaps converted in water soluble salts, catalysts, glycerol and other water soluble substances (Gomes et al, 2011). In water, glycerol forms greater droplets that are retained during the microfiltration step, been the biodiesel obtained by this way with glycerol content lower than 0.02% wt, the limit of free glycerol specified by Brazilian regulation. Mesoporous silica catalyst was used in a heterogeneous catalysis of soybean transesterification with methanol. $La_{50}SBA$-15 is used to obtain an ethyl biodiesel in mild conditions after 6 hours. The main advantage of this technique is the use of a heterogeneous catalysis to obtain ethyl biodiesel in mild condi-tions, don't use the usual high temperatures 473K and lower amounts of the ratio Oil/ Alco-hol, from 36 to 20. This lower amount of alcohol facilitates the phase separation organic/ water, less amount of alcohol causes an easier purification process, because diminishes the possibility of emulsion formation (Quintela et al, 2012). The heterogeneous catalysis of bio-diesel has reached an advance with the development of methodologies using membranes.

These membranes can be prepared in a simple way by the use of clays as hydrotalcite and poly (vinyl alcohol). The biodiesel is prepared by transesterification and can be obtained in mild conditions, 60ºC in a volume ratio oil/methanol of 5:60. The methyl biodiesel by this methodology can be obtained in 90%. The catalyst can be reused at least by 7 times (Guerreiro et al, 2010). An alternative method to obtain biodiesel is the enzymatic-catalytic way. In general this methodology has as main advantage the enzymatic selectivity, is a reusable catalytic and facilitates the separation, purification process. Methodology using a immobilized lipase onto a nanostructure has been developed due to the good transesterification activity of lipase and the use of electrospinning method to obtain nanofibrous membranes, which have larger surface area and porous structure that can lower the substrate resistance and facilitate enzyme immobilization, generating a reusable catalyst for biodiesel synthesis. The use of this methodology simplifies the separation process, where after the reaction, glycerol can be removed by centrifugation and the biodiesel obtained in the 90% range (Li et al, 2010).

Variable	Base Catalyst	Acid Catalyst	Lipase Catalyst	Supercritical Alcohol	Heterogeneous Catalyst
Reaction temperature (ºC)	60 - 70	55 - 80	30 - 40	239 - 385	180 - 220
Free fatty acid in raw material	Saponified products	Esters	Methyl esters	Esters	Non sensitive
Water in raw materials	Interfere with reaction	Interfere with reaction	No influence		Non sensitive
Yields of methyl esters	Normal	Normal	Higher	Good	Normal
Recovery of glycerol	Difficult	Difficult	Easy		Easy
Purification of methyl esters	Repeated washing	Repeated washing	None		Easy
Production cost of catalyst	Cheap	Cheap	Relatively expensive	Medium	Potentially cheaper

Table 3. Example of different technologies to produce biodiesel.

8. Non usual methods of soybean biodiesel analysis of cold properties and oxidation state

Several methods have been used to characterize biodiesel, and each methodology analyses some aspects of biodiesel as cold properties and oxidation process. Most legislation assumes a small group of tests to determination of biodiesel quality. Eighteen percents of Brazilian biodiesel production uses soybean as oil sources. Several nations have been establishing standards and legislation about biodiesel. Mainly determinations include iodine value, acid content, specific mass, esters content among others. The aim of this work are the availability of methodologies that wasn't includes on official methodologies.

8.1. Cold properties

Official methodologies of biodiesel are: Cold Filter Plugging Point (CFPP), Cloud Point and Pour Point. Pour point indicates a moment of initial crystallization, but this methodology has low accuracy. For studies with more complexity other methodologies has better performance.

8.2. Differential scanning calorimetry

One of most usual and versatility methodology is a differential scanning calorimetry, these methodology are based on monitoring the difference in energy provided/released to/ by the sample (reagent system) in relation to a reference system (inert) as a function of temperature when both systems are subjected to a controlled temperature program. The changes in the temperature of the sample are caused by phase rearrangements, dehydration reaction, dissociation or decomposition reactions, oxidation or reduction reaction, gelatinization and other chemical reactions. DSC evaluates absorption or energy liberation to determine the initial of the reaction. A typical curve of biodiesel is presented at figure 1:

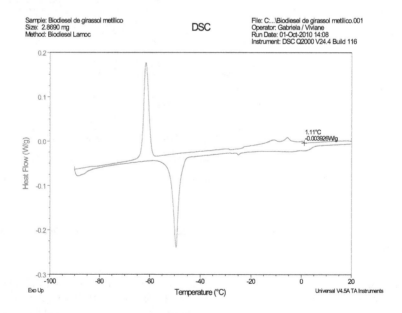

Figure 5. Differential Scanning Calorimetry of biodiesel.

Small signal at temperature of -1.1°C is due a crystallization of saturated ester notably palmitic and stearic methyl esters, shaped signal of -60°C is caused by crystallization of unsaturated esters (oleic, linoleic and linolenic methyl esters). These temperatures can change by cooling rate, so is important promoting standardization for independent analysis.

DSC trying to be associates with pour point results, because, a priori, both methods analyze a formation of firsts crystals, Formation of crystals initiate with nucleation of crystals that precedes crystallization is dependent on the formation and growth of aggregates or clusters of molecules. These aggregates must overcome a critical size in order to keep a steady growth and become a crystal of detectable dimensions [Avrami, 1940; Avrami, 1941]. At the stage of crystal growth, molecules of solute adsorb on the crystal surface and the process depends on the diffusion of material from the liquid phase to the solid phase which is being formed. Any of these stages can control crystal growth. The added substance must be capable to interfere with one of these stages: either avoiding or delaying the growing of aggregates to a critical size or reducing crystal growth rate [Mullin,2001].

In this point it's crucial to understand the difference of results obtained by pour point analyses and DSC. Calorimetric method is very sensible and detect energy associate of crystallization phenomena under of critical crystal size. While pour point detection occur only when crystal reached a minimum size.

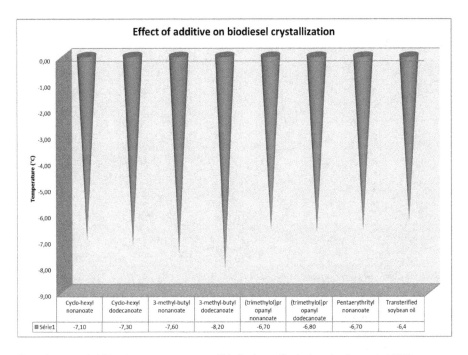

Figure 6. Impact of additives into onset temperature of biodiesel crystallization based on Soares et al, 2009

Soares et al (2009) present a work in which several esters derived from branched chain, cyclic monohydroxylated alcohols or polyhydroxylated alcohols were added to methyl transesterified soybean oil (biodiesel like) to investigate their effect on the transesterified soybean oil crystallization. In this work were added kinetic studies, were conducted in order to detect differences in crystallization mechanisms due to differences in additive structures.

They obtained a depression of onset crystallization point about 2ºC when used an additive (0,08mol/100g of transesterifed soybean oil)

Still about this work were determined induction times of crystallization. Induction time indicates how many time initiate crystallization at determined temperature. This time tends to reduce while temperature decreases. This information is important because permitted association with limit time to storage before crystallization.

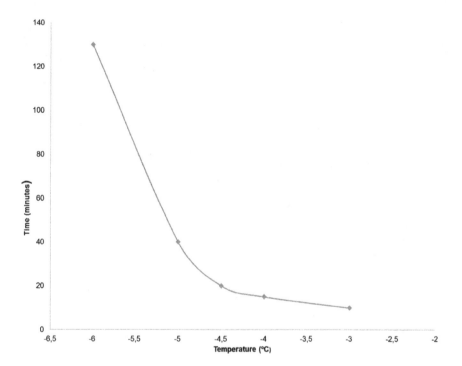

Figure 7. Induction time of crystallization based on Soares et al (2009)

These kinetic models describe how the extension of the phase transformation of a given material occurs as a function of time and temperature. Their equation is based on the suppositions of isothermal condition of crystallization, aleatory homogeneous or heterogeneous nucleation and that the new phase growth rate is temperature dependent. Avrami admitted that a number of tiny nuclei (aggregates of subcritical size) are already present in the phase to be transformed and that these aggregates must grow to a critical size to start a steady growth. By simplifying his statistical treatment presented in the calculation of transformed matter, he came to the generalized expression:

$$\alpha = 1 - \exp(-kt^n) \tag{1}$$

Where α is the volume fraction transformed (crystallized mass); k is dependent on a shape factor, on nucleation probability, on nucleation and growth rates and on the dimensionality of crystal growth while n reflects the mechanism of nucleation and growth and the crystal morphology [Avrami, 1940; Avrami, 1941]. Several works has been using DSC as assessment cold properties (ref), with better accuracy and precision.

8.3. Oxidative properties

Oxidative properties of biodiesel commonly assessment by EN 14112 called rancimat and Iodine value, but several methods have been used by analysis of oxidation state of oil and biodiesel. Oxidation products from these compounds as Petrooxy, differential scanning calorimetry (DSC), Pressure Differential Scanning Calorimetry (PDSC) (Dufaure et al, 1999) iodine value (IV) and mainly Rancimat Method. Each method is based at one step, intermediate compounds or reactants of biodiesel oxidation.

At the PetroOxy, the sample is inducted to oxidation through an intense oxygen flow, manipulating by this way the stability conditions through a specific apparatus. The analysis time is recorded as the required time to the sample absorbs 10% of oxygen pressure. Analysis is based on oxygen consumption (reactant) but not detect products.

The differential scanning calorimetry (DSC) monitors the difference in energy provided/released between the sample (reagent system) and the reference system (inert) as a function of temperature when both, the system are subjected to a controlled temperature program. Changes in temperature sample are caused by rearrangements of induced phase changes, dehydration reaction, dissociation or decomposition reactions, oxidation or reduction reaction, gelatinization and other chemical reactions. DSC evaluated absorption or energy liberation for determining initial reaction. This process can present some problems because of formation of lipid alkyl radical is an endotermic process and others reactions are exotermic (Santos et al, 2011). The time for secondary product formation from the primary oxidation product, hydroperoxide, varies with different oils. Secondary oxidation products are formed immediately after hydroperoxide formation in olive and rapeseed oils. However, in sunflower and safflower oils, secondary oxidation products are formed when the concentration of hydroperoxides is appreciable (Guillen and Cabo 2002).

At the Rancimat technique, oxidative stability is based at the electric conductivity increase (Hadorn & Zurcher, 1974.). The biodiesel is prematurely aged by the thermal decomposition. The formed products by the decomposition are blown by an air flow (10L/ 110 ºC) into a measuring cell that contains bi-distilled, ionized water. The induction time is determined by the conductivity measure and this can be totally automated. Rancimat is the most used technique to determine finalized biodiesel stability, under oxidative accelerated conditions, according to standard EN14112. This technique evaluated final products of thermal decomposition.

The differential scanning calorimetry (DSC) monitors the difference in energy provided/released between the sample (reagent system) and the reference system (inert) as a function of temperature when both in the system are subjected to a controlled temperature program. Changes in temperature sample are caused by rearrangements of induced phase changes, dehydration reaction, dissociation or decomposition reactions, oxidation or reduction reaction, gelatinization and other chemical reactions. DSC evaluated consumption or energy liberation for determining initial reaction.

The Pressure Differential Scanning Calorimetry (P-DSC) is a thermo analytical technique that measures the oxidative stability using a differential heat flow between sample and reference thermocouple under variations of temperatures and pressure. This technique differs from the Rancimat for being a fast method and presents one more variable - the pressure, allowing to work at low temperatures and using a small amount of sample (Candeia, 2009). All of these methods evaluated one aspects of oxidation process. But to predict behavior or design an adequate biodiesel its necessary to associate a oxidation process with structural properties or composition.

Iodine Value (IV) has been used for a long time to quantify unsaturated bonds on vegetable oil and, actually, biodiesel. Iodine Value is considering mainly structural method to assessment oxidation stability. Although currently some authors agree that this method is not necessarily the better method to evaluate stability.

Agreement with oxidation mechanism of fatty acids its very common associated presence of unsaturated with tendency of low stability, but Jain and Sharma (2010) presents weak relationship ($R^2= 0,4374$) between unsaturated esters content and induction period for several biodiesels, its associate this discrepancy with differ technology productions or presence of impurities.

Knothe (2007) bring an important discussion about relevance of Iodine value, its associate iodine value with some others structural index as APE and BAPE and associate to several properties of biodiesel in his work; he gives examples of how different mixtures of methyl esters of the three most common unsaturated FA—oleic, linoleic, and linolenic—can achieve nearly identical IV just slightly below the value of 115. This is an important fact because if oxidative stability depends of how many times one biodiesel can resist to be oxidized, and Linolenic acid reacts more fast then linoleic and oleic acid, this mixtures should present a differing oxidation behavior independently to have same iodine Value.

One of innovative uses of Thermogravimetric analysis is based on gum formation during biodiesel storage. Figure 4 shows a impact of storage at high temperature storage (85°C) of biodiesel. High resistance of biodiesel after four weeks is probably due formation of polymeric species as trimmers or tetramers of unsaturation esters.

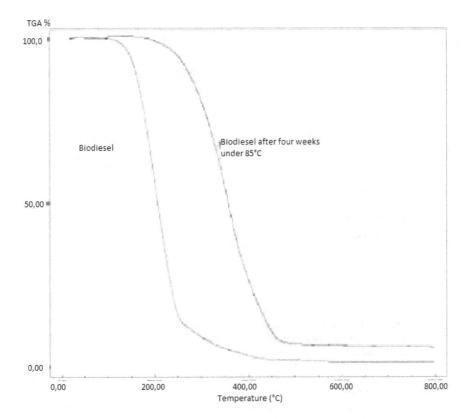

Figure 8. Formation of high molecular level species tends to increase oil viscosity

In presence of oxygen molecule polymer are forming by C-O-C linkages (Jonhson et al,1957; Wexler, 1964; Formo et al, 1979) and C-C linkages, while under an inert atmosphere only polymers with C-C linkages primordially are founded. High contents of polyunsaturated fatty acid chains enhance oxidative polymerization in fatty oils (Korus et al, 1983).Trimers and other fatty acid polymers presents higher thermal stability enhanced Termogravimetric curves of biodiesel, therefore this methodology can be used to determine biodiesel oxidation stage.

9. Conclusion

Many advanced techniques are obtained throughout the world to obtain and characterize bi-odiesel. To follow all these methodologies is necessary a constant research in many different science fields as chemistry, theoretical chemistry and even physics methodologies. This chapter contains only a simple updated tool to help to verify all the nowadays latest news, but never forget the most important tool to use is your brain, the determination and interest in the studied subject to unravel the science frontiers.

Author details

Mauricio G. Fonseca[1*], Luciano N. Batista[1], Viviane F. Silva[1], Erica C. G. Pissurno[1], Thais C. Soares[1], Monique R. Jesus[1] and Georgiana F. Cruz[2]

1 INMETRO – National Institute of Metrology, Quality and Technology, Metrological Chemistry Division Xerém, Duque de Caxias, Rio de Janeiro, Brasil

2 UENF – North Fluminense University, Engineering and exploitation of petroleum laboratory, Macaé, Rio de Janeiro, Brasil

References

[1] Agarwal, A. K., & Das, L. M. (2000). Biodiesel development and characterization for use as fuel in compression ignition engines. *Am. Soc. Mech. Eng. J., Eng. Gas Turbine Power*, 123, 440-447.

[2] Albuquerque, M. C. G., Jimenez-Urbistondo, I., Santamaria-Gonzales, J., Merida-Robles, J. M., Moreno-Tost, R., Rodriguez-Castellom, E., Jimenez-Lopez, A., Azevedo, D. C. S., Cavalcante Jr, , , C. L., Maireles-Torres, P., et al. (2008). CaO supported on mesoporous sílica as basic catalysts for transesterifications reactions. *Appl. Catal. A: Gen*, 334, 35-43.

[3] Anitescu, G., Deshpande, A., & Tavlarides, L. L. (2008). Integrated technlogy for supercritical biodiesel production and Power cogeneration. *Energy Fuels*, 22, 1391-1399.

[4] Antunes, W. M., Veloso, C. O., & Henriques, C. A. (2008). Transesterification of soybean oil with methanol catalyzed by basic solids. *Catal. Today*, 133-135, 548 EOF-554 EOF.

[5] Avrami, M. (1940). Kinetics of phase change. II: transformation-time relations for random distribution of nuclei. *J Chem Phys.*, 8, 212-224.

[6] Avrami, M. (1941). Granulation, phase change, and microstructure: kinetics of phase change. III. *J Chem Phys.*, 9, 177-184.

[7] Arzamendi, G., Campoa, I., Arguinarena, E., Sanchez, M., Montes, M., & Gandia, L. M. (2007). Synthesis of biodiesel with heterogeneous NaOH/Alumina catalysts: Comparison with homogeneous NaOH. *Chem. Eng. J.*, 134, 123-130.

[8] Barakos, N., Pasias, S., & Papayannakos, N. (2008). Transesterification of triglycerides in high and low quality oil feeds over an HT2 hydrotalcite catalyst. *Bioresource Technol*, 99, 5037-5042.

[9] Batistella, L., Lerin, L. A., Brugnerotto, P., Danielli, A. J., Trentin, C. M., Popiolski, A., Treichel, H., Oliveira, J. V., & Oliveira, D. (2012). Ultrasound-assisted lipase-catalyzed transesterification of soybean oil in organic solvent system. *UltrasonicsSonochemistry*, 19, 452-458.

[10] Candeia, R. A., Silva, M. C. D., Carvalho, Filho. J. R., Brasilino, M., Bicudo, T. C., Santos, I. M. G., & Souza, A. G. (2009). Influence of soybean biodiesel content on basic properties of biodiesel diesel blends. *Fuel*, 88, 738-743.

[11] Dufaure, C., Thamrin, U., & Moulongui, Z. (1999). Comparison of the thermal behaviour of some fatty esters and related ethers by TGA-DTA analysis. *Thermochim. Acta,,* 368, 77-83.

[12] Berchmans, H. J., & Hirata, S. (2008). Biodiesel production from crude Jathropa Curcas L. Seed oil with a high content of free fatty acids. *Bioresour. Technol.,,* 99, 1716-1721.

[13] Cao, W., Han, H., & Zhang, J. (2005). Preparation of biodiesel from soybean oil using supercritical methanol and co-solvent. *Fuel*, 84, 347-351.

[14] El -Mashad, H. M., Zhang, R., & Avena-Bustillos, R. J. (2008). A two steps process for biodiesel production from salmon oil. *Bisosys. Engin.,,* 99, 220-227.

[15] Encinar, J. M., González, J. F., Martínez, G., Sánchez, N., & Pardal, A. (2012). Soybean oil transesterification by the use of a microwave flow system. *Fuel*, 95, 385-393.

[16] Fan, M., Zhang, P., & , Q. (2012). Enhancement of biodiesel synthesis from soybean oil by potassium fluoride modification of a calcium magnesium oxides catalyst. *Bioresource technology*, 104, 447-450.

[17] Formo, M. W., Jungermann, E., Noris, F., & Sonntag, N. O. (1979). *Bailey's Industrial Oil and Fat Products,,* 1(4), Daniel Swern, Editor: John Wiley and Son,, 698-711.

[18] Fukuda, H., Konda, A., & Noda, H. (2001). Bioidesel fuel production by transesterification of oils. *J. Biosc. Bioeng.,,* 92, 405-416.

[19] Furata, S., Matsuhasbi, H., & Arata, K. (2004). Biodiesel fuel production with solid superacid catalysis in fixed bed reactor under atmospheric pressure. *Catal. Commun*, 5, 721-723.

[20] Gomes, M. C. S., Arroyo, P. A., & Pereira, N. C. (1999). Biodiesel production from degummed soybean oil and glycerol removal using ceramic membrane. *Journal of Membrane Science*, 378, 453-461.

[21] Gryglewicz, S. (1999). Rapeseed oil methyl esters preparation using heterogeneous catalysts. *BioresourTechnol*, 70, 249-253.

[22] Guerreiro, L., Pereira, P. M., Fonseca, I. M., Martin-Aranda, R. M., Ramos, A. M., Dias, J. M. L., Oliveira, R., & Vital, J. (2010). PVA embedded hydrotalcite membranes as basic catalysts for biodiesel synthesis by soybean oil methanolysis. *Catalysis Today*, 156, 191-197.

[23] Guillen, M. D., & Cabo, N. (2002). *Food Chemistry,*, 77(4), 503-510.

[24] Hadorn, H., & Zurcher, K. (1974). Zur bestimmung der oxydationsstabilitat von olen und fetten. *Deutsche Lebensmittel Rundschau.*, 70, 57-65.

[25] Han, H., Cao, W., & Zhang, J. (2005). Preparation of biodiesel from soybean oil using supercritical methanol and CO_2 as co-solvent. *Process Biochem,.*, 40, 3148-3151.

[26] Harding, K. G., Dennis, J. S., von, Blottnitz. H., & Harrison, S. T. L. (2008). A life-cycle comparision between inorganic and biological catalyse for the production of biodie-sel. *J. Clean. Produc.*, 16(13), 1368-1378.

[27] He, H., Wang, T., & Zhu, S. (2007). Continuous production of biodiesel fuel from vegetable oil using supercritical methanol process. *Fuel,*, 86, 442-447.

[28] Hegel, P., Mabe, G., Pereda, S., & Brignole, E. A. (2007). Phase transitions in a biodie-sel reactor using supercritical methanol. *Ind. Eng. Chem. Res.,*, 46, 6360-6365.

[29] Helwani, Z., Othman, M. R., Aziz, N., Fernando, W. J. N., & Kim, J. (2009). Technolo-gies for production of biodiesel focusing on green catalytic techniques: A review. *Fuel Processing Technologies*, 90, 1502-1514.

[30] Hsiao, M., , C., Lin, C., , C., Chang, Y., , H., Chen, L., & , C. (2010). Ultrasonic mixing and closed microwave irradiation-assisted transesterification of soybean oil. *Fuel*, 89, 3618-3622.

[31] Hsiao-C, M., Lin-C, C., & Chang-H, Y. (2011). Microwave irradiation-assisted trans-esterification of soybean oil to biodiesel catalyzed by nanopowder calcium oxide. *Fuel*, 90, 1963-1967.

[32] Huaping, Z., Zongbin, W., Yuanxiao, C., Ping, Z., Shije, D., Xiaohua, L., & Zong-qiang, M. (2006). Preparation of biodiesel catalyzed by solid super base of calcium oxide and its refining process. *Chinese J. Catalysis*, 27, 391-396.

[33] Johnson, O. C., & Kummerow, F. A. (1957). Chemical Changes Which Take Place in an Edible Oil During Thermal Oxidation. *JAOCS*, 34, 407-409.

[34] Kaieda, M., Samukawa, T., Matsumoto, T., Ban, K., Kondo, A., Shimada, Y., Noda, H., Nomoto, F., Obtsuka, K., Izumoto, E., & Fukuda, H. (1999). Biodiesel fuel produc-tion from plant oil catalyzed by rhizopus orizae lipase in a water containing system without an organic solvent. *J. Biosc. Bioeng.,*, 88(6), 627-631.

[35] Kaieda, M., Samukawa, T., Kondo, A., & Fukuda, H. (2001). Effect of methanol and water contents on production of biodiesel fuel from plant oil catalyzed by various li-pases in a solvent free system. *J. Biosc. Bioeng.,*, 91(1), 12-15.

[36] Knothe, G. H. (2007). Some aspects of biodiesel oxidative stability. *Fuel Processing Technology*, 88, 677-699.

[37] Korus, R. A., Mousetis, T. L., & Lloyd, L. (1982). Polymerization of Vegetable Oils. *American Society of Agricultural Engineering,,* Fargo, ND,, 218-223.

[38] Leclercq, E., Finiels, A., & Moreau, C. (2001). Transesterification of rapessed oil in the presence of basic zeolites and related solid catalysts. *J. Am. Oil Chem. Soc,* 78, 1161-1165.

[39] Lee, S. B., Lee, J. D., & Hong, I. K. (2011). Ultrasonic energy effect on vegetable oil based biodiesel synthetic process. *Journal of Industrial Engineering Chemistry,* 17, 138-143.

[40] Lee, S., Posarac, D., & Ellis, N. (2012). An experimental investigation of biodiesel synthesis from waste canola oil using supercritical methanol. *Fuel,* 91, 229-237.

[41] Li, S., , F., Fan, Y., , H., Hu, R., , F., Wu, W., & , T. (2011). Pseudomonas cepacia lipase immobilized onto electrospun PAN nanofibrous membranes for biodiesel production from soybean oil. *Journal of Molecular Catalysis B: Enzymatic,,* 72, 40-45.

[42] Li, E., & Rudolph, V. (2008). Transesterification of vegetable oil to biodiesel over MgO-functionally mesoporous catalysts,. *Energy and Fuels,* 22, 143-149.

[43] Liu, X., He, H., Wang, Y., & Zhu, S. (2008). Transesterification of soybean oil to biodiesel using CaO as a solid base catalyst. *Fuel,,* 87, 216-221.

[44] Liu, X., He, H., Wang, Y., Zhu, S., & Piao, X. (2007). Transesterification of soybean oil to biodiesel using CaO as a solid base catalyst. *Cat. Commun,* 8, 1107-1111.

[45] Marulanda, V. F. (2012). Biodiesel production by supercritical methanol transesterification : process simulationand potential environmental impact assessment. *Journal of Cleaner Production,,* 33, 109-116.

[46] Meher, L. C., Dharmagada, V. S. S., & Naik, S. N. (2006). Optimization of Alkaly-catalyzed transesterification of Pongamia pinnata oilfor production of biodiesel. *Bioresour. Technol.,,* 97, 1392-1397.

[47] Meher, L. C., Sagar, D. V., & Naik, S. N. (2008). Technical aspects of biodiesel production by transesteriofication- Review. *Renew. Sustain Energy Ver.,* 10, 248-268.

[48] Miao, X., & Wu, Q. (2006). Biodiesel production from heterotrophic microalgal oil. *Bioresour. Technol.,* 97, 841-846.

[49] Modi, M. K., Reddy, J. R. C., Rao, B. V. S. K., & Prasad, R. B. N. (2002). Lipase mediated conversion of vegetables oils in to biodiesel using ethyl acetate as acyl acceptor. *Bioresource Technology,* 98(6), 1260-1264.

[50] Mohamad, I. A. W., & Ali, O. A. (2002). Evaluation of the transesterification of waste palm oil into biodiesel. *BioresourTechnol,* 85, 225-256.

[51] Mullin, J. W. (2001). *Crystallization,* 4, Boston: Butterworth-Hetnemann,, 181.

[52] Pinto, A. C., Guarieiro, L. L. N., Rezende, M. J. C., Ribeiro, N. M., Torres, E. A., Lopes, A. W., Pereira, P. A. P., & Andrade, J. B. (2005). Biodiesel: An overview. *J. Braz. Chem. Soc*, 16(6B), 1313-1330.

[53] Quintella, S. A., Saboya, R. M. A., Salmin, D. C., Novaes, D. S., Araujo, A. S., Albuquerque, M. C. G., Cavalcante Jr, , & , C. L. (2012). Transesterification of soybean oil using ethanol and mesoporous silica catalyst. *Renewable Energy*, 38, 136-140.

[54] Raganathan, S. V., Narasiham, S. L., & Muthukumar, K. (2008). An overview enzymatic production of biodiesel. *Bioresour. Technol*, 99(10), 3975-3981.

[55] Salis, A., Pinna, M., Manduzzi, M., & Solinas, V. (2008). Comparision among immobilised lipases on macroporous polypropilene toward biodiesel synthesis. *J. Molec. Catal. B. Enzim.*, 54(1), 19-26.

[56] Santos, V. M. L., Silva, J. A. B., Stragevitch, L., & Longo, R. (2011). *Fuel*, 90(2), 811-817.

[57] Schumaker, J. L., Crofcheck, C., Tackett, S. A., Santillan-Jimenez, E., Morgan, T., Crocker, M., Ji, Y., & Toops, T. J. (2008). Biodiesel synthesis using calcined layered double hydroxide catalysts. *App. Cat. B: Env*, 82, 120-130.

[58] Seong-J, P., Jeon, B. W., Lee, M., Cho, D. H., Kim-K, D., Jung, K. S., Kim, S. W., Han, S. O., Kim, Y. H., & Park, C. (1993). Enzymatic coproduction of biodiesel and glycerol carbonate from soybean oil and dimethyl carbonate. *Enzyme and Microbial Technology*, 48, 505-509.

[59] Shay, E. G. (1993). Diesel fuel from vegetable oil: Status and opportunities. *Biomass Bioenergy*, 4(4), 227-242.

[60] Shrivastava, A., & Presad, R. (2002). Triglycerides-based diesel fuels. *Renewable Sustainable Energy Rev*, 4, 111-133.

[61] Shu, Q., Yang, B., Yuan, H., Qing, S., & Zhu, G. (2007). Synthesis of biodiesel from soybean oil methanol catalyzed by zeolite beta modified with La^{3+}. *Catal. Commun.*, 8, 2159-2165.

[62] Siddharth, J., & Sharma, J. P. (2010). Stability of Biodiesel and its Blends: A Review. *Renewable and Sustainable Energy Reviews*, 14(2), 667-678.

[63] Silva, G. F., Camargo, F. L., & Ferreira, A. L. O. (2011). Application of response surface methodology for optimization of biodiesel production by transesterification of soybean oil with ethanol. *Fuel Processing Technology*, 92, 407-413.

[64] Singh, Couhan. A. P., & Sarma, A. K. (2011). Modern heterogeneous catalysts for biodiesel production: A comprehensive review. *Renewable and Sustainable Energy Reviews*, 15(9), 4378-4399.

[65] Soares, V. L., Nascimento, R. S. V., & Albinante, S. R. (2009). Ester-additives as inhibitors of the gelification of soybean oil methyl esters in biodiesel. *Journal of Thermal Analysis and Calorimetry*, 97, 621-626.

[66] Stamenkovic, O. S., Lazic, Z. B., Todorovic, M. L., Veljkovic, V. B., Skala, D. U., & (2004, . (2004). The effect of agitation intensity on alkali-catalyzed methanolysis of sunflower oil. *Bioresour. Tech.*, 98(14), 2688-2699.

[67] Suppes, G. J., Dasari, M. A., Doskocil, E. J., Mankidy, P. J., & Goff, M. J. (2004). Transesterification of soybean oil with zeolite and metal catalyst. *Appl. Catal. A: Gen*, 257, 213-223.

[68] Suppes, G. J., Bockwinkel, K., Lucas, S., Botts, J. B., Mason, M. H., & Heppert, J. A. (2001). Calcium carbonate catalyzed alcoholysis of fats and oils. *J. AM. Oil Chem. Soc.*, 78(2), 139-146.

[69] Tomasevic, A. V., & Marinkovic, S. S. (2003). Methanolysis of used frying oils. *Fuel Process Technol*, 81, 1-6.

[70] Trakarnpruk, W., & Porntangjitlikit, S. (2008). Palm oil biodiesel synthesized with potassium loaded calcined hydrotalcite and effect of biodiesel blend on elastomers properties. *Renew Energy*, 33, 1558-1563.

[71] Usta, N. (2005). Use of tobacco seed oil methyl esterin a turbocharged indirect injection diesel engine. *Biomass and Bioenergy*, 28, 77-86.

[72] Wang, L., & Yang, J. (2007). Transesterification of soybean oil with nano-MgO or not in supercritical and subcritical methanol. *Fuel*, 86(3), 328-333.

[73] Wang, Y., Zhang, F., Yu, S., Yang, L., Li, D., Evans, D. G., & Duan, X. (2008). Preparation of macrospherical magnesia-rich magnesium aluminate spinel catalysts for methanolysis of soybean oil. *Chem. Eng. Sci*, 63(17), 4306 EOF-4312 EOF.

[74] Wang, Y. D., Al-Shemmeri, T., Eames, P., Mcnullan, J., Hewitt, N., Huang, Y., et al. (2006). An experimental investigation of the performance of gaseous exhaust emissions of a diesel engine using blends of a vegetable oil. *Appl. Thermal Eng.*, 26, 1684-1691.

[75] Wexler, H. (1964). Polymerization of Drying Oils. *Chemical Reviews*, 64, 591-611.

[76] Yin-Z, J., , Z., Shang-Y, Z., Hu-P, D., & Xiu-L, Z. (2012). Biodiesel production from soybean oil transesterification in subcriticalmethanol with K3PO4 as a catalyst. *Fuel*, 93, 284-287.

[77] Yu, D., Tian, L., Wu, H., Wang, S., Wang, Y., , D., & Fang, X. (2010). Ultrasonic irradiation with vibration for biodiesel production from soybean oil by Novozym 435. *Process Biochemistry*, 45, 519-525.

[78] Zeng, H., , Y., Feng, Z., Deng, X., Li, Y., & , Q. (2008). Activation of Mg-Al hydrotalcite catalyst for transesterification of rape oil,. *Fuel*, 87(13-14), 3071-3076.

[79] Zhou, W., Konar, S. K., & Boocock, D. G. V. (2003). Ethyl esters from the single-phase base-catalyzed ethanolysis of vegetable oils. *J. Am. Oil Chem. Soc.*, 80(4), 367-371.

Chapter 2

Critical Evaluation of Soybean Role in Animal Production Chains Based on the Valorization of Locally Produced Feedstuff

Stefano Tavoletti

Additional information is available at the end of the chapter

1. Introduction

Commodities such as soybean and maize are respectively protein and energy concentrates that represent the basic raw materials used by the animal feeding industry and their prices influence the overall market of agricultural products. In animal feeding, soybean is mostly used as soybean meal which is a by-product of oil seed extraction industry and the availability of this raw material on the international market has led to its worldwide diffusion as the main source of protein for animal feed formulation. Soybean can also be used as raw seed due to its high fat content that makes this grain legume a valid feed to increase both protein and energy concentration of animal diets [1]. The close relationship between oil extraction industry and feed industry together with its high nutritional value for human consumption, as reported by several articles included in the present book, made soybean a perfect crop to be treated as a commodity in the world trade of raw materials.

However, despite all these positive characteristics, the diffusion of soybean and its by-products, together with the overall increased importance of the commodities trade, has triggered in agriculture several downstream effects that had deep consequences on the evolution of agricultural practices. After World War II agriculture initiated a course of progressive structural changes toward the implementation of more intense production processes due to both the need of increasing world food supply and to the progressive reduction of people employed in agriculture. The diffusion of improved varieties, fertilizers, pesticides, advanced agricultural machineries, intensive systems of animal rearing, efficient systems for the storage and transformation of agricultural products led to the abandonment of traditional cropping and animal farming systems [2]. Therefore, important agronomic practices such as crop rotations including a cereals and legumes, cultivation of forage crops for animal feeding, a close link between animal farming and field productions useful to ensure an adequate content of organic matter in the soil were almost completely abandoned due to the diffusion of monocultures [3]. As a consequence, agricultural soil fertility decrease dramatically, as indicated by the dangerous low levels of organic matter content that at present are generally recorded in most countries that have been characterized by such an intensification of agricultural practices, and the use of chemical fertilizers became an indispensable necessity to reach economically valuable productions [4-7]. At the same time animal farming was based on the use of by-products that were available on the market to reduce the costs of production and simplify the animal production system.

At present, globalization together with the rapid economic development actually under way in eastern countries and the world economic crisis are jeopardizing the economic feasibility of many agricultural activities, mainly in the European Union. Those farms that restructured their production processes in order to simply satisfy the demand for raw materials by the food and feed industries and by multiple retailers that manage marketing and commercialization have recently experienced the negative effects of increased costs of production followed by the low prices of agricultural products paid to the farmers. In particular, the increased costs of commodities such as soybean seed and meal that recently happened several times together with the low price of animal products paid to the farmers, mainly in the beef, pork and dairy production chains, made unprofitable the economic activities of animal farms, especially small or medium size farms.

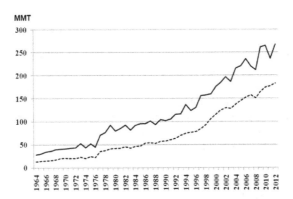

Figure 1. World soybean seed (continuous line) and meal (dashed line) production.

Recently new strategies for agricultural development have emerged due to the interest of consumers toward high quality products and production chains, the increased request of non-standardized food and the attention given to the impact of agricultural activities on the environment, human and animal health [8-10]. Therefore, an increasing number of animal farms adopted more sustainable instead of intensive production systems that were closely linked to the area of production by using locally produced animal feed and reducing their dependence from commodities. Moreover, these farms developed direct commercialization systems trying to make their business more profitable with an emphasis on the quality of both the final products and the production system.

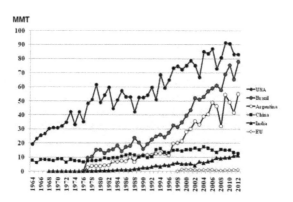

Figure 2. Soybean seed production of USA, Brazil, Argentina, China, India and European Union (EU).

Aim of the present article is to critically evaluate the effects of the large diffusion of soybean on the international market concerning aspects related to the evolution of soybean trade, the effects on agricultural systems where soybean cannot be cultivated, the present dependence from soybean of intensive animal farming systems and the consequences on small or medium farms applying not-intensive animal production chains. An experience under way in the Marche Region (Central Italy) will also be illustrated as an attempt to make agricultural activity both profitable and integrated within local soil and climatic characteristics.

2. Soybean seed and meal trade

Data on soybean seed and meal (oil was not considered) production, import, export and prices were obtained at the Index Mundi website [11] (data source:USDA) where information on all commodities trade is available. Soybean data were usually available from 1964 to 2011 together with estimates referred to the current year 2012, even though for some countries and also for the European Union data availability covered a shorter period of time. Data

were organized in an excel data sheet and elaborated to obtain information concerning single countries involved in the soybean international trade. The total worldwide value of seed
and meal production, import and export, expressed as Million Metric Tons (MMT), were
then calculated by summing the data available for each country for each year. Results were
summarized by graphics concerning the overall soybean trade, the characteristics of single
countries significantly involved in the international market of soybean and the comparisons
among different countries.

Soybean seed production progressively increased from 1964 (28,3 MMT) to 2010 (264,7
MMT) and, although followed by a slight decrement in 2011 (236,4 MMT), soybean seed US-
DA 2012 estimated production (referred to june 2012) is 266,8 MMT confirming the positive
trend for this commodity (Figure 1).

USDA data identified 42 countries characterized by an estimated soybean seed production
of at least 0.001 MMT. However since 1964 more than 90% of total soybean seed production
was concentrated in 5 countries (USA, Brazil, Argentina, China and India) and USA, Brazil
and Argentina covered about 80% of worldwide soybean production (Figure 2).

European Union (Figure 2) produced between 0.6 MMT (year 2008) and 1,4 MMT (year
1999) of soybean seeds and in 2011 EU27 ranked twelfth with a production of 1,1 MMT
(0.47% of world production). These data confirm the almost complete dependence of Europe
from non-EU and mainly American countries to satisfy the needs of protein concentrates of
European animal production chains.

Figure 1 also shows that the production of soybean meal had almost the same trend of
world soybean seed production, with a steady increase from 1964 (13,5 MMT) to 2011 (177,4
MMT). However, the relative contribution of each country is different than what has been
described for seed production. USA was the highest soybean meal producer until 2009 when
it was exceeded by China that, based on the 2012 estimate, at present seems to be the world
leader in soybean meal production. China showed a relatively low meal production until
1997 when it started a progressive increase in the production of this by-product of oil extraction (Figure 3).

It is interesting to compare soybean seed and meal amounts produced over time in China;
soybean seed production in this country was always lower than 20 MMT, ranging from 6.14
MMT in 1995 to 17.4 MMT in 2004 (Figure 2), whereas soybean meal production was lower
than 10 MMT until 1997 but increased from 10 MMT in 1998 to 46.9 MMT in 2011 with an
estimated 50.2 MMT for the year 2012 (Figure 3). Starting from 1997 also Brasil and Argentina began to increase their soybean meal production reaching about 28 MMT in 2011, whereas USA, after a constant increase from 1964 to 1996, was characterized by and almost
constant meal production of about 35 MMT/year in the 1997-2011 time period (Figure 3).

India, which was the fifth producer of soybean seeds, also showed a constant increase in
soybean meal production from 1987 to 2011 although remaining below the 10 MMT level of
meal production (Figure 3). On the contrary in the 2001-2011 time period the European Union was characterized by a negative trend of soybean meal production that decreased from
14 MMT in 2001 to less than 10 MMT after 2008 (Figure 3).

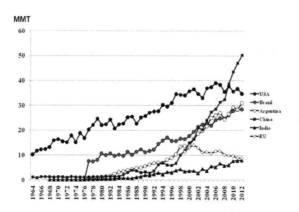

Figure 3. Soybean meal production of USA, Brazil, Argentina, China, India and European Union (EU).

Moreover, in the 1994-2011 time period the relative contribution to the worldwide soybean meal production of Japan decreased from 8% (1994) to 0.82% (2011) and Canada was also characterized by the same negative trend (from 3.12% in 1994 to 0.62% in 2011); this behaviour could be attributed to the low increase in meal production over time that characterized these two countries (from 1.07 to 1.45 MMT for Japan and from 0.42 to 1.10 MMT for Canada) compared to the progressive overall world increase of soybean meal production.

Therefore, the international scenario concerning soybean clearly shows that 3 countries (USA, Brasil and Argentina) handle almost all the world production of soybean seed, whereas China must be added to USA, Brasil and Argentina concerning the production of soybean meal. Conversely European Union has a negligible level of soybean seed production and a low level of soybean meal production as a by-product of oil extraction from imported soybean seeds. This determines that European Union animal production chains rely completely on imported soybean and this situation generates an almost total dependence of European farms from the international trade of these commodities. Data on the import/export of soybean seeds and meal also confirm that EU animal farming is suffering the effects of globalization of the markets, mostly because the dynamics of the international market of soybean is changing as a consequence of the new scenario due to the increased interest toward this commodity by several new countries.

Figure 4 and Figure 5 summarize the world scenario of soybean seed and meal trade, respectively. World seed production increase was also followed by an increase in the amount, expressed as percentage of total world production value, of overall exported seed that was about 25% until 1995, then raised to about 30% between 1996 and 2005 and finally reached the value of 35 % in 2010 with an estimated amount of 36% for year 2012 (Figure 4).

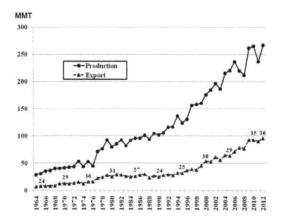

Figure 4. Soybean seed: comparison between world production and export; every five year export values expressed as percentage of total world production are shown (2012 data are estimated values).

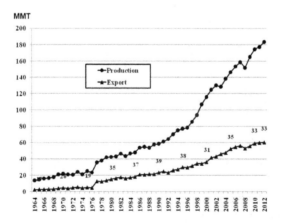

Figure 5. Soybean meal: comparison between world production and export; every five year export values expressed as percentage of total world production are shown (2012 data are estimated values).

A similar trend was shown by global soybean meal production. An increase from 17% to 35% in the percentage of exported meal, expressed as percentage the total world meal production, was registered in the 1965-1980 time period, a further slight increase until 38% was observed between 1980 and 1995 and subsequently the percentage of exported meal went back to the value of 33% for year 2010, the same value estimated for year 2012 (Figure 5). Therefore, about one third of total world soybean seed and meal production is exported to countries that show a deficit in the internal production of these commodities.

The relative importance of the import/export of soybean seed and meal compared with the internal production is also shown separately for each of the 4 most important countries (USA, China, Brazil and Argentina) together with the situation characterizing the European Union (Figures 6-10).

At present, USA is the highest producer of soybean seed in the world and in 2010 45% of total USA internal production was exported, this level being maintained in 2012 estimate. About 20-25% of internal USA soybean meal has always been exported (Figure 6).

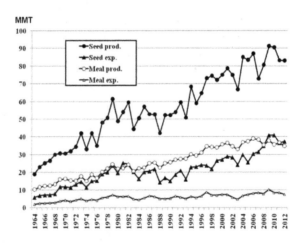

Figure 6. USA: Soybean seed and meal production and export; imported amounts are negligible and therefore are not shown.

Together with USA, also Brazil and Argentina (data available from 1978 to 2012) are characterized by a high amount of internal production which is exported followed by a negligible import of soybean products. In particular, Brazil is getting close to USA levels of production for both soybean seed and meal (Figures 2 and 3) and seed export, that ranged between 12 and 20% in the 1980-1995 time period, steadily increased from 30% in 1996 to 55% in 2011 with an estimate of 45% for year 2012 (Figure 7). A different trend characterized soybean meal Brazilian export that was about 76% from 1978 to 1990 and then decreased to 50% in 2011 with an estimated level of 48% for year 2012. On the whole, about 50% of both seed and meal internal Brazilian productions is exported (Figure 7).

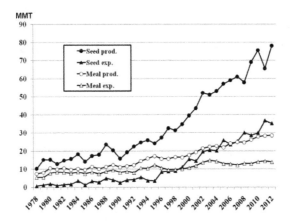

Figure 7. Brazil: Soybean seed and meal production and export; imported amounts are negligible and therefore are not shown.

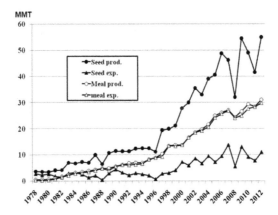

Figure 8. Argentina: Soybean seed and meal production and export; imported amounts are negligible and therefore are not shown.

A large amount of Argentina's seed production in 1978-1980 period was exported, but since 1981 the amount of exported seed dropped drastically to 33-55% until 1985 and thereafter it decreased even more reaching 19% in 2011 with an estimate of 20% in 2012 (Figure 8). On the other end the amount of soybean meal internally produced increased approximately four times and almost 100% of the meal was exported, as clearly shown by Figure 9. Therefore these data suggest that Argentina is essentially growing soybean to export soybean meal.

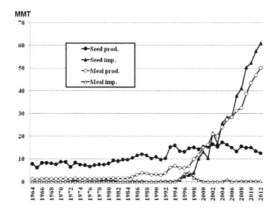

Figure 9. China: soybean seed and meal production and import; exported amounts are negligible and therefore are not shown.

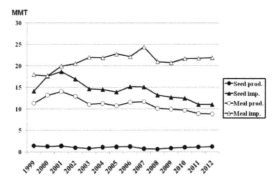

Figure 10. European Union: soybean seed and meal production and import; exported amounts are negligible and therefore are not shown.

On the other hand, an important soybean importer is China (Figure 9) whose inner soybean seed production only slightly increased over time whereas this country has become the largest importer of soybean seed (61 Mt estimated for 2012). All the internal production and import of soybean seed is used for oil extraction and meal by-product production. The large amount of soybean meal produced is almost completely used within the country to support internal animal productions, since China export of meal is negligible. Due to this large volume of import China is getting a predominant role influencing the worldwide exchange of soybean products, sometimes competing with other strong importers such as the European Union (Figure 10). As a matter of fact estimated levels of 2012 EU import of soybean seed and meal are 11 and 21.9 MMT respectively, against an internal production of 1.2 and 8.8 MMT of seed and meal, respectively.

Therefore, costs of production of European animal farms strictly depended on soybean prices that are set by the international market. As shown in Figure 11 monthly price of soybean seeds and meal in the 1983-2011 time period showed a marked change in part due to the trend of world soybean production. In particular, after 2007 seed and meal prices showed a clear average increase compared to the previous years. The unpredictable variation in the price of the basic protein concentrate for animal feeding strongly influenced the incomes of European farmers since it was also related to a general increase in the prices of other commodities and production factors (fertilizers, pesticides, seed, fuels etc.) whereas the prices of animal products to the farmers did not follow the same positive trend.

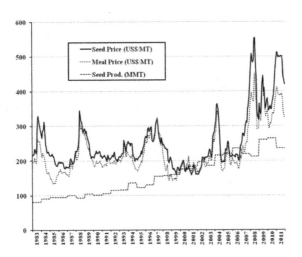

Figure 11. Trend of soybean seed and meal prices (US$ per Metric Ton) compared to the world total seed production (MMT) of each year.

Furthermore, the worldwide diffusion of soybean as protein feed component has almost completely replaced any other protein source for animal feeding and this happened mainly in highly intensive farms trying to maximize animal growth rate, to simplify the management of feedstuff and to standardize meat or milk production systems. The same trend occurred for maize which is currently the most important cereal for animal production chains. Moreover, most of the soybean is produced by using new varieties obtained by genetic engineering [12] whose acceptance by European consumers is strongly debated.

On the whole, globalization of the market of animal products favoured the development of animal farms rearing large numbers of animals determining a high concentration of animals in a reduced number of farms which is typical of intensive farming systems, followed by a progressive reduction in the number of small farms that were not able to compete in such a situation of high costs of production coupled with low prices of products for the farmers.

3. Strategies for partial or total soybean replacement.

Therefore, even though both soybean and maize have extremely interesting nutritional characteristics as feed intended for almost all farm animals, an alternative strategy was necessary to reduce this complete dependence of European animal production chains from the market of the commodities. However, environmental and climatic conditions in most European countries hinder the cultivation of soybeans and maize because they are warm-season crops growing during the spring-summer season making irrigation a fundamental need for the success of these crops. Therefore the limited amount of lands suitable for cultivation of these crops, the high water request making irrigation a fundamental component of the production cost, the low levels of soybean grain production per hectare suggested that attention should have been addressed toward other sources of proteins to reduce dependence on imported soybean.

This is particularly true for European countries characterized by a Mediterranean climate, where rainfall is mainly concentrated during winter until late spring whereas summer is usually characterized by high temperatures coupled with very low and often irregular rainfall. Moreover, the use of water for irrigation is often very expensive and therefore, where it is available, it is devoted to other crops such as orchards and vegetables. As a consequence, in these areas agriculture is essentially based on cool-season crops that could be sown in autumn-end of winter and harvested at the end of spring or during summer. Therefore, these agricultural systems are typically based on rotations between winter cereals and grain legumes such as faba bean (*Vicia faba* var. *minor* L.), field pea (*Pisum sativum* L.), chick pea (*Cicer arietinum* L.) and white lupin (*Lupinus albus* L.). For animal feeding the most commonly used grain legume was faba bean. Recently, genetic selection of field pea new varieties stimulated the diffusion of this crop as protein source for animal feeding. Also chickpea has been proposed as a possible alternative to soybean together with sweet lupin, which is characterized by the highest protein content among the grain legumes alternative to soybean. However, since chickpea is mainly cultivated for human consumption and lupin can be cultivated only in locations with specific soil pH conditions, the main grain legume crops on which to focus attention as possible replacements of soybean are field pea and faba bean.

As a consequence, scientific research was directed to increase knowledge on cultivation, plant breeding and utilization as animal feed of these grain legumes [13-23]. Results showed that in ruminant, monogastric and avian animals at least a partial replacement of soybeans is feasible in intensive animal farming systems, whereas in low input or organic farming systems soybean can be completely replaced by grain legumes that can be grown on farm where soybean cultivation is not feasible [24]. However, despite these encouraging results, grain legume cultivation has suffered a clear reduction in Europe. This trend was related to the development of feed industry that, due to the large quantity of raw materials handled, rely mostly on commodities available on the international market and on the use of industrial protein reach by-products of oil extraction such as rapeseed meal, cottonseed meal, sunflower meal and others.

Therefore, the identification of grain legumes such as faba bean, field pea, chickpea and lupin that could at least partially replace soybean in animal feeding systems and their introduction of in crop rotation systems targeted at supplying animal production chains with locally produced protein concentrates could have several positive effects on European agricultural systems. This set of crops could guarantee, together with forage crops and pastures, the development of animal production chains that were fully integrated with local environmental characteristics. Moreover, the close link between animal farming and field productions supported the maintenance of good soil fertility and organic matter content. Finally, new strategies for the commercialization of final animal products must be undertaken for the full valorisation of the whole production chain.

4. Consequences of simplified cropping systems

The negative trend shown by grain legume diffusion was therefore a consequence of agriculture evolution toward highly specialized intensive production systems that determined the progressive gap between animal productions and field agriculture. This trend led to well-known agricultural and economical drawbacks such as lost of soil fertility, dramatic decrease of soil organic matter content, increased need of inputs (fertilizers and pesticides) to reach the highest agronomic performances, search for high productions to counterbalance the lowering prices of raw materials on both global and local markets. Moreover, a clear separation between farmers that progressively became simple producers of raw materials and industry that managed commercialization and transformation of agricultural products, over time led farmers to lose any possibility of market control. These aspects determined a deep crisis in the agricultural sectors of countries where agriculture was characterized by small or medium sized farms that lost their ability to compete on the market since they were confined to the role of simple low value raw material producers. The diffusion of monoculture, the reduction of forage crops due to the intensive feeding systems mainly based of protein and energy concentrates, the trend toward part-time agriculture, the massive use of chemicals to maximize crop production characterized agriculture for several decades after World War II. As a consequence, market of agricultural products was invaded by standardized

products that replaced most of the typical and local productions that previously characterized agricultural systems strictly integrated with the areas of production.

Recently, the need for a more environmental friendly agriculture together with the increasing request by the consumers of high quality products, stimulated farmers to recapture the
market of agricultural products [25]. These farms progressively abandoned standardized
and intensive agriculture and dedicated to animal and crop productions following the voca-
tion of their own local region.

As far as animal production chains are concerned, the reconstitution of local production systems led to the valorisation of animal feeding systems based on locally produced raw materials, both forages and protein (grain legumes) or energy (cereals) concentrates. This allowed
these farms to reintroduce rational crop rotations, that were abandoned due to the diffusion
of monocultures, by alternately cultivating those cereals and legumes that could also be intended for animal feeding. Most of these farms started to commercialize by their own the
products of their farms by an action aimed at informing the consumers about all aspects of
their production system, receiving economical and professional satisfactions.

This approach led to partly or completely replace soybean in animal diets, to reintroduce
forage crops both in field crop rotations and in animal feeding systems, to develop less intensive animal farming systems, to stimulate creation of local networks among farmers
which could represent a further stimulus for local farms to reintroduce grain legumes and
forage crops for animal feeding by making agreements with local animal farms that would
withdraw their legume products.

5. Experiences on soybean replacement carried out in the Marche Region (Central Italy)

On the feasibility of local animal production chains a research, funded by the Marche Region (Central Italy), has been carried out by our research group since year 2000 to evaluate
the technical and economical possibilities of soybean replacement with grain legumes such
as *faba* bean and field pea. At the same time research evaluated the possibility of a total or
partial replacement of maize with barley or sorghum grain, since both these crops are valuable energy crops for areas characterized by high temperatures and low rainfall during
summer, since barley is a cool season cereal whereas sorghum is a worm season but drought
resistant cereal crop.

The first objective of the project was to test the agronomic feasibility of both faba bean and
field pea in different areas of the Marche Region [26-27]. Very favourable lands, where
these crops have been evaluated in optimal agronomic conditions, and more marginal fields
were included in field trials. Moreover, several farms including both conventional and organic farms were involved as partners of the project to carry out "on farm" field trials
based on large plots that were managed by the farmers themselves. The "on farm" approach allowed the evaluation of the real potential of these crops in the areas under exami-

nation, that were mainly located in the inner part of the Marche region were irrigation is not feasible (Figure 12).

Figure 12. Field experimental trial including faba bean and field pea carried out in typical agricultural landscape of inner areas of the Marche Region (Province of Ancona).

Results showed that both faba beans and field pea could be effectively reintroduced in crop rotations with winter cereals such as wheat and barley, reconstituting a correct alternation between nitrogen fixing legumes and cereals. However, the grain productions obviously varied based on the environmental and soil conditions and the seasonal climatic conditions that varied from year to year. This experience allowed the creation of a useful data set indicating that faba beans showed, on the average, a range between 1.0 and 3.5 tons/hectare of grain, the lowest productions being obtained when very dry growing seasons occurred with extremely low values of rainfall during end of winter and spring. Farmers know very well these characteristics of faba beans since this crop has been traditionally used across the whole region mainly as protein grain for beef cattle of the Marchigiana breed. However, plant breeding efforts are requested to stabilize grain production in the variable environmental conditions characterizing the inner areas of Central Italy.

Field pea was characterized by a higher average seed production than faba beans, showing a range between 1.5 and 4.5 tons/hectare. For this crop low production can be due to environmental adverse conditions but also grain loss due to seed shattering during harvesting is a primary cause of production losses mainly when the crop is grown on soils with an irregular surface that makes threshing difficult.

To compare faba bean and field pea with soybean, few farms where irrigation was feasible were asked to try cultivation of soybean. The results showed an average seed production between 2.5 and 3 tons/hectare, similar to a good faba bean or field pea harvest, but costs of irrigation made this crop unprofitable. Moreover, chemical weed control was necessary in order to obtain acceptable seed production because watering also favoured the development of weeds. Herbicides are used also to protect faba beans and field pea against weed competition. However, our field trials carried out in organic farms showed that multi-year rotations including forage crops such as alfalfa, highly competitive cereals against weeds such as barley or wheat, and a higher crop density (number of plants/m^2) than conventional cultivation can avoid the use of herbicides on crops such as faba beans and field pea.

Results of the field trials showed that these grain legumes can be effectively produced in inner areas of the Marche Region where soybean cannot be cultivated and animal farming is traditionally an economic source of income for local farmers. After gathering information on the agronomic feasibility, research has been addressed to verify the possibility of total or partial substitution of soybean with faba beans and/or field pea in beef cattle, dairy cattle and swine feeding systems. Therefore, feeding trials were conducted in one large dairy farm, located in the Province of Ancona, 4 organic farms rearing beef cattle of the Marchigiana breed and located in the Provinces of Macerata and Fermo, one conventional farm located in the Province of Pesaro-Urbino, and one conventional small familiar swine farm, located in the Province of Fermo, rearing pigs using a non-intensive farming system.

Concerning the dairy sector, our experimental trials were carried out while farmers were experiencing the continuous fluctuation of the prices of raw materials, mainly commodities such as soybean meal and corn grain, coupled with the crisis of the dairy sector across all Europe that determined low milk prices despite the increasing costs of production. Therefore, we were able to evaluate both the potential and the limits of soybean and corn partial replacement in a very critical agricultural sector such as dairy farming. Due to the peculiarities of dairy production, based on the animal physiology and on the nutritional characteristics of the raw materials under examination, soybean meal cannot be totally replaced by faba bean and field pea. Based on the results of the "on farm" feeding trial carried out at this dairy farm about 50% of soybean meal present in the ration was replaced by faba bean and field pea. To understand the potential effect of this partial substitution on the local agricultural system it can be considered that the farm was initially using 3kg/cow of soybean meal. Therefore, 1.5 kg of soybean meal were substituted by about 2 kg/cow of faba bean/field pea mix, considering that also part of the corn grain was also partially replaced by these grain legumes due to their starch content. Having the farm an average of 450 lactating cows per day, daily feeding requested about 9 tons/day of faba bean and field pea, which was about 330 tons/year. Based on the results of the field trials, assuming an average field production of 2.5 tons/hectare (2 tons/ha for faba beans and 3 tons/ha for field pea) it can be estimated that this farm could support about 130 hectares cultivated with grain legumes for animal feeding. Assuming a multi-rotation such as wheat-grain legume-barley- 3 years of alfalfa we can roughly estimate that partial substitution of soybean in this case could support an overall agricultural system covering about 790 hectares. Moreover, the presence of a 3 years forage crop such as alfalfa would reduce drastically the use of pesticides and fertilizers and also would increase nitrogen fixation and soil organic matter content, the introduction of organic farming practices to manage cereal and grain legume crops would avoid or at least reduce the use of chemicals and in particular of herbicides also in conventional farms, the milk production would be closely linked to the area of production. However, to make this system working it is necessary to make it economically profitable for the farm. When soybean meal prices increased the use of grain legumes produced by local farms helped the dairy farm to compensate for the increased costs of production. Despite these encouraging results the crisis of the dairy sector is hampering the implementation of this integrated production system that is based on close relationships between the dairy farm and farmers producing raw materials used as animal feed. The identification of different commercialization systems able to fully valorise the quality of the overall production chains is becoming a fundamental step in order to counteract the continuous decrease of milk price on the national and international market.

Beef cattle field trials were carried out at four organic farms located in the inner areas of Provinces of Macerata and Fermo. Differently from the experience previously described in the dairy sector, the small size of these farms, the high quality of both final products and production system and a different approach to the product valorisation based on the direct sales of meat by the producers themselves, allowed research results to be transferred to the final step of the production chain: the product marketing and commercialization.

Feeding trials on organic beef cattle showed that for not intensive production systems soybean can be totally replaced by faba beans and field pea. Moreover, these farms are characterized by self-producing almost all the forage and a high amount of protein (grain legumes) and energy (mainly barley) needed for animal feeding. It is relatively simple for these farms to make arrangements with neighbouring farmers to secure themselves the supply of the raw materials they are not able to self produce. This production system is extremely interesting because, as shown in Figure 13, it is based on rational agronomic crop rotations maintaining both a high level of crop diversification together with a low if not positive environmental impact due to the organic farming practices. The lower animal daily growth rate characterizing organic or non-intensive animal farming (1-1.2 kg/day), compared to growth rate of intensive systems (higher than 1.6 kg/day), is also an aspect of the production system which is valorised in the final product.

Figure 13. Organic animal farm located at Monte San Martino (Province of Macerata) where both experimental field and feeding trials were carried out.

Aim of the research project was also to verify the technical and economic feasibility of GM*free* production chains, that is production systems that do not include genetically modified (GM) feed in the feeding system. Among the commodities, soybean show the highest amount of worldwide production obtained from GM varieties, mainly cultivated in USA or South America [12]. Therefore, since almost soybean seed or meal used in the European Union is imported by American countries, the risk of GM soybean contamination is very high. This is confirmed by the introduction of a threshold of 0.9% technically unavoidable contamination also in organic feedstuff. Therefore, our results demonstrated that also organic farming systems could avoid GM contamination whenever a strict control of the raw material production or origin is made directly from the farmers. Feed composition therefore becomes an index of the raw materials used in the production chains and can be used as further information for the consumers to valorise the value of the final product. For these reason a DNA method has been developed as further result of the project aimed at the identification of the presence of faba beans and/or field pea within feed samples [28]. This could be a simple and not expensive approach to certify the use of local raw materials as feedstuff together with the absence of soybean from the ration. Moreover, an attempt to increase consumers' information about the characteristics of GM *free* organic production has been started as part of dissemination of the project activities and this increased the number of consumers interested in purchasing GM *free* products. On the whole, results on organic beef cattle showed that GM *free* production chains based on feeding systems that rely on locally produced raw materials can be an efficient alternative to intensive production chains. This approach could also be useful to maintain or increase economically effective agricultural systems in inner areas of Central Italy by reducing the dependence from the international market of commodities.

Encouraging results have also been obtained concerning the swine production chain. The farm where feeding trials were carried out had the possibility of rearing pigs both indoor and outdoor (open air). Therefore a feeding trial was conducted to compare one conventional feed (Control) based on the use of soybean meal and corn with an experimental feed where soybean meal was replaced completely by faba beans and field pea and corn was partially replaced with barley. Both feeds were formulated respecting the differences requested between the growth and the finishing phases. No differences in animal growth rate (600 g/ day) were detected between the two feeds (Control vs Experimental). At the same time a group of pigs was reared outdoor and fed with the experimental feed. Average daily growth rate was slightly lower than observed in the indoor trial. The same experimental feed was subsequently tested in one organic and one conventional farm and results confirmed that regular growth rates can be obtained when soybean is not included in the feed, with slightly higher average daily gains obtained in the conventional farm (750 g/day). Therefore, non intensive swine production chains could represent another animal farming system that could stimulate the development of production systems linked to the production area, the networking among local farms concerning the exchange of raw materials for animal feeding, the reintroduction of rational not intensive agricultural systems. Commercialization of the final products is again fundamental to guarantee profitability for all the actors of the production chain and for this purpose direct selling is showing to be an effective marketing strategy to reach this objective.

6. Conclusions

The main objective of this paper was to stimulate a critical evaluation of soybean impact on agricultural systems where soybean cannot be cultivated. Notwithstanding soybean positive nutritional characteristics, this commodity may not be the only solution for animal production chains for those countries that may suffer from a complete dependence on import of raw materials.

Field and feeding trials carried "on farm" furnished information on the feasibility of local animal production chains using feeding systems based on locally produced raw materials that could partially or completely replace soybean. This possibility would be a stimulus to recreate networks among local farmers in order to develop local production chains that could restore an economically feasible agricultural systems to farms that are unable to compete on the global market. This also represents an attempt to reintroduce sustainable agricultural systems characterized by a reduced use of chemicals and pesticides due to the cultivation of low input cereals and legumes in rationale crop rotations.

Moreover, the interest toward research results on feedstuff that do not include soybean is also related to the risk of GM contamination due to this commodity. GM free animal production chains are strongly exposed to GM contamination when soybean is included as the main protein source in the feed. The decision to include the 0,9% threshold also for organic farming production chains underlines the real risk of GM contamination and the difficulties to create GM free production chains when the feed is based on the use of the same raw materials characterizing GM animal products.

Our results however showed that for large animal farms, that carryout intensive production systems, it is more difficult than for small farms, characterized by not intensive production chains, to manage soybean replacement. This aspect confirms the almost complete dependence from imported commodities that has been reached in the time by agricultural production sectors aimed at the mass production of large amounts of standardized products. The large volume of raw materials requested followed by the low internal availability of feedstuff that can be used as an alternative to soybean exposes these farms to the risks of international market variations both in the availability and in the price of this commodity.

On the other hand, the implementation of animal production chains based on the use of locally produced feedstuff is a valid approach for small farms producing high quality products using not intensive animal farming systems. These farms can in this way gain a market space for products that can be an alternative to standardized products and at the same time activate agricultural systems well integrated with local environmental features, that make less use of intensive production techniques, reduce the use of fertilizers and pesticides, restore soil fertility. However, new marketing strategies are necessary to make consumers aware of the importance of the overall characteristics of local production chains in defining the quality of a final product and to ensure at the same time a profitable price for the producers.

Author details

Stefano Tavoletti *

Address all correspondence to: s.tavoletti@univpm.it

Dipartimento di Scienze Agrarie, Alimentari ed Ambientali – Università Politecnica delle Marche - Via Brecce Bianche, 60131 ANCONA, Italy

References

[1] Masuda, T., & Goldsmith, P. D. (2009). World Soybean Production: Area Harvested, Yield, and Long-Term Projections. *International Food and Agribusiness Management Association Review*, 12(4), 143-162.

[2] Borsari, B. (2011). Agroecology as a Needed Paradigm Shift to Insure Food Security in a Global Market Economy. *International Journal of Agricultural Resources, Governance and Ecology*, 9(1/2) 1-14.

[3] Altieri MA. (1995). Agroecology: the Science of Sustainable Agriculture. *Westview Press Inc.*

[4] Reicosky D.C, Kemper W.D, Langdale C.W, Douglas C.LJ., 1995. Rasmussen PE. Soil organic matter changes resulting from tillage and biomass production. *Journal of Soil and Water Conservation* 50(3) 253-261.

[5] Tilman, D., Cassman, K. G., Matson, P. A., Naylor, R., & Polasky, S. (2002). Agricultural sustainability and intensive production practices. *Nature*, 418(6898), 671-677.

[6] Montgomery, DR. (2007). Soil erosion and agricultural sustainability. *Proceedings of the National Academy of Sciences of the United States*, 104(33), 13268-13272.

[7] Tilman, D., Balzer, C., Hill, J., & Befort, B. L. (2011). Global food demand and sustainable intensification of agriculture. Proceedings of National Academy of Science; , 108(50), 20260-20264.

[8] Sleper D.A., Barker T.C., Bramel-Cox P.J. editors. 1991. Plant Breeding and Sustainable Agriculture: Considerations for Objectives and Methods. *CSSA Special Publication N.18.*

[9] Jackson, D.L., & Jackson, L.L. (2002). The Farm as Natural Habitat. Reconnecting Food Systems with Ecosystems. *Island Press, Washington DC.*

[10] Jordan, N., Boody, G., Broussard, W., Glover, Keeney. D., Mc Cown, B. H., Mc Isaac, G., Muller, M., Murray, H., Neal, J., Pansing, C., Turner, R. E., Warner, K., & Wyse, D. (2007). Sustainable Development of the Agricultural Bio-Economy. *Science*, 316(5831), 1570-1571.

[11] Index Mundi.

[12] International Service fort he Acquisition og Agri-Biotech Applications. *ISAAA: executive summary*.

[13] Yu, P., Goelema, J. O., Leury, B. J., Tamminga, S., & Egan, A. R. (2002). An analysis of the nutritive value of heat processed legume seeds for animal production using the DVE/OEB model: a review. *Animal Feed Science and Technology*, 99(1), 141-176.

[14] Froidmont, E., & Bartiaux-Thill, N. (2004). Suitability of lupin and pea seeds as a substitute for soybean meal in high-producing dairy cow feed. *Animal Research*, 53(6), 475-487.

[15] Prandini, A., Morlacchini, M., Moschini, M., Fusconi, G., Masoero, F., & Piva, G. (2005). Raw and extruded pea (Pisum sativum) and lupin (Lupinus albus var. Multitalia) seeds as protein sources in weaned piglets' diets: effect on growth rate and blood parameters. *Italian Journal of Animal Scince*, 4(4), 385-394.

[16] Stein, H. H., Everts, A. K. R., Sweeter, K. K., Peters, D. N., Maddock, R. J., Wulf, D. M., & Pedersen, C. (2006). The influence of dietary field peas (Pisum sativum L.) on pig performance, carcass quality, and the palatability of pork. *Journal of Animal Science*, 84(11), 3110-3117.

[17] Gilbery T.C., Lardy G.P., Soto-Navarro S.A., Bauer M.L. and Anderson V.L. 2007 Effect of field peas, chickpeas, and lentils on rumen fermentation, digestion, microbial protein synthesis and feedlot performance in receiving diets for beef cattle. *J Anim Sci* 85(11) 3045-3053.

[18] Perella, F., Mugnai, C., Dal, Bosco. A., Sirri, F., Cestola, E., & Castellini, C. (2009). Faba bean (Vicia faba var. minor) as a protein source for organic chickens: performance and carcass characteristics. *Italian Journal of Animal Science*, 8(4), 575-584.

[19] Volpelli, L. A., Comellini, M., Masoero, F., Moschini, M., Lo, Fiego. D. P., Scipioni, R., & Pea, . (2009). (Pisum sativum) in dairy cow diet: effect on milk production and quality. *Italian Journal of Animal Science*, 8(2), 245-257.

[20] Crépon, K., Marget, P., Peyronnet, C., Carrouée, B., Arese, P., & Duc, G. (2010). Nutritional value of faba bean (Vicia faba L.) seeds for feed and food. *Field Crops Research*, 115(3), 329-339.

[21] Jensen ES, Peoples MB,. (2010). Hauggaard-Nielsen H. Faba bean in cropping systems. *Field Crops Research*, 115(3), 203-216.

[22] Jezierny, D., Mosenthin, R., & Bauer, E. (2010). The use of grain legumes as a protein source in pig nutrition: A review. *Animal Feed Science and Technology*, 157(3), 111-128.

[23] Masey, O'Neill. H. V., Rademacher, M., Mueller-Harvey, I., Stringano, E., Kightley, S., & Wiseman, J. (2012). Standardised ileal digestibility of crude protein and amino acids of UK-grown peas and faba beans by broilers. *Animal Feed Science and Technology*, 175(3), 158-167.

[24] Blair, R. (2011). Nutrition and Feeding of organic cattle. *CABI, UK*, 13978184593758.

[25] Jarosz, L. (2012). Growing inequality: agricultural revolutions and the political ecology of rural development. *International Journal of Agricultural Sustainability*, 10(2), 192-199.

[26] Tavoletti, S., Mattii, S., Pasquini, M., & Trombetta, M. F. (2004). La reintroduzione delle colture proteiche nelle filiere agrozootecniche marchigiane. *L'Informatore Agrario*, 21-31.

[27] Migliorini, P., Tavoletti, S., Moschini, V., & Iommarini, L. (2008). Performance of grain legume crops in organic farms of central Italy. *In: Neuhoff D, Halberg N, Alföldi T, Lockeretz W, Thommen A, Rasmussen IA, Hermansen J, Vaarst M, Lueck L, Caporali F, Jensen HH, Migliorini P, Willer H (eds.) Cultivating the future based on science. Proc. Of 16th IFOAM Organic World Congress*, 16-20, 978-3-03736-023-1.

[28] Tavoletti, S., Iommarini, L., & Pasquini, M. (2009). A DNA method for qualitative identiWcationof plant raw materials in feedstuff. *European Food Research Technology*, 229(3), 475-484.

Chapter 3

Phytoestrogens and Colon Cancer

B. Pampaloni, C. Mavilia, E. Bartolini, F. Tonelli,
M.L. Brandi and Federica D'Asta

Additional information is available at the end of the chapter

1. Introduction

Colorectal carcinoma (CRC) represents the most frequent malignancy of the gastrointestinal tract in the Western world in both genders. There is a wide variation of incidence rate for both colonic and rectal cancer among the populations of different countries: up to a 30-40-fold difference is seen between North America (Canada, Los Angeles, San Francisco), New-Zealand (non–Maori), Northern Italy (Trieste), Northern France (Haut- and Bas-Rhin) in which the rate of CRC is around 50/100,000 inhabitants, and India (Madras, Bangalore, Trivadrum, Barshi, Paranda, Bhum, Karunagappally), Algeria (Setif), and Mali (Bamako) in which the rate is around 3/100,000 [1]. It is estimated that approximately 6% of the United States population will eventually develop a CRC, and that 6 million of American citizens who are living will die of CRC [2].

The geographic differences in CRC incidence are due more to environment, life-style, and diet than to racial or ethnic factors. Demonstration of this fact is that migrants from low to high incidence areas have the same incidence as the host country within one generation, having assimilate western lifestyle and diet [3].

Colonoscopy to screen asymptomatic adults older than 50 years allows an estimation of the prevalence of adenomatous polyps or CRC: in North America CRC is found in 2%, and advanced adenoma (more than 1 cm in diameter) in 10% [4[.

Population-based studies have investigated several environmental factors as contributors to the initiation of sporadic colorectal carcinogenesis. High-calorie diet, high red meat consumption, overcooked red meat consumption, high saturated fat consumption, excess alcohol consumption, cigarette smoking, sedentary lifestyle, and obesity are considered to increase the incidence of CRC, while consumption of fiber, fresh fruit and vegetables, and a high-calcium diet could have a protective effect [5]. A recent review [6] provided an overview of the epidemiological evidence supporting the roles of diet, lifestyle, and medication in reducing the risk of colorectal cancer. Similarly, many studies that implicate effects of dietary agents in various types of cancers are available and suggest that much of the suffering and death from cancer could be prevented by consuming a healthy diet, reducing tobacco use, performing regular physical activity, and maintaining an optimal body weight [7]. Even if several epidemiological and experimental studies support the role of these factors in the genesis of CRC, other well-designed prospective and randomized clinical trials conducted in recent years report conflicting evidence, in particular on the role of the diet component in the etiology of CRC [8, 9].

Meanwhile, the great majority of CRC are sporadic, with 2 to 6% of them related to a hereditary disease due to mutations of highly penetrant autosomic dominant genes. Mutations of *APC* tumor suppressor gene is responsible for familial adenomatous polyposis (FAP), and mutations of the mismatch repair (*MMR*) genes are related to hereditary non polypoid colorectal cancer (HNPCC or Lynch's syndrome). Mutations of *MLH1* and *MSH2* are responsible for more than 90% of the family affected by HNPCC. In these familial events, the onset of CRC is greatly anticipated in comparison to the sporadic counterpart which is usually diagnosed after 50 years of age. However, an increasing incidence rate of CRC not clearly related to the presence of inheritable or predisposing colonic diseases was observed in individuals less than 40 years of age in recent decades [10]. Furthermore, an enhanced risk for CRC and colonic adenomas is present in individuals whose first-degree relatives are affected by CRC, especially if the tumor occurs before the age of 60 [8]. Possible factors of this inherited susceptibility to CRC are polymorphisms of genes deputed to glutathione synthesis such as *GSTP1*, *GSTM1* and *GSTT1* genes [11].

Prognosis of CRC is in relationship to local and distant tumor progression. Deep penetration of carcinogenic cells in the colonic wall, invasion of adjacent organs, diffusion in lymph nodes or peritoneum, and distant metastases must be evaluated for staging of the disease and correct therapeutic planning. One third of all colorectal tumors are located in the rectum: prognosis of distally sited rectal cancer is worse than that of proximally sited rectal cancer or of colonic cancer. Despite great advances in population screening, early diagnosis, surgical interventions, and complementary therapies, long-term survival for CRC remains in the range of 50-60%.

Tumor formation in humans is a multistage process involving a series of events, and generally occurs over an extended period. During this process, several genetic and epigenetic alterations lead to the progressive transformation of a normal cell into a cancer cell. These cells acquire various abilities that transform them into malignant cells: they become resistant to growth inhibition, proliferate without dependence on growth factors, replicate without limit, evade apoptosis, and invade, metastasize, and support angiogenesis. Mechanisms by which cancer cells acquire these capabilities can vary considerably, but most of the physiological changes associated with these mechanisms involve alteration of signal transduction pathways [7].

It is commonly agreed that the first step of colorectal tumorigenesis is the shift of the proliferative zone in the glandular crypts, accompanied by the development of aberrant crypt foci, and followed by the formation of an adenomatous polyp. These pathological features are considered the precursor of the carcinoma in a temporal sequence that also can be completed in several years. However, CRC is not a homogenous disease: several histological types can be distinguished such as tubular or villous, mucinous, serrated, medullary, signet-ring, squamous cell, adenosquamous, small cell, and undifferentiated, and different molecular basis can also be recognized in histologically similar tumors. In recent years, the identification of the genetic mutations of hereditary forms of CRC has clarified two fundamentals types of carcinogenesis. The first is similar to that described for the development of the FAP, and is characterized by a progressive accumulation of genetic changes starting from a biallelic inactivation of *APC*. Additional mutations either of oncogenes *KRAS* and *p53* or of oncosuppressor genes (*DCC* and *DPC4*) are necessary for the neoplastic progression and invasivity [12]. The genetic alterations are responsible for an increased mucosal proliferation and a reduced apoptosis, causing a clonal cellular expansion. The second, similar to the CRC arising in the HNPCC, is due to inactivation of *MLH1* or of other *MMR* genes. Repetitive sequences of DNA, sited in non-encoding microsatellite regions throughout the genome, are specifically found in this type of CRC, hence, the definition of micro satellite instability (MSI). The mechanism responsible for the carcinogenesis is epigenetic due to an extensive DNA methylation. Rarely in this type of CRC both proto-oncogenes (*KRAS*, *p53*) and oncosuppressor genes (*APC, TGFBRII, IGF2R, BAX*) are mutated or inactivated [13]. The former genetic mechanism explains the most frequent form of sporadic CRC characterized by the sequence adenoma-carcinoma and a long period for the formation of cancer; vice versa, the last mechanism is only present in 15% of sporadic CRC, and can have the character of an accelerated carcinogenesis.

Improved knowledge of the molecular mechanisms of colorectal carcinogenesis allows a rationale chemopreventive use in individuals who have an increased risk of developing colorectal adenomas or cancer. Both natural or synthetic agents have been employed to prevent or suppress the colorectal tumorigenesis. In particular, in experimental animals, cohort and clinical case-control studies have shown inverse association between the use of either anti-inflammatory non steroidal drugs (NSAIDs), estrogens or phytoestrogens, and incidence of both colonic adenomas and CRC. NSAID use appears to prevent the occurrence of carcinogen-induced animal colonic tumors [14] and to decrease the number and size of colo-rectal polyps in FAP (Familial Adenomatous Polyposis) patients [15]. Randomized placebo controller trials showed that aspirin reduced the risk of colorectal adenomas in populations with an intermediate risk of developing adenomas [16]. Furthermore, NSAIDs or selective COX-2 inhibitors reduce the in vitro growth of human colon cancer cell [17]. The effect of NSAIDs is mediated by cell cycle arrest due to inhibition of the Wnt-signaling pathway that favors the phosphorilation of beta-catenin and by induction of apoptosis [18, 19].

The fact that estrogens have an effect in decreasing the risk of colo-rectal cancer is shown by the following data:

1. Several epidemiologic studies show a smaller incidence of sporadic CRC in the female gender. Also the occurrence of CRC in HNPCC is lower in females than in males;

2. women who are multipare are a reduced risk of CRC in confront to nullipare;

3. epidemiologic studies of postmenopausal women show that users of HRT have a significant reduction of CRC development in respect to women who had never used HRT. The risk appears to be halved with 5-10 years of HRT use [20, 21];

4. use of non-contraceptive hormones for more than 5 years reduces by (OR = 0.47, 95 percent CI = 0.24-0.91) the risk of colon cancer [22].

2. Nutrition and colon cancer

It is now believed that 90–95% of all cancers are attributed to lifestyle, with the remaining 5–10% attributed to faulty genes [7]. Almost 30 years ago epidemiological research suggested that appropriate nutrition could prevent approximately 35% of cancer deaths, and up to 90% of certain cancers could be avoided by dietary enhancement [23, 24].

Colon cancer is a multifactorial disease that results from the interaction of different factors such as aging, family history, and dietary style. Identifying modifiable factors associated with colorectal cancer is of importance, the ultimate goal being primary prevention, and particularly the role of diet in the aetiology, initiation, and progression of colorectal cancer remains an area of important research. Moreover, several components of food can exert a potent activity also in the later stages of cancer. Several studies have indicated that inhibition of metastasis by genistein, one of the most important constituents of soy foods, represents an important mechanism by which it is possible to reduce mortality associated with solid organ cancer.

Many plant-derived dietary agents have multitargeting properties and are therefore called nutraceuticals. A nutraceutical (a term formed by combining the words "nutrition" and "pharmaceutical") is simply any substance considered to be a food or part of a food that provides medical and health benefits. During the past decade, a number of nutraceuticals have been identified from natural sources. Nutraceuticals are chemically diverse and target various steps in tumor cell development [7].

Several epidemiological studies have consistently shown an inverse association between consumption of vegetables and fruits and the risk of human cancers at many sites. Wickia & Hagmannc (2011) recently reported that many case-control and cohort studies are dealing with the effect of fruits and vegetables on cancer incidence [25]. Early data indicated a beneficial effect [26] and, as recently as 2008, Freedman et al. found a reduced occurrence of head and neck cancers with increased fruit and vegetable consumption [27].

The concept that a diet that is high in fiber, especially from fruits and vegetables, lowers risk of colorectal cancer has been in existence for more than 4 decades. The majority of case-control studies have shown an association between higher intake of fiber, vegetables, and possibly fruits, and lower risk of colon cancer [28]. A meta-analysis of six case-control studies found that a high intake of vegetables or fiber was associated with an approximate 40%–50% reduction in risk for colon cancer [29]. Similarly, a pooled analysis of 13 case-control studies reported an approximately 50% lower risk of colon cancer associated with higher intake of fiber [30].

Increasing intake of fruits, vegetables, or fiber is unlikely to prevent a large proportion of colorectal cancers, particularly among the US population, which has a food supply already fortified with folate and other dietary factors that might protect against colorectal neoplasia. There is also little evidence that concentrated sources of one type of fiber are efficacious, although fiber-rich diets have health benefits for other gastrointestinal conditions, such as diverticular disease and constipation, and possibly other chronic diseases [6].

All evidence supporting the decreased risk include results from a few studies of adenomatous polyps (which may progress to colorectal carcinomas). Fruit and grain intake also appears to be inversely related to risk of colorectal cancer and polyps, although less consistently than vegetables. These potentially protective associations may result from the high levels of dietary fibres, antioxidants (e.g., beta-carotene, vitamin C), or other anticarcinogenic constituents (e.g., protease inhibitors, phytoestrogens) in these vegetables, fruits, and grains. However, the association of adenomatous polyps of the large bowel with intake of vegetables, fruits, and grains has not been studied to any great degree, and existing data on these associations are not entirely consistent. Because adenomatous polyps are precursors to colorectal cancer, studying polyps instead of cancer might allow one to measure the diet of relatively asymptomatic subjects closer to the time of the initial neoplastic process. [31].

A recent meta analysis and data review, conducted by Magalhães B. [32], substantiates that the risk of colon cancer was increased with patterns characterized by high intake of red and processed meat, and decreased with those labelled as 'healthy.'

There are many plausible mechanisms by which intake of vegetables, fruits and "healthy foods" may prevent carcinogenesis.

Plant foods contain a wide variety of anticancer phytochemicals with many potential bioactivities that may reduce cancer susceptibility [7,33, 34].

3. Soyfoods and colon cancer

Many epidemiologic studies evidence a lower rate of hormone-related cancers among Asian populations which are characterized by regular consumption of soy based foods. Soy is a major plant source of dietary protein for humans. A review of epidemiologic studies (most of which were case-control studies published before 2000) suggested an inverse association between high soy intake and colon cancer risk in humans [35]. Moreover, migration studies show that Japanese immigrants in the United Status have incidence rates of colorectal cancers very near to the rates among the whites in the country [6]. Thus the protective effect of soy foods and isoflavones is a matter of interest in the etiology of colorectal cancer.

Soy and soy foods contain a wide variety of chemical compounds, biologically active, that may contribute, individually or synergistically, to the health benefits of this plant; in particular, polyphenols are considered to possess chemopreventive and therapeutic properties against cancer.

Among these compounds, certainly, there are isoflavones, the most important and abundant of which is genistein, which also have estrogenic properties. In fact, in recent decades, there have been several studies showing that isoflavones are promising candidates for cancer prevention [36, 37, 38, 39].

Data associating soybean consumption with reduced cancer rates have been used as evidence for a role of isoflavones in cancer prevention. However, soybeans are also a rich source of trypsin inhibitor, other proteins with health benefits, phosphatidyl inositol, saponins, and sphingolipids, all of which have potential health benefits. All of these soybean constituents demonstrate tumor preventive properties in animal models. Research by Birt et al. demonstrated that 20% by weight of dietary soy protein significantly reduced rat intestinal mucosa levels of polyamine, a biomarker of cellular proliferation for colorectal cancer risk [39].

Surely, soy foods are complex foods, and it is difficult to assume that associations which suggest protective properties of soy foods are due only to a single constituent. Because of the association between diets in Japan and China and lower rates of cancers, such as those of the breast, prostate, and colon, than in Europe and the United States, many investigators have assumed that this is due to soy food consumption in Japan and China.

Other factors in the Asian diet may be responsible, and it's important to evaluate the possible confounding dietetic factors in the studies.

Several studies suggest that soy foods, the predominant source of isoflavones, are associated with reductions in cancer rate, but they do not consistently appear to be the primary protective component of the Asian diet.

Wu et al. noted the difficulties in assessing the relationship between the level of intake and protection. Case control and prospective epidemiological investigations that have provided a suggestion of protection against cancer by soy foods have not provided adequate information on the bioactive constituents in the soy foods, the portion size, or other components that may be protective in the diets of people who eat soy foods [40].

Isoflavones and flavonoids may be rapidly and predominately glucuronidated in the GI mucosa, if genistein can be considered a model for all of these phenolic compounds [41]. Further, glucuronidation occurs in the liver. Genistein undergoes biliary excretion, with more than 70% of a dose recovered in bile within 4 hr after dosing in rats. Although genistein may be absorbed well initially, a maximum of 25% of an oral genistein dose would be eliminated in rat urine. About 20–25% of an oral dose of genistein (predominantly as its glucoside from soy foods) is recovered in human urine [42, 43].

The presence of hydroxylated and methylated genistein metabolites correlated positively with inhibition of cancer cell proliferation, but genistein sulfates were not associated with antiproliferative effects of genistein, suggesting that some types of metabolism of the isoflavones may be crucial for their action [44].

Witte, et al, showed that higher consumption of tofu (or soybeans) was inversely associated with polyps. Tofu (or soybeans) contain a number of potentially anticarcinogenic constituents, including isoflavones, saponins, genistein, and phytosterols. They were able to look at tofu (or soybeans) as a single food item (i.e., separate from legumes) because almost 15 percent of our multiethnic study population reported consuming tofu (or soybeans) at least once a week. The strongest association observed was for vegetables—including those high in carotenoids, cruciferae, and broccoli—as well as garlic and tofu (or soybeans), and these associations were found even after adjusting for dietary fiber, folate, beta-carotene, vitamin C, and other commonly measured antioxidants [31].

Men tend to have a slightly higher incidence of colorectal cancer than women of similar age (American Cancer Society, 2007), and oestrogen seems to be implicated for this decreased risk in women. Epidemiological studies and results of a Women's Health Initiative (WHI) clinical trial provide strong evidence that colorectal cancer is hormone sensitive because the cancer risk is reduced by post-menopause hormone therapy [35]

In effect, many epidemiological and experimental studies suggest a protective role of estrogens against colorectal cancer. The decrease in the number of deaths from large bowel carcinoma observed in the United States in the last 40 years was significantly higher in women (30%) as compared to men (7%). A link was observed between oral contraceptive use and a reduction of colorectal cancer, whereas there was a higher than expected frequency of colorectal tumors among non users [45].

Interestingly, as reported by Barone et al., although several experimental studies have confirmed a protective role of estrogens for CRC, few studies have been conducted, and with conflicting results, on the possible protective effect of estrogens against the development of adenomatous polyps in the colon, although it is well known that the development of adenocarcinoma mostly involves polyp formation [46].

Gender differences in the incidence and behavior of colorectal cancer (CRC), as well as epidemiologic data indicating a protective effect of hormone replacement therapy in women, have further supported the concept of hormonal influence on the development of CRC. It has been suggested that the protective effect of estrogens (or phytoestrogens) may be mediated through activation of ERβ, which has been shown to be the predominant subtype of ER in the gastrointestinal tract [47].

ERs are nuclear receptors belonging to the steroid hormone receptor superfamily which have the characteristic of being activated upon binding of the ligand. If the ligand is not present, ERs bind to a shock protein. Otherwise, when the ligand is present, the ERs make a stable dimer and initiate the specific estrogenic response, with transcription of the target genes. Two main types of ER have been identified: alfa (ERα) and beta (ERβ). They are the so-called ligand-activated transcriptional factors through which estrogens exert their effects on various tissues and have a different tissue distribution. ERα is mainly present in the mammary glands and in the utterus; ERβ is mainly present in endothelial cells, the urogenital tract, the central nervous system, and the colonic mucosa. Experimental data have demonstrated that CRC express an elevated number of estrogen receptors (ERs), but while ERα is detected in very low levels either in normal or pathological colonic mucosa (adenoma and carcinoma), ERβ expression is high in the normal colonic mucosa, and progressively decreased in the pathological mucosa in relationship to the cellular differentiation and CRC stage.

The observation that the level of ERβ protein is lower in malignant tumors than in normal tissue of the same organ has fostered the hypothesis that ERβ may function as a tumor suppressor, protecting cells against malignant transformation and uncontrolled proliferation.

ERβ is present in various isoforms: studying different types of colonic tumoral cells, isoform 1 of ERβ is found in the Lo-Vo, HCT8, HCT116, DLD-1 and isoform 2,3,4 and 5 only in the HCT8 and HCT116. It has not been well investigated whether the function of the various isoforms of ERβ, but loss of the expression of isoform 1 of ERβ, is accompanied by undifferentiated proliferation, mucinous histological type, and tumor progression [48]. It is accepted that the binding of estrogens to the ERβ blocks the activity of AP-1 on the genes involved in the cellular proliferation and provokes an activation of p53. Conversely, SERM, such as tamoxifene and raloxifene, induce an antiproliferative effect in human colorectal cell lines by a citostatic or cytotoxic effect [49]. Several observations on the CRC cellular cultures and on the experimental mouse with germinal mutation of APC have clarified the role of the ER and estrogenes for colorectal cancerogenesis: 17β estadiol decreases the proliferation in vitro of the HCT116, Lo-Vo and DLD1 cells, but increases the proliferation of the HCT8 cells. However, the effect on the last type of cells is completely changed by increasing the level of RRb by transfection with ERβ. The overexpression of ERβ can have an inhibitory effect on the proliferation. In the transfected HCT8 cells the levels of CD4 and CP21, which are onco-suppressor genes, are significantly increased, and the level of cyclinE, which have oncogenic activity, significantly decreased, in respect to normal HCT8 [50].

ERβ is lower in the adenomatous polyps of FAP patients and in the intestinal adenomas which develop in APC Min+/- mouse than in the colonic normal mucosa. The restoration of normal levels of ERβ obtained with dietary phytoestrogenes is accompained by regression or disappearance of the polyps in the experimental animal. Patients with sporadic adenomas in the colon show an increase of apoptoic activity, and ERβ expression of the colonic mucosa, if their diet is supplemented by phytoestrogenes [45]. These data strongly support a pivotal role of ERβ in a protective action against the initiation and progression of colorectal carcinogenesis.

Many epidemiologic studies evidence a lower rate of hormone-related cancers among Asian populations which are characterized by regular consumption of soy based foods. Soy is a major plant source of dietary protein for humans. Among other components, soy contains large amounts of phytoestrogens.

As proposed for estrogens, genomic and non-genomic mechanisms have also been suggested for phytoestrogens to explain their biological activities

As reported by several authors in the past, genomic pathways are mediated through the ability of phytoestrogens to interact with enzymes and receptors, and cross the plasma membrane. In this way, they bind ERs and induce the transcription of estrogen-responsive genes, stimulate cell growth in the breast, and modify ER transcription itself. However, some of their effects are not due to interaction with ERs, and are therefore denominated non-genomic effects. For example: inhibition of tyrosine kinase and DNA topoisomerase, suppression of angiogenesis, and antioxidant effects [33, 36, 46].

The bioavailability of phytoestrogens (determined by: absorption, distribution, metabolism (bioconversion in the gut and biotransformation in the liver) and escretion) and their activity is highly variable and changes with respect to several factors, such as administration rules, dosage, metabolism and interaction with other pharmacological substances. Moreover, their biological effect is influenced by the type of target tissue, the number and type of ERs expressed in the tissue, their serum concentration, and sex steroid hormone concentration [51, 52].

Phytoestrogens, present in soy and soy-based food, may act through hormonal mechanisms to reduce cancer risk by binding to estrogen receptors (ER) or interacting with enzymes involved in sex steroid biosynthesis and metabolism [53].

Although cancer incidence in women is much lower than in men in both countries, there is also a difference when the 2 countries are compared. Japanese men as well as women have a lower colorectal cancer incidence than their American counterparts, although mortality is quite similar when related to specific incidence data. In hormone-dependent cancers such as those of the breast and prostate, incidence is exceedingly low in Japan (and was even lower in earlier decades) compared with that in the United States. Mortality, again in proportion to incidence, is rather similar. Numerous reports have suggested that this difference in tumor incidence is probably due to consumption of soy as a staple food in Asian countries in contrast to Western industrialized countries. These substances, through their potential to act as selective estrogen receptor modulators, may affect vitamin D–related inhibition of tumor growth by upregulating extrarenal synthesis of 1,25- D3. Genistein, the most prominent phytoestrogen in soy, is known to regulate other P450 enzymes, such as 5-reductase and 17-hydroxysteroid dehydrogenase, which are essential for metabolism of sex hormones [54].

In vitro studies of DLD1 colon adenocarcinoma cells have linked the effects of soy with estrogen receptor beta. Experiments conducted on this cell line, with or without ER-β gene silencing by RNA interference (RNAi), have shown that soy isoflavones decreased the expression of proliferating cell nuclear antigen (PCNA), extracellular signal-regulated kinase (ERK)-1/2, AKT, and nuclear factor (NF)-κB. Soy isoflavones dose-dependently caused G2/M cell cycle arrest and downregulated the expression of cyclin A. This was associated with inhibition of cyclin dependent kinase (CDK)-4 and upregulation of its inhibitor p21 expressions. ER-β gene silencing lowered soy isoflavone-mediated suppression of cell viability and proliferation. ERK-1/2 and AKT expressions were unaltered and NF-κB was modestly upregulated by soy isoflavones after transient knockdown of ER-β expression.

Soy isoflavone-mediated arrest of cells at G2/M phase and upregulation of p21 expression were not observed when ER-β gene was silenced. These findings suggest that maintaining the expression of ER-β is crucial in mediating the growth-suppressive effects of soy isoflavones against colon tumors. Thus, upregulation of ER-β status by specific foodborne ER-ligands such as soy isoflavones could potentially be a dietary prevention or therapeutic strategy for colon cancer [55].

4. Genistein and isoflavones: Other mechanisms of action

In addition to estrogenic/antiestrogenic activity, some mechanisms of action have been identified for isoflavone/flavone prevention of cancer: antiproliferation, induction of cell cycle arrest and apoptosis, prevention of oxidation, induction of detoxification enzymes, regulation of host immune system, and changes in cellular signaling [39, 56, 57]. It is expected that also combinations of these mechanisms may contribute to cancer prevention.

Gene silencing due to the promoter methylation provides an opportunity for clinical intervention, as gene-re-expression can be induced by a variety of DNA demethylating agents.

Recent studies show that genistein may affect DNA methylation, serves as a natural demethylation agent, and that it is specifically effective on colon cancer cells from early-stage colon cancer [58]. WNT family members are highly conserved, secreted signaling molecules that play important roles in both tumorigenesis and normal development and differentiation. Study of Hibi et al. evidences that genistein treatment affected the DNA methylation of WNT5a, and that WNT5a downregulation is correlated with hypermethylation of its promoter in human colon cancer patients [60, 59].

Moreover, genistein may inhibit cancer progression by inducing apoptosis or inhibiting proliferation, and the mechanisms by which genistein exerts its anti-tumor effects has been the subject of considerable interest [61, 62, 63].

Genistein has been shown to induce epigenetic changes in several cancer cell lines and in the in vivo animal models. [64].

The presence of hydroxylated and methylated genistein metabolites correlated positively with inhibition of cancer cell proliferation, but genistein sulfates were not associated with antiproliferative effects of genistein, suggesting that some types of metabolism of the isoflavones may be crucial for their action.

Genistein is a known inhibitor of protein-tyrosine kinase (PTK), which may attenuate the growth of cancer cells by inhibiting PTK-mediated signaling mechanisms [65]. Sakla et al. (2007) recently reported that genistein inhibits the protooncogene HER-2 protein tyrosine phosphorylation in breast cancer cells as well as delaying tumor onset in transgenic mice that overexpress the HER-2 gene. These data support its potential anti-cancer role in chemotherapy of breast cancer. However, effects independent of this activity have also been demonstrated [66, 67].

Soy isoflavone supplemented diets also prevented the development of adenocarcinomas in the prostate and seminal vesicles in a rat carcinogenesis model [68].

Phytoestrogens, present in soy based food, may act through hormonal mechanisms to reduce cancer risk by binding to estrogen receptors (ER) or interacting with enzymes involved in sex steroid biosynthesis and metabolism [53]. Moreover, genistein may inhibit cancer progression by inducing apoptosis or inhibiting proliferation, and the mechanisms by which genistein exerts its anti-tumor effects have been the subject of considerable interest [61, 62, 63].

Studies demonstrate that ERβ is highly expressed in superficial and crypt epithelium of the normal colon in both genders. ERβ expression was highly correlated among all cell types in both genders, and the strongest correlation was observed between surface and crypt ERβ expression. This finding suggests that there may be an intersubject difference in ERβ expression that is manifested in all cell types. ERβ expression was significantly lower in colon cancer cells compared with normal colonic epithelium, and there was a progressive decline in ERβ expression that paralleled the loss of cancer cell differentiation. The present findings are consonant with previous results reported by Foley and colleagues [69], who also detected a loss of ERβ protein expression in malignant colon tissue by western immunoblotting. Another immunohistochemical study of ERβ in 55 patients with colorectal adenocarcinomas showed that 32% of all tumors in both genders were ERβ-negative; the 10% cut-off threshold was used to distinguish ERβ-positive from negative tumors [70].

Studies conducted with ER subtype-specific ligands and those performed with estrogen receptor b-knockout mice (ERβKOs) have illustrated the involvement of ERβ in cellular anti-inflammatory pathways and tissue homeostasis in the colon. These results suggest that ERβ-specific ligands may be promising targets in the pharmaceutical and therapeutical treatment of inflammatory bowel disease and the prevention of CRC. ERβKOs suggest that ERβ-specific agonists and ERβ-selective phytoestrogens like genistein (GEN) and coumestrol may serve as potential regulators of intestinal tissue homeostasis [71, 72, 73].

Schleipen et al. investigate the influence of ERα and ERβ-specific agonists, and of genestein on cell proliferation and apoptosis of the small intestine and the colon. Recent data indicate that ERβ-specific agonists and GEN inhibit epithelial proliferation of the prostate and mammary gland, and can even impede prostate cancer development [74, 76, 75]. It can therefore be assumed that ERβ-specific agonists may also inhibit the proliferation of the intestinal epithelium. To prove this hypothesis in the study, ovariectomized rats were treated with 17β-Estradiol (E2), the phytoestrogen GEN and ER subtype-specific agonists for ERα and ERβ for 3 weeks.

Genistein has been shown to induce epigenetic changes in several cancer cell lines and in *in vivo* animal models [64]. Recent studies show that genistein may affect DNA methylation, serves as a natural demethylation agent, and is specifically effective on colon cancer cells from early-stage colon cancer [58].

WNT family members are highly conserved, secreted signaling molecules that play important roles in both tumorigenesis and normal development and differentiation. Study of Hibi *et al.* evidence that genistein treatment affected the DNA methylation of *WNT5a*, and that *WNT5a* down-regulation is correlated with hypermethylation of its promoter in human colon cancer patients [59, 60]. Aberrant WNT signaling is considered one of the most correlated factors in over 90% of both benign and malignant colorectal tumors [77].

Many epigenetic silencing and activating events have been discovered in the WNT pathway that are also related to aberrant WNT signaling, including aberrant expression of *sFRP1*, *DKK1*, and *APC* [78, 79]. Therefore, Wang and Chen investigate the effect of genistein on WNT pathway regulation in colon cancer development [58]. This study showed that: genistein treatment selectively induced *WNT5a* expression in specific colon cancer cell lines; *WNT5a* showed the lowest expression compared to other more advanced tumor cell lines; and the novel finding that *WNT5a* mRNA expression was upregulated by genistein in this early-stage colon cancer cell line.

These results support the notion that genistein serves as a natural demethylation agent and that it is specifically effective on colon cancer cells from early-stage colon cancer. Genistein treatment affected the DNA methylation of *WNT5a*. It has been shown that *WNT5a* downregulation is correlated with hypermethylation of its promoter in human colon cancer patients [59, 60].

Wang and Chen studies showed that the time dependent induction of WNT5a by genistein in colon cancer cell line SW 1116 was correlated with decreased methylation of a CpG island within its promoter, as determined by bisulfate sequencing [58].

Demethylation of CpGs inhibition of Dnmt and MBD2 activity, and activation of the histones by acetylation and demethylation at the BTG3 promoter followed by genistein treatment, were observed in renal cancer cells [80]. Using the mouse differential methylation hybridization array, alteration of DNA methylation in specific genes in mice was observed following feeding of a diet containing genistein compared to that in mice fed a control casein diet [81].

Other direct evidence that genistein affected DNA methylation was that maternal exposure to dietary genistein altered the epigenome of offspring in viable yellow agouti (Avy/a) mice. Overall, the potential of genistein as an effective epigenome modifier, which may greatly impact CRC metastasis, highlights the potential ability of dietary genistein to improve CRC prognosis [82].

Downregulation by promoter hypermethylation occurs in cell lines from earlier stages of colon cancer but not in cell lines from later stages.

These findings suggest that maintaining the expression of ER-β is crucial in mediating the growth-suppressive effects of soy isoflavones against colon tumors. Upregulation of ER-β by specific foodborne ER-ligands, such as soy isoflavones, could potentially be a dietary prevention strategy for colon cancer. [55].

Genistein has been shown to inhibit cancer metastasis through its ability to regulate nearly every step of the metastatic cascade, including cell adhesion, migration invasion, and angiogenesis. The effect of genistein on the metastaic cascade involves many metastasis suppressor or related signaling pathways, such as NFKappaB. Genistein can affect both of these processes, as well as modulate key regulatory protein such as Akt and nuclear factor κB (NF-κB). In general, low-to-mid micro molar concentrations of genistein are required for these effects in cell-culture-based models, although, interestingly, effects in animal models have been observed at lower concentrations. Genistein inhibits critical pathways in cancer invasion and can specifically target MEK4. This inhibition results in inactivation of the MEK4 pathway, decreased MMP-2 production, and decreased cell invasion. Genistein also activates Smad1, which is activated by the endoglin signaling pathway, and causes decreased cell invasion. Additionally, genistein inhibits FAK activation, resulting in increased cell adhesion. At this time, it is unclear whether the activation of Smad1 and FAK are due to genistein's inhibition of MEK4 or via a different signaling mechanism [83].

Several reports have demonstrated that genistein can induce cell cycle arrest and that it can therapeutically modulate key regulator cell cycle proteins at concentrations ranging from 5 to 200 μM [84]. It is important to note that these concentrations are greater than the blood levels that are observed with dietary consumption, indicating that this is likely not the primary mechanism by which genistein inhibits metastasis. However, it is theoretically possible to achieve these levels in humans, and various animal studies have also demonstrated that genistein can reduce the primary tumor size in certain contexts.

Studies by Wentao et al. show that genistein inhibits EGF induced loss of FOXO3 activity by targeting the PI3K/ Akt pathway. Downstream, genistein inhibits EGF induced FOXO3 disassociation from p53(mut), which further promotes FOXO3 activity and leads to increased expression of the p27kip1 cell cycle inhibitor, which inhibits proliferation in colon cancer cells. The author demonstrated that one of the anti-proliferative mechanisms of genistein in colon cancer cells is to promote FOXO3 activity by inhibiting EGF-induced FOXO3 phosphorylation (inactivation) via the PI3K/Akt pathway. Active FOXO3 negatively regulates proliferation of colon cancer cells and shows that its inactivation is an essential step in EGF-mediated proliferation [85, 86].

5. Conclusion

Several studies shown that consumption of fiber, fresh fruit and vegetables, a high-calcium diet could have a protective effect on the increased risk of colorectal cancer, and suggest that much of the suffering and death from cancer could be prevented by consuming a healthy diet, reducing tobacco use, performing regular physical activity, and maintaining an optimal body weight [5].

Soy is one of the most consumed foods worldwide. Soy foods contain larger amounts of phytoestrogens of which the isoflavon genistein is surely the biologically most important.

This compound, in recent years, has received much attention in the field of oncology research, as it exerts a wide range of biological effects of direct relevance to cancer.

Phytoestrogens and in particular genistein, have shown to be an important tool for the inhibition of cancer metastasis, exerting effects on both the initial steps of primary tumor growth as well as the later steps of the metastatic cascade.

The international literature suggests that phytoestrogens have potentially a high clinical impact and the expansion of knowledge on soy, soy foods, and soy products will lead to novel future developments in the field of cancer treatment.

Phytoestrogens in soy foods	
Foods	Total isoflavons (mg/100 g)
Miso	41.45
Natto	82.29
Roasted soybeans	148.50
Soy beans	154.53
Soy cheese american	17.95
Soy flour (textured)	172.55
Soy milk	10-200
Soy milk curd, dried	83.30
Soy milk fortified or unfortified	10.73
Soy milk skin or film (Foo jook or yuba), cooked	44.67
Soy milk skin or film (Foo jook or yuba), raw	196.05
Soy protein concentrate	94.65
Soy protein drink	81.65
Soy protein isolate	91.05
Soy yogurth	33.17
Tempeh	60.61
Tofu (dried frozen)	83.20
Tofu raw regular with calcium and sulphate	22.73
Tofu yogurt	16.30

Table 1. Isoflavone Content of Selected Soy Foods (USDA Database 2008)

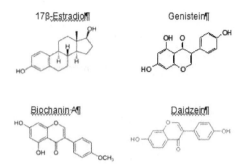

Figure 1. Chemical structures of soy phytoestrogens are similar to17 beta estadiol

Author details

B. Pampaloni[1], C. Mavilia[1], E. Bartolini[1], F. Tonelli[2], M.L. Brandi[1] and Federica D'Asta[3]

1 Department of Internal Medicine, School of Medicine, University of Florence, Florence, Italy

2 Department of Clinical Physiopathology, School of Medicine, University of Florence, Florence, Italy

3 Department of Neurosciences, Psychology, DrugArea and Child Health, School of Medicine, University of Florence, Florence, Italy

References

[1] Parkin DM, Whelan SL, Ferlay L et al Cancer incidence in five continents, vol VII. IARC Scientific Publications No.143. Lyon: IARC,1997

[2] Jemal A, Siegel R, Ward E et al: Cancer statistics, 2006. CA Cancer J Clin 2006;56:107

[3] Whittemore As, Wu-Williams AH, Lee M et al. Diet, physical activity, and colorectal cancer among Chinese in North America and China. J Natl Cancer Inst 1990, 82:915

[4] Lieberman DA, Weiss DG, Bond JH et al. Use of colonoscopy to screen asymptomatic adults for colorectal cancer. Veterans Affairs Cooperative Study Group 380. N Engl J Med 2000;343:162

[5] Libutti SK, Saltz LB, Rustgi AK, Tepper JE. Cancer of the colon in DeVita VT, Hellman S, Rosenberg SA eds Cancer: Principles & Practis of Oncology. 7th Edition, Lippincott &W, Philadelphia, 2005, pag 1062

[6]   Chan Andrew T., Giovannucci Edward L. "Primary Prevention of Colorectal Cancer" Gastroenterology 2010;138:2029–2043

[7]   Gupta Subash C., Ji Hye Kim, Sahdeo Prasad & Bharat B. Aggarwal "Regulation of survival, proliferation, invasion, angiogenesis, and metastasis of tumor cells through modulation of inflammatory pathways by nutraceuticals" Cancer Metastasis Rev (2010) 29:405–434

[8]   Fuchs Cs, Giovannucci EL, Colditz GA et al. A prospective study of family history and the risk of colorectal cancer. New Engl J Med 1994;331:1669

[9]   Alberts DS, Martinz ME, Roe DE et al. Lack of effect of high-fiber cereal supplement on the recurrence of colorectal adenomas. Phoenix Colon Cnacer Prevention Physicians'Network. N Engl J Med 2000;342:1156

[10] O'Connell JB, Maggard MA, Liu JH et al. Rates of colon and rectal cancers are increasing in young adults. Am Surg 2003; 69: 866-872

[11] Welfare M, Monesola Adeokun A, Bassendine MF, Daly AK. Cancer Epidemiol Biomarkers Prev 1999;8:289

[12] Vogelstein B, Fearon ER, Hamilton SR et al. Genetic alterations during colorectal tumor development. N Engl J Med 1988; 319: 525

[13] Grady WM. Genomic instability and colon cancer. Cancer Metast Rev 2004;23:11

[14] Jacoby RF, Seibert K, Cole CE et al. The cyclooxygenase-2 inhibitor celecoxib is a potent preventive and therapeutic agent in the Min mouse model of adenomatous polyposis. Cancer Res 2000;60:5040-44

[15] Giardiello FM, Hamilton SR, Krush AJ et al. Treatment of colonic and rectal adenomas with sulindac in familial adenomatous polyposis. N Engl J Med 1993;328:1313-16

[16] Baron JA, Cole BF, Sandler RS et al. A randomized trial of aspirin to prevent colorectal adenomas N Engl J Med 2003; 348:891-99

[17] Sheng H, Shao J, Kirkland SC et al. Inhibition of human colon cancer cell growth by selective inhibition of cyclooxygenase-2 J Clin Invest. 1997 May 1;99(9):2254-9.

[18] Picariello L, Brandi Ml, Formigli L et al. Apoptosis induced by sulindac sulfide in epithelial and mesenchymal cells frm human abdominal neoplasms. Eur J Pharmacol 1998;360:105-112

[19] Boon EMJ, Keller JJ, Wormhoudt TAM et al. Sulindac targets nuclear beta-catenin accumulation and Wnt signaling in adenomas of patients with familial adenomatous polyposis and in human colorectal cancer cell lines Br J Cancer. 2004 Jan 12;90(1): 224-9

[20] Calle EE, Miracle-McMahill HL, Thorn MJ, Heath CW. Estrogen replacement therapy and risk of fatal colon cancer in a prospettive color of post-menopausal women. J Natl cancer Inst 1995;87:517-523

[21] Koo JH, Leong RWL. Sex differences in epidemiologic, clinical and pathological characteristics of colorectal cancer. J Gastroenterol Hepatol 2010; 25:33-42

[22] Jacobs EJ, White E, Weiss NS. Exogenous hormones, reproductive history and colon cancer. Cancer Causes Contr 1994;5:359-366

[23] Doll, R., & Peto, R. (1981). The causes of cancer: Quantitative estimates of avoidable risks of cancer in the United States today. Journal of the National Cancer Institute, 66, 1191–1308. 10.

[24] Hardy, G., Hardy, I., & Ball, P. A. (2003). Nutraceuticals—A pharmaceutical viewpoint: Part II. Current Opinion in Clinical Nutrition and Metabolic Care, 6, 661–671

[25] Wickia A. & Hagmannc J., "Diet and cancer" Swiss Med Wkly. 2011;141:w13250

[26] Block G, Patterson B, Subar A. Fruit, vegetables, and cancer prevention: a review of the epidemiological evidence. Nutr Cancer. 1992;18:1–29.

[27] Freedman ND, Park Y, Subar AF, et al. Fruit and vegetable intake and head and neck cancer risk in a large United States prospective cohort study. Int J Cancer. 2008;122:2330–6.

[28] Levi F, Pasche C, Lucchini F, La Vecchia C. Dietary fibre and the risk of colorectal cancer. Eur J Cancer 2001;37:2091–2096.

[29] Trock B, Lanza E, Greenwald P. Dietary fiber, vegetables, and colon cancer: critical review and meta-analyses of the epidemiologic evidence. J Natl Cancer Inst 1990;82:650–61.

[30] Howe GR, Benito E, Castelleto R, et al. Dietary intake of fiber and decreased risk of cancers of the colon and rectum: evidence from the combined analysis of 13 case-control studies. J Natl Cancer Inst 1992;84:1887–1896.

[31] Witte JS., Matthew PL, Cristy LB, Eric R.L, Harold D.F and Robert W. H. Relation of Vegetable, Fruit, and Grain Consumption to Colorectal Adenomatous Polyps; Am J Epidemiol. 1996 Dec 1;144(11):1015-25.

[32] Magalhães B, Peleteiro B, Lunet N. Dietary patterns and colorectal cancer: systematic review and meta-analysis.Eur J Cancer Prev. 2012 Jan;21(1):15-23.

[33] Adlercreutz, H. (1990). Western diet and Western diseases: some hormonal and biochemical mechanisms and associations. Scand J Clin Lab Invest 50 (Suppl. 201), 3– 23.

[34] Waladkhani, A. R., & Clemens, M. R. (1998). Effect of dietary phytochemicals on cancer development. Int J Med 1, 747– 753.

[35] Spector D, Anthony M, Alexander D, Arab L. "Soy consumption and colorectal cancer". Nutr Cancer 2003;47:1–12

[36] Adlercreutz, H. (1995). Phytoestrogens: epidemiology and a possible role in cancer protection. Environ Health Perspect 103 (suppl. 7), 103– 112.

[37] Knight, D. C., & Eden, J. A. (1996). A review of the clinical effects of phytoestrogens. Obstet Gynecol 87, 897–904.

[38] Knekt P, Järvinen R, Seppänen R, Hellövaara M, Teppo L, Pukkala E, Aromaa A. Dietary flavonoids and the risk of lung cancer and other malignant neoplasms. Am J Epidemiol. 1997 Aug 1;146(3):223-30.

[39] Birt DF, Hendrich S, Wang W. Dietary agents in cancer prevention: flavonoids and isoflavonoids. Pharmacol Ther. 2001 May-Jun;90(2-3):157-77.

[40] Wu, A. H., Ziegler, R. G., Nomura, A. M. Y., West, D. W., Kolonel, L. N., Horn-Ross, P. L., Hoover, R. N., & Pike, M. C. (1998). Soy intake and risk of breast cancer in Asians and Asian Americans. Am J Clin Nutr 68, 1437S–1443S.

[41] Zhang Y, Wang GJ, Song TT, Murphy PA, Hendrich SUrinary disposition of the soybean isoflavones daidzein, genistein and glycitein differs among humans with moderate fecal isoflavone degradation activity. J Nutr. 1999 May;129(5):957-62.

[42] Watanabe S, Yamaguchi M, Sobue T, Takahashi T, Miura T, Arai Y, Mazur W, Wähälä K, Adlercreutz H. Pharmacokinetics of soybean isoflavones in plasma, urine and feces of men after ingestion of 60 g baked soybean powder (kinako). J Nutr. 1998 Oct; 128(10):1710-5.

[43] Zhang Y, Song TT, Cunnick JE, Murphy PA, Hendrich S. Daidzein and genistein glucuronides in vitro are weakly estrogenic and activate human natural killer cells at nutritionally relevant concentrations. J Nutr. 1999 Feb;129(2):399-405

[44] Peterson TG, Ji GP, Kirk M, Coward L, Falany CN, Barnes S. Metabolism of the isoflavones genistein and biochanin A in human breast cancer cell lines. Am J Clin Nutr. 1998 Dec;68(6 Suppl):1505S-1511S.

[45] Barone M., Katia Lofano, Nicola De Tullio, Raffaele Licino, Francesca Albano, Alfredo Di Leo. "Dietary Endocrine and metabolic factors in the development of colorectal cancer" J Gastrointest Canc 2012, 43: 13-19

[46] Barone M., Sabina Tanzi, Katia Lofano, Maria Principia Scavo, Raffaella Guido, Lucia Demarinis, Maria Beatrice Principi , Antongiulio Bucci and Alfredo Di Leo Estrogens, phytoestrogens and colorectal neoproliferative lesions, 2008, Genes Nutr (2008) 3:7–13

[47] Nüssler NC, Reinbacher K, Shanny N, Schirmeier A, Glanemann M, Neuhaus P, Nussler AK, Kirschner M. Sex-specific differences in the expression levels of estrogen receptor subtypes in colorectal cancer. Gend Med. 2008 Sep;5(3):209-17.

[48] Fiorelli G, Picariello L, Martineti V et al. Functional estrogen receptor beta in colon cancer cells. Biochem Biophys Res Communic 1999; 261: 521-527

[49] Picariello L, Fiorelli G, Martineti V et al. Anticancer Res 2003; 23:2419-24

[50] Martineti V, Picariello L, Tognarini I et al ERB is a potent inhibitor of cell proliferation in the HCT8 human colon cancer cell line through regulation of cell cycle components. Endocrine-Related Cancer 2005; 12:455-469

[51] Wiseman H (1999) The bioavailability of non-nutrient plants factors: dietary flavonoids and phytoestrogens. Proc Nutr Soc 58(1):139–146

[52] Xu X, Harris KS, Wang HJ, Murphy PA, Hendrich S (1995) Bioavailability of soybean isoflavones depends upon gut microflora in women. J Nutr 125(9):2307–2315

[53] Cotterchio Michelle, Beatrice A. Boucher, Michael Manno, Steven Gallinger, Allan Okey, and Patricia Harper. "Dietary Phytoestrogen Intake Is Associated with Reduced Colorectal Cancer Risk". J. Nutr. 136: 3046–3053, 2006

[54] Cross Heide S., Eniko° Ka' llay, Daniel Lechner, Waltraud Gerdenitsch, Herman Adlercreutz, H and H. James Armbrecht Phytoestrogens and Vitamin D Metabolism: A New Concept for the Prevention and Therapy of Colorectal, Prostate, and Mammary Carcinomas 2004 American Society for Nutritional Sciences.

[55] Bielecki A, Roberts J, Mehta R, Raju J. "Estrogen receptor-β mediates the inhibition of DLD-1 human colon adenocarcinoma cells by soy isoflavones". Nutr Cancer. 2011;63(1):139-50

[56] Chrzan BG, Bradford PG. (2007). Phytoestrogens activate estrogen receptor β1 and estrogenic responses in human breast and bone cancer cell lines. Mol Nutr Food Res 51: 171–177.

[57] Murata IC, Itoigawa M, Nakao K, Kumagai M, Kaneda N, Furukawa H. (2006). Induction of apoptosis by isofl avonoids from the leaves of Millettia taiwaniana, in human leukemia HL-60 cells. Planta Med 72 (5): 424–429.

[58] Wang Z. and H. Chen, "Genistein Increases Gene Expression by Demethylation of WNT5a Promoter in Colon Cancer Cell Line SW1116" Anticancer Research 30: 4537-4546 (2010)

[59] Ying, J, Li, H, Yu, J, Ng, KM, Poon, FF, Wong, SCC et al. "WNT5A exhibits tumor-suppressive activity through antagonizing the Wnt/β-catenin signaling, and is frequently methylated in colorectal cancer". Clinical Cancer Res 14(1): 55-61, 2008 54,

[60] Hibi K, Mizukami H, Goto T, Kitamura Y, Sakata M, Saito M et al: WNT5A gene is aberrantly methylated from the early stages of colorectal cancers. Hepatogastroenterology 56(93): 1007- 1009, 2009.

[61] Wang HK. "The therapeutic potential of flavonoids". Expert Opinion on Investigational Drugs. 2000;9(9):2103–2119

[62] Barnes S, Peterson TG. "Biochemical targets of the isoflavone genistein in tumor cell lines". Proceedings of the Society for Experimental Biology and Medicine Society for Experimental Biology and Medicine New York, NY. 1995; 208(1):103–108

[63] Sarkar FH, Li Y. "Soy isoflavones and cancer prevention". Cancer Investigation. 2003; 21(5):744–757)

[64] Zhang Yukun and Hong Chen "Genistein attenuates WNT signaling by up-regulating sFRP2 in a human colon cancer cell line" Experimental Biology and Medicine 2011; 236: 714–722).

[65] Szkudelska K., L. Nogowski, Genistein – a dietary compound inducing hormonal and metabolic changes, J. Steroid. Biochem. Mol. Biol. 105 (2007) 37–45.

[66] Sakla M.S., N.S. Shenouda, P.J. Ansell, R.S. Macdonald, D.B. Lubahn, Genistein affects HER2 protein concentration, activation, and promoter regulation in BT-474 human breast cancer cells, Endocrine 32 (2007) 69–78.

[67] Abler A., J.A. Smith, P.A. Randazzo, P.L. Rothenberg, L. Jarett, Genistein differentially inhibits postreceptor effects of insulin in rat adipocytes without inhibiting the insulin receptor kinase, J. Biol. Chem. 267 (1992) 3946–3951

[68] . Onozawa M., T. Kawamori, M. Baba, K. Fukuda, T. Toda, H. Sato, et al., Effects of a soybean isoflavone mixture on carcinogenesis in prostate and seminal vesicles of F344 rats, Jpn. J. Cancer Res. 90 (1999) 393–398.

[69] Foley EF, Jazaeri AA, Shupnik MA, Jazaeri O, Rice LW. Selective loss of estrogen receptor beta in malignant human colon. Cancer Res 2000, 60, 245–248.

[70] Witte D, Chirala M, Younes A, Li Y, Younes M. Estrogen receptor beta is expressed in human colorectal adenocarcinoma. Hum Pathol 2001, 32, 940–944

[71] Harris,H.A. et al. (2003) Evaluation of an estrogen receptor-beta agonist in animal models of human disease. Endocrinology, 144, 4241–4249.

[72] Wada-Hiraike,O. et al. (2006) New developments in oestrogen signalling in colonic epithelium. Biochem. Soc. Trans., 34, 1114–1116. ,Weige,C.C. et al. (2009) Estradiol alters cell growth in nonmalignant colonocytes and reduces the formation of preneoplastic lesions in the colon. Cancer Res., 69, 9118–9124.

[73] Weige,C.C. et al. (2009) Estradiol alters cell growth in nonmalignant colonocytes and reduces the formation of preneoplastic lesions in the colon. Cancer Res., 69, 9118–9124.

[74] Lechner,D. et al. (2005) Phytoestrogens and colorectal cancer prevention. Vitam. Horm., 70, 169–198;

[75] Mak,P. et al. (2010) ERβeta impedes prostate cancer EMT by destabilizing HIF-1alpha and inhibiting VEGF-mediated snail nuclear localization: implications for Gleason grading. Cancer Cell, 17, 319–332

[76] 76. Sarkar,F.H. et al. (2006) The role of genistein and synthetic derivatives of isoflavone in cancer prevention and therapy. Mini Rev. Med. Chem., 6, 401–407

[77] Giles RH, van Es JH and Clevers H: Caught up in a Wnt storm: Wnt signaling in cancer. Biochim Biophys Acta 1653(1): 1-24, 2003.

[78] Aguilera O, Fraga MF, Ballestar E, Paz MF, Herranz M, Espada J et al: Epigenetic inactivation of the Wnt antagonist DICKKOPF-1 (DKK-1) gene in human colorectal cancer. Oncogene 25(29): 4116-4121, 2006.

[79] Suzuki H, Watkins DN, Jair KW, Schuebel KE, Markowitz SD, Chen WD et al: Epigenetic inactivation of SFRP genes allows constitutive WNT signaling in colorectal cancer. Nat Genet 36(4): 417-422, 20?.

[80] Shahana Majid, Altaf A.Dar1, Ardalan E.Ahmad, Hiroshi Hirata, Kazumori Kawakami, Varahram Shahryari, Sharanjot Saini, Yuichiro Tanaka, Angela V.Dahiya, Gaurav Khatri and Rajvir Dahiya BTG3 tumor suppressor gene promoter demethylation, histone modification and cell cycle arrest by genistein in renal cancer. Carcinogenesis vol.30 no.4 pp.662–670, 2009

[81] Kevin J. Day et al Genistein Alters Methylation Patterns in Mice 2002 American Society for Nutritional Sciences]

[82] Qian Li and Hong Chen "Epigenetic modifications of metastasis suppressor genes in colon cancer metastasis" Epigenetics 6:7, 849-852; July 2011

[83] Pavese Janet M. & Rebecca L. Farmer & Raymond C. Bergan: Inhibition of cancer cell invasion and metastasis by genistein. Cancer Metastasis Rev (2010) 29:465–482

[84] Ramos, S. (2007). Effects of dietary flavonoids on apoptotic pathways related to cancer chemoprevention. The Journal of Nutritional Biochemistry, 18(7), 427–442

[85] Wentao Qi1, Christopher R Weber2*, Kaarin Wasland1 and Suzana D Savkovic1: Genistein inhibits proliferation of colon cancer cells by attenuating a negative effect of epidermal growth factor on tumor suppressor FOXO3 activity. BMC Cancer 2011, 11:219

[86] Dijkers PF, Medema RH, Pals C, Banerji L, Thomas NS, Lam EW, Burgering BM, Raaijmakers JA, Lammers JW, Koenderman L, Coffer PJ: Forkhead transcription factor FKHR-L1 modulates cytokine-dependent transcriptional regulation of p27(KIP1). Molecular and Cellular Biology 2000, 20(24):9138-914 36

Bowman-Birk Protease Inhibitor as a Potential Oral Therapy for Multiple Sclerosis

Farinaz Safavi and Abdolmohamad Rostami

Additional information is available at the end of the chapter

1. Introduction

1.1. The Bowman-Birk Protease Inhibitor (BBI)

Legume seeds contain different kinds of proteins and protease inhibitors. Serine proteases are a large sub-group of the protease family [1] and they play a role in various pathological conditions such as cancer and thrombotic and inflammatory diseases [2]. Thus they are excellent targets for treatment of many disorders.

Various plant species and, in particular, legumes contain a great number of serine protease inhibitors. The Bowman-Birk protease inhibitor belongs to a family of serine protease inhibitors that has been widely studied for the past 60 years [3, 4].

The soybean-derived Bowman-Birk protease inhibitor (BBI) is a small protein consisting of 71 amino acids and 7 disulfide bonds [4]. BBI is a double-headed serine protease inhibitor, with two functional active sites at opposite sides of the molecule, which inhibits both trypsin and chymotrypsin-like proteases [1,3] (Figure 1). It is a water-soluble protein that is resistant to acidic conditions and proteolytic enzymes [3]. These characteristics make it a good candidate for use as an oral agent for therapeutic purposes.

Crude soybean contains a small amount of BBI and may have components that counter some of the beneficial effects of BBI. Bowman-Birk Inhibitor Concentrate (BBIC) is a soybean extract enriched in BBI [5]. Researchers prefer to use BBIC in their studies because a smaller amount of BBIC contains the proposed dose of BBI compared to crude soybean.

In rodents, BBI is detectable in the blood, tissue and urine after ingestion [6]. Interestingly, BBI can be detected in the central nervous system (CNS) of animals even when the blood-brain barrier is intact. In human studies, although the BBI level could not be detected in blood after oral BBIC dosing, it could be measured in urine [6].

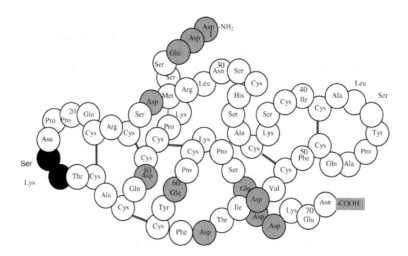

Figure 1. Crystal structure of soybean-derived Bowman-Birk protease inhibitor

The ability of certain serine protease inhibitors to prevent the malignant transformation of cells was shown two decades ago [7, 8]. BBI prevents/suppresses carcinogenesis in a variety of in vitro and in vivo systems [8, 9].

Several human clinical trials to evaluate the effect of BBIC have been completed or are in progress [10-12]. To date, in completed clinical trials, neither toxicity nor neutralizing anti-bodies against BBIC have been reported in patients receiving BBIC [9].

2. Multiple Sclerosis (MS)

MS is the second cause of disability in young adults and is considered to be a demyelinating disease of the central nervous system (CNS) along with chronic inflammation, demyelina-tion and gliosis [13]. Lesions are characterized by periventricular cuffing and infiltration consisting mainly of T lymphocytes and macrophages, leading to myelin destruction. Re-cently neuronal degeneration and axonal involvement have also been shown in MS lesions [14]. Current findings therefore raise some doubts about the original assumption that MS is exclusively a white matter disease.

Based on the MS inflammatory phenotype, it has been considered an autoimmune disorder in which peripherally activated myelin-reactive T cells enter the CNS and begin an immuno-logic cascade that subsequently causes myelin damage.

Activated antigen presenting cells (APCs) and auto-reactive T cells produce pro-inflamma-tory cytokines, including IL-23, IFN-γ, TNF-α, IL-17, that enhance cell-mediated immunity in the CNS [15-17]. Conversely, other cytokines, such as IL-10, IL-27, IL-4 and TGF-β, play an immunoregulatory role and may be protective in MS [17-20].

Despite extensive research, only a few pharmacotherapeutic agents (e.g., IFN-β, glatiramer acetate, and mitoxantrone) are available, all of which are administered by injection, demon-strate mild to moderate efficacy and have potential side effects [21,22]. A new oral therapeu-tic agent (Fingolimod) was approved by the FDA and shows potential benefits in MS patients [23].

Recently, in two phase three clinical trials, BG-12 (dimethyl fumarate), a newly proposed or-al drug for the treatment of multiple sclerosis, showed a significant reduction in relapse rate and number of MRI lesions in treated patients compared to the placebo group [68, 69].

Development of a new, effective and oral therapy for MS, with fewer side effects, is there-fore desirable.

3. Experimental Autoimmune Encephalomyelitis (EAE)

Experimental Autoimmune Encephalomyelitis (EAE) is an autoimmune animal model of MS. Immunization with myelin peptides in different strains of mice induces chronic or re-lapsing types of the disease, which makes EAE a good tool for studying disease mechanisms and testing therapeutic agents [24] To date, three of the four therapies currently approved for MS were first tested in this animal model [24].

After immunization with myelin protein, APCs present myelin on the surface of MHC II and produce pro-inflammatory cytokines. Dendritic cell-derived IL-12 and IL-23 lead to de-velopment of myelin-specific Th1 and Th17 cells, respectively. Th1 and Th17 cells are the two main culprits in pathogenesis of EAE and MS [25]. Auto-reactive T cells enter the CNS and facilitate recruitment of other immune cells such as monocytes and neutrophils. Accu-mulation of inflammatory cells within the CNS promotes myelin damage, axonal loss and clinical manifestations in affected animals [24].

Recently it has been shown that dendritic cells are also able to produce another cytokine from the IL-12 family called IL-27. Compared with IL-12 and IL-23, IL-27 elicits different im-munoregulatory effects. IL-27 inhibits encephalitogenicity of T cells and suppresses EAE disease [26]. In addition, it stimulates IL-10 production in T cells and induces Tr1 cells [17]. IL-10 is a widely studied immunoregulatory cytokine, which virtually all immune cells are able, in different conditions, to release and which suppresses inflammatory response [27]. IL-10 also plays a significant role in suppression of EAE [28-30].

In general, if a therapeutic agent is able to stimulate IL-10 production and Tr1 cells, it could be an excellent candidate for MS therapy.

4. Proteases in inflammation

Several proteases are associated with the pathogenesis of inflammatory disorders [24, 31]. Proteolytic enzymes are involved in activation and migration of immune cells, cytokine and chemokine activation/inactivation and complement function [32].

Various studies demonstrate that neutrophil serine proteases induce proinflammatory activity of both IL-32 and IL-33 cytokines [33, 34]. They are also able to convert inactive forms of IL-1 and IL-18 to the active form of these cytokines [35]. Cytotoxic T cell-derived proteases called granzymes are also involved in inflammation. Granzymes promote T cell entry into the site of inflammation. In addition, they stimulate B cell proliferation [36].

The complement cascade contains different enzymes that activate each other and proteases that play a role in initiation of the cascade, which results in formation of the membrane attack complex [37].

In general, proteases are involved in all aspects of the immune response and play a significant role in inflammation.

5. Proteases in pathogenesis of EAE and MS

Modulators of neuronal and endogenous proteolysis show a different pattern in spinal cords of EAE rats compared to control animals. This finding indicates higher activity of some proteases in EAE than in control groups, which makes specific proteases good potential biomarkers for disease activity or therapeutic targets in the EAE model and MS [38]. Various types of proteases, including lysosomal proteases and matrix metalloproteinases (MMPs), are highly expressed in MS lesions [24, 39-42]. Serine proteases such as plasmin, cathepsin G, chymase and trypsin activate inert MMP proenzymes to their active forms [24, 41, 42].

GelatinaseB (MMP-9] increases the number of leukocytes entering the site of inflammation and promotes myelin breakdown [39, 43]. Plasmin is a serine protease that mainly participates in the coagulation cascade. It has been demonstrated that plasmin directly induces myelin destruction and demyelination [44].

Levels of gelatinase and tissue plasminogen activator (t-PA) are also increased in MS lesions and in the cerebrospinal fluid (CSF) of active MS patients [46, 47]. Reactive astrocytes and infiltrating lymphocytes, macrophages and microglia express MMP-2, MMP-9 and t-PA in early active MS plaques [24, 41, 45, 47].

6. Anti-inflammatory effects of BBI

BBI suppresses the function of several proteases such as leukocyte elastase, trypsin and human cathepsin G released from human inflammatory cells. [48-50]. Mast cell chymase stimulates migration of lymphocytes and purified T cells, and BBI inhibits this enzyme quite efficiently [49]. In addition, BBI significantly suppresses the chemotactic activity of chymase, thus suppressing lymphocyte migration [51].

Stimulated human polymorphonuclear leukocytes produce reactive oxygen species (super-oxide and hydrogen peroxide) that may damage cell membranes by reacting with phoso-pholipids to form peroxides [52]. BBI is able to suppress the production of reactive oxygen species and inhibits their destructive effects [53]. Macrophage-derived proteases and free radicals are also associated with inflammation. BBI down-regulates NO and PGE2 inflam-matory pathways in LPS-activated macrophages [54]. Activated macrophages also induce neurotoxicity in the CNS. Anti-inflammatory effects of BBI prevent macrophage-induced neurotoxicity [55].

Serine protease inhibitors can prevent conversion of pro-MMPs to enzymatically active forms [56, 57]. BBI inhibits generation of active MMP-1 and MMP-9 in vitro, and BBIC re-duces MMP-2 and -9 activity in supernatants of spleen cells [58].

The aforementioned mechanisms may be particularly relevant in the context of the patho-genesis of multiple sclerosis and myelin destruction in the CNS.

BBI may have significant immunomodulatory effects and can be an excellent potential can-didate for treatment of inflammatory and autoimmune diseases.

7. BBI and other protease inhibitors in treatment of inflammatory disease

The role of proteases in inflammation has been reviewed in previous sections. Based on the fact that proteases are actively involved in inflammation, they can be a good therapeutic tar-get in suppression of inflammatory response and treatment of inflammatory diseases.

RWJ-355871 is a synthetic protease inhibitor that effectively suppresses allergic inflammato-ry diseases of the respiratory system [59]. 4-(2-Aminoethyl) benzenesulfonyl fluoride (AEBSF) is another protease inhibitor that attenuates ovalbumin-induced allergic airway in-flammation in its animal model [60].

Several studies have reported that protease inhibitors diminish inflammatory response in in-flammatory bowel diseases [1]. Nafamostat is a serine protease inhibitor that suppresses dextran sulfate sodium-induced colitis and diminishes inflammatory infiltration in the colon [61]. BBI is able to suppress gland inflammation in the gastrointestinal tract and shows a strong anti-inflammatory effect in the acute colitis model [62]. In addition, in a completed clinical trial [12], BBI demonstrates anti-inflammatory effects and a degree of amelioration of clinical disease and remission rate in patients with ulcerative colitis. We have also shown that administration of oral BBIC significantly inhibits experimental autoimmune neuritis (EAN) in rats [63, 64].

All of the above findings show the potential immunomodulatory and therapeutic effect of BBI in autoimmune diseases.

8. Immunoregulatory effect of BBI in the EAE model

We have shown that oral treatment of BBIC in MBP-induced EAE in rats, reduces disease severity from clinical score 3 (complete hind limb paralysis) to less than 1 (flaccid tail) compared to control animals. In addition, BBIC treatment significantly diminished demyelination in the peripheral nerve tissue of treated animals [58]. We have also shown that both BBI and BBIC suppress clinical and pathologic manifestations of chronic and relapsing EAE in B6 and SJL mice. In addition, the therapeutic effect of oral BBI is dose-dependent, and oral administration of higher amounts of BBI inhibits EAE more efficiently [65] (Figure 1).

BBI treatment also decreased pathogenicity of myelin-reactive T cells and induced milder disease in the adoptively transferred EAE model (unpublished data).

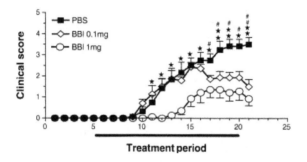

Figure 2. Effect of oral BBI compared to PBS treatment in EAE. Mice that received BBI showed significantly less severe disease compared to control group. The therapeutic effect of BBI is dose-dependent. Elsevier Publications Ltd. has kindly granted us permission to reproduce this figure from Touil et al., 2008.

BBI inhibits invasion of immune cells through the blood-brain barrier (BBB). BBI-treated mice showed dramatically lower numbers of CNS-infiltrating MNCs than control animals [58, 65, 66]. In addition, BBI suppresses generation of active MMP-1 and MMP-9 in vitro, and BBIC reduces MMP-2 and -9 activity in supernatants of spleen cells [58]. Consistent with other findings, BBI decreased migration of splenocytes in Boyden's chamber assay [65].

However, BBI may inhibit release of active MMP-2 and MMP-9 at the blood-brain barrier and prevent immune cell infiltration into the CNS; it might decrease expression of adhesion molecules on immune cells or invasiveness of immune cells, resulting in an altered cytokine pattern of inflammatory cells that hinders their migration from peripheral immune organs to the site of inflammation.

In order to clarify immunoregulatory mechanisms of BBI, the direct effect of BBI on immune cells was evaluated, and it was shown that splenocytes produce a higher amount of IL-10 following BBI treatment [65, 66]. Several reports have demonstrated the immunoregulatory effect of IL-10 in the EAE model of multiple sclerosis. To determine whether the immuno-modulatory effect of BBI depends on IL-10, we have compared the therapeutic effect of oral

BBI in EAE in WT and IL-10 KO mice. Although BBI-treated WT mice showed less severe disease, BBI treatment did not affect clinical disease in BBI-treated IL-10 KO mice compared to the control group [66] (Figure 3).

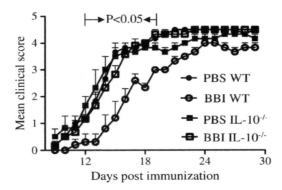

Figure 3. Effect of oral BBI compared to PBS treatment in EAE in WT and IL-10 KO mice. Although BBI-treated WT mice showed significantly less severe disease compared to the control group, there was no significant difference in treated and control IL-10 KO mice. Figure reproduced from Dai et al., 2012 with the kind permission of Elsevier Publications Ltd.

Different types of immune cells can release IL-10 cytokine [27]. However, BBI treatment induces IL-10 mainly in CD4$^+$ T cells [66]; it increases IL-10 production in CD8$^+$ T cells (unpublished data), demonstrating that BBI has a strong ability to activate IL-10 producing pathways in T cells. Exploring these underlying mechanisms will be a major focus of our future studies.

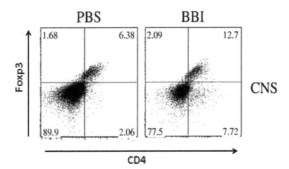

Figure 4. Higher expression of Foxp3$^+$ Treg cells in CNS infiltrating cells after oral treatment with BBI, Figure reproduced from Dai et al., 2012 with the kind permission of Elsevier Publications Ltd.

Treg cells are a subgroup of CD4$^+$ T cells that expresses the Foxp3$^+$ transcription factor. They produce IL-10 in the CNS and can suppress EAE disease [67]. Oral administration of BBI also induces Treg cells in the CNS, which might be one of the underlying mechanisms of the therapeutic effect of BBI in EAE [66] (Figure 4)

BBI also induces IL-10 in other types of effector T cells, and the immunomodulatory effect of BBI might be related to an increase in Tr1 cells. Should this be the case, BBI can be used to induce regulatory T cells and for treatment of autoimmune diseases such as multiple sclerosis.

9. Conclusion

BBI is a soybean-derived serine protease inhibitor. It can be administered orally with several immunomodulatory characteristics and no major side effects. Our observations have shown that BBI dramatically decreases severity of EAE and that its therapeutic effect is mediated through IL-10. In addition, BBI decreases infiltration of inflammatory cells across the BBB and inflammation in the CNS. BBI has potential as a safe and effective oral therapy for multiple sclerosis and other autoimmune diseases.

Acknowledgments

This study was supported by a grant from the National Institutes of Health (NIH 1R01AT005322). We are very grateful to Katherine Regan for editing the chapter.

Author details

Farinaz Safavi and Abdolmohamad Rostami

Department of Neurology, Thomas Jefferson University, Philadelphia, PA, USA

References

[1] Clemente A, Sonnante G, Domoney C. Bowman-Birk inhibitors from legumes and human gastrointestinal health: current status and perspectives. Curr Protein Pept Sci. 2011;12:358-373.

[2] Losso JN. The biochemical and functional food properties of the bowman-birk inhibitor. Crit Rev Food Sci Nutr. 2008;48:94-118. doi: 10.1080/10408390601177589.

[3] Birk Y. The Bowman-Birk inhibitor. Trypsin- and chymotrypsin-inhibitor from soybeans. Int J Pept Protein Res. 1985;25:113-131.

[4]   Odani S, Ikenaka T. Studies on soybean trypsin inhibitors. 8. Disulfide bridges in soybean Bowman-Birk proteinase inhibitor. J Biochem. 1973;74:697-715.

[5]   Kennedy, A. R. 1993. Overview: anti-carcinogenic activity of protease inhibitors. In Protease inhibitors as cancer chemopreventive agents. W. Troll, Kennedy A. R., ed. Plenum Press, New York. 9-64.

[6]   Wan XS, Lu LJ, Anderson KE, Ware JH, Kennedy AR. Urinary excretion of Bowman-Birk inhibitor in humans after soy consumption as determined by a monoclonal antibody-based immunoassay. Cancer Epidemiol Biomarkers Prev. 2000;9:741-747.

[7]   Kennedy AR, Little JB. Protease inhibitors suppress radiation-induced malignant transformation in vitro. Nature. 1978;276:825-826.

[8]   Yavelow J, Collins M, Birk Y, Troll W, Kennedy AR. Nanomolar concentrations of Bowman-Birk soybean protease inhibitor suppress x-ray-induced transformation in vitro. Proc Natl Acad Sci U S A. 1985;82:5395-5399.

[9]   Kennedy AR. Chemopreventive agents: protease inhibitors. Pharmacol Ther. 1998;78:167-209.

[10] Malkowicz SB, McKenna WG, Vaughn DJ, Wan XS, Propert KJ, Rockwell K, et al. Effects of Bowman-Birk inhibitor concentrate (BBIC) in patients with benign prostatic hyperplasia. Prostate. 2001;48:16-28. doi: 10.1002/pros.1077.

[11] Armstrong WB, Kennedy AR, Wan XS, Taylor TH, Nguyen QA, Jensen J, et al. Clinical modulation of oral leukoplakia and protease activity by Bowman-Birk inhibitor concentrate in a phase IIa chemoprevention trial. Clin Cancer Res. 2000;6:4684-4691.

[12] Lichtenstein GR, Deren JJ, Katz S, Lewis JD, Kennedy AR, Ware JH. Bowman-Birk inhibitor concentrate: a novel therapeutic agent for patients with active ulcerative colitis. Dig Dis Sci. 2008;53:175-180. doi: 10.1007/s10620-007-9840-2.

[13] Peterson LK, Fujinami RS. Inflammation, demyelination, neurodegeneration and neuroprotection in the pathogenesis of multiple sclerosis. J Neuroimmunol. 2007;184:37-44. doi: 10.1016/j.jneuroim.2006.11.015.

[14] Geurts JJ, Kooi EJ, Witte ME, van der Valk P. Multiple sclerosis as an "inside-out" disease. Ann Neurol. 2010;68:767-8; author reply 768. doi: 10.1002/ana.22279.

[15] Cua DJ, Sherlock J, Chen Y, Murphy CA, Joyce B, Seymour B, et al. Interleukin-23 rather than interleukin-12 is the critical cytokine for autoimmune inflammation of the brain. Nature. 2003;421:744-748. doi: 10.1038/nature01355.

[16] Langrish CL, Chen Y, Blumenschein WM, Mattson J, Basham B, Sedgwick JD, et al. IL-23 drives a pathogenic T cell population that induces autoimmune inflammation. J Exp Med. 2005;201:233-240. doi: 10.1084/jem.20041257.

[17] Fitzgerald DC, Zhang GX, El-Behi M, Fonseca-Kelly Z, Li H, Yu S, et al. Suppression of autoimmune inflammation of the central nervous system by interleukin 10 secret-

ed by interleukin 27-stimulated T cells. Nat Immunol. 2007;8:1372-1379. doi: 10.1038/ni1540.

[18] Bettelli E, Das MP, Howard ED, Weiner HL, Sobel RA, Kuchroo VK. IL-10 is critical in the regulation of autoimmune encephalomyelitis as demonstrated by studies of IL-10- and IL-4-deficient and transgenic mice. J Immunol. 1998;161:3299-3306.

[19] Lenz DC, Swanborg RH. Suppressor cells in demyelinating disease: a new paradigm for the new millennium. J Neuroimmunol. 1999;100:53-57.

[20] Zhao Z, Yu S, Fitzgerald DC, Elbehi M, Ciric B, Rostami AM, et al. IL-12R beta 2 promotes the development of CD4+CD25+ regulatory T cells. J Immunol. 2008;181:3870-3876.

[21] Hafler DA, Slavik JM, Anderson DE, O'Connor KC, De Jager P, Baecher-Allan C. Multiple sclerosis. Immunol Rev. 2005;204:208-231. doi: 10.1111/j. 0105-2896.2005.00240.x.

[22] Boster A, Edan G, Frohman E, Javed A, Stuve O, Tselis A, et al. Intense immunosuppression in patients with rapidly worsening multiple sclerosis: treatment guidelines for the clinician. Lancet Neurol. 2008;7:173-183. doi: 10.1016/S1474-4422[08]70020-6.

[23] Devonshire V, Havrdova E, Radue EW, O'Connor P, Zhang-Auberson L, Agoropoulou C, et al. Relapse and disability outcomes in patients with multiple sclerosis treated with fingolimod: subgroup analyses of the double-blind, randomised, placebo-controlled FREEDOMS study. Lancet Neurol. 2012;11:420-428. doi: 10.1016/S1474-4422[12]70056-X.

[24] Cuzner ML, Opdenakker G. Plasminogen activators and matrix metalloproteases, mediators of extracellular proteolysis in inflammatory demyelination of the central nervous system. J Neuroimmunol. 1999;94:1-14.

[25] Chen SJ, Wang YL, Fan HC, Lo WT, Wang CC, Sytwu HK. Current status of the immunomodulation and immunomediated therapeutic strategies for multiple sclerosis. Clin Dev Immunol. 2012;2012:970789. doi: 10.1155/2012/970789.

[26] Fitzgerald DC, Ciric B, Touil T, Harle H, Grammatikopolou J, Das Sarma J, et al. Suppressive effect of IL-27 on encephalitogenic Th17 cells and the effector phase of experimental autoimmune encephalomyelitis. J Immunol. 2007;179:3268-3275.

[27] Saraiva M, O'Garra A. The regulation of IL-10 production by immune cells. Nat Rev Immunol. 2010;10:170-181. doi: 10.1038/nri2711.

[28] Santambrogio L, Crisi GM, Leu J, Hochwald GM, Ryan T, Thorbecke GJ. Tolerogenic forms of auto-antigens and cytokines in the induction of resistance to experimental allergic encephalomyelitis. J Neuroimmunol. 1995;58:211-222.

[29] Cash E, Minty A, Ferrara P, Caput D, Fradelizi D, Rott O. Macrophage-inactivating IL-13 suppresses experimental autoimmune encephalomyelitis in rats. J Immunol. 1994;153:4258-4267.

[30] Rott O, Fleischer B, Cash E. Interleukin-10 prevents experimental allergic encephalomyelitis in rats. Eur J Immunol. 1994;24:1434-1440. doi: 10.1002/eji.1830240629.

[31] Vaday GG, Baram D, Salamon P, Drucker I, Hershkoviz R, Mekori YA. Cytokine production and matrix metalloproteinase (MMP)-9 release by human mast cells following cell-to-cell contact with activated T cells. Isr Med Assoc J. 2000;2 Suppl:26.

[32] Scarisbrick IA. The multiple sclerosis degradome: enzymatic cascades in development and progression of central nervous system inflammatory disease. Curr Top Microbiol Immunol. 2008;318:133-175.

[33] Novick D, Rubinstein M, Azam T, Rabinkov A, Dinarello CA, Kim SH. Proteinase 3 is an IL-32 binding protein. Proc Natl Acad Sci U S A. 2006;103:3316-3321. doi: 10.1073/pnas.0511206103.

[34] Lefrancais E, Roga S, Gautier V, Gonzalez-de-Peredo A, Monsarrat B, Girard JP, et al. IL-33 is processed into mature bioactive forms by neutrophil elastase and cathepsin G. Proc Natl Acad Sci U S A. 2012;109:1673-1678. doi: 10.1073/pnas.1115884109.

[35] Guma M, Ronacher L, Liu-Bryan R, Takai S, Karin M, Corr M. Caspase 1-independent activation of interleukin-1beta in neutrophil-predominant inflammation. Arthritis Rheum. 2009;60:3642-3650. doi: 10.1002/art.24959.

[36] Mullbacher A, Waring P, Tha Hla R, Tran T, Chin S, Stehle T, et al. Granzymes are the essential downstream effector molecules for the control of primary virus infections by cytolytic leukocytes. Proc Natl Acad Sci U S A. 1999;96:13950-13955.

[37] Walport MJ. Complement. First of two parts. N Engl J Med. 2001;344:1058-1066. doi: 10.1056/NEJM200104053441406.

[38] Jain MR, Bian S, Liu T, Hu J, Elkabes S, Li H. Altered proteolytic events in experimental autoimmune encephalomyelitis discovered by iTRAQ shotgun proteomics analysis of spinal cord. Proteome Sci. 2009;7:25. doi: 10.1186/1477-5956-7-25.

[39] Bever CT,Jr, Rosenberg GA. Matrix metalloproteinases in multiple sclerosis: targets of therapy or markers of injury? Neurology. 1999;53:1380-1381.

[40] Halonen T, Kilpelainen H, Pitkanen A, Riekkinen PJ. Lysosomal hydrolases in cerebrospinal fluid of multiple sclerosis patients. A follow-up study. J Neurol Sci. 1987;79:267-274.

[41] Hartung HP, Kieseier BC. The role of matrix metalloproteinases in autoimmune damage to the central and peripheral nervous system. J Neuroimmunol. 2000;107:140-147.

[42] Kieseier BC, Seifert T, Giovannoni G, Hartung HP. Matrix metalloproteinases in inflammatory demyelination: targets for treatment. Neurology. 1999;53:20-25.

[43] Leppert D, Lindberg RL, Kappos L, Leib SL. Matrix metalloproteinases: multifunctional effectors of inflammation in multiple sclerosis and bacterial meningitis. Brain Res Brain Res Rev. 2001;36:249-257.

[44] Romanic AM, White RF, Arleth AJ, Ohlstein EH, Barone FC. Matrix metalloproteinase expression increases after cerebral focal ischemia in rats: inhibition of matrix metalloproteinase-9 reduces infarct size. Stroke. 1998;29:1020-1030.

[45] Cuzner ML, Gveric D, Strand C, Loughlin AJ, Paemen L, Opdenakker G, et al. The expression of tissue-type plasminogen activator, matrix metalloproteases and endogenous inhibitors in the central nervous system in multiple sclerosis: comparison of stages in lesion evolution. J Neuropathol Exp Neurol. 1996;55:1194-1204.

[46] Gijbels K, Masure S, Carton H, Opdenakker G. Gelatinase in the cerebrospinal fluid of patients with multiple sclerosis and other inflammatory neurological disorders. J Neuroimmunol. 1992;41:29-34.

[47] Maeda A, Sobel RA. Matrix metalloproteinases in the normal human central nervous system, microglial nodules, and multiple sclerosis lesions. J Neuropathol Exp Neurol. 1996;55:300-309.

[48] Larionova NI, Gladysheva IP, Tikhonova TV, Kazanskaia NF. Inhibition of cathepsin G and elastase from human granulocytes by multiple forms of the Bowman-Birk type of soy inhibitor. Biokhimiia. 1993;58:1437-1444.

[49] Gladysheva IP, Larionova NI, Gladyshev DP, Tikhonova TV, Kazanskaia NF. The classical Bowman-Birk soy inhibitor is an effective inhibitor of human granulocyte alpha-chymotrypsin and cathepsin G. Biokhimiia. 1994;59:513-518.

[50] Tikhonova TV, Gladysheva IP, Kazanskaia NF, Larionova NI. Inhibition of elastin hydrolysis, catalyzed by human leukocyte elastase and cathepsin G, by the Bowman-Birk type soy inhibitor. Biokhimiia. 1994;59:1739-1745.

[51] Tani K, Ogushi F, Kido H, Kawano T, Kunori Y, Kamimura T, et al. Chymase is a potent chemoattractant for human monocytes and neutrophils. J Leukoc Biol. 2000;67:585-589.

[52] Halliwell B. Free radicals, reactive oxygen species and human disease: a critical evaluation with special reference to atherosclerosis. Br J Exp Pathol. 1989;70:737-757.

[53] Frenkel K, Chrzan K, Ryan CA, Wiesner R, Troll W. Chymotrypsin-specific protease inhibitors decrease H_2O_2 formation by activated human polymorphonuclear leukocytes. Carcinogenesis. 1987;8:1207-1212.

[54] Dia VP, Berhow MA, Gonzalez De Mejia E. Bowman-Birk inhibitor and genistein among soy compounds that synergistically inhibit nitric oxide and prostaglandin E2 pathways in lipopolysaccharide-induced macrophages. J Agric Food Chem. 2008;56:11707-11717. doi: 10.1021/jf802475z.

[55] Li J, Ye L, Cook DR, Wang X, Liu J, Kolson DL, et al. Soybean-derived Bowman-Birk inhibitor inhibits neurotoxicity of LPS-activated macrophages. J Neuroinflammation. 2011;8:15. doi: 10.1186/1742-2094-8-15.

[56] Bawadi HA, Antunes TM, Shih F, Losso JN. In vitro inhibition of the activation of Pro-matrix Metalloproteinase 1 (Pro-MMP-1] and Pro-matrix metalloproteinase 9

(Pro-MMP-9] by rice and soybean Bowman-Birk inhibitors. J Agric Food Chem. 2004;52:4730-4736. doi: 10.1021/jf034576u.

[57] Losso JN, Munene CN, Bansode RR, Bawadi HA. Inhibition of matrix metalloproteinase-1 activity by the soybean Bowman-Birk inhibitor. Biotechnol Lett. 2004;26:901-905.

[58] Gran B, Tabibzadeh N, Martin A, Ventura ES, Ware JH, Zhang GX, et al. The protease inhibitor, Bowman-Birk Inhibitor, suppresses experimental autoimmune encephalomyelitis: a potential oral therapy for multiple sclerosis. Mult Scler. 2006;12:688-697.

[59] Maryanoff BE, de Garavilla L, Greco MN, Haertlein BJ, Wells GI, Andrade-Gordon P, et al. Dual inhibition of cathepsin G and chymase is effective in animal models of pulmonary inflammation. Am J Respir Crit Care Med. 2010;181:247-253. doi: 10.1164/rccm.200904-0627OC.

[60] Saw S, Kale SL, Arora N. Serine protease inhibitor attenuates ovalbumin induced inflammation in mouse model of allergic airway disease. PLoS One. 2012;7:e41107. doi: 10.1371/journal.pone.0041107.

[61] Cho EY, Choi SC, Lee SH, Ahn JY, Im LR, Kim JH, et al. Nafamostat mesilate attenuates colonic inflammation and mast cell infiltration in the experimental colitis. Int Immunopharmacol. 2011;11:412-417. doi: 10.1016/j.intimp.2010.12.008.

[62] Ware JH, Wan XS, Kennedy AR. Bowman-Birk inhibitor suppresses production of superoxide anion radicals in differentiated HL-60 cells. Nutr Cancer. 1999;33:174-177. doi: 10.1207/S15327914NC330209.

[63] Fujioka T, Purev E, Kremlev SG, Ventura ES, Rostami A. Flow cytometric analysis of infiltrating cells in the peripheral nerves in experimental allergic neuritis. J Neuroimmunol. 2000;108:181-191.

[64] Olee T, Powell HC, Brostoff SW. New minimum length requirement for a T cell epitope for experimental allergic neuritis. J Neuroimmunol. 1990;27:187-190.

[65] Touil T, Ciric B, Ventura E, Shindler KS, Gran B, Rostami A. Bowman-Birk inhibitor suppresses autoimmune inflammation and neuronal loss in a mouse model of multiple sclerosis. J Neurol Sci. 2008;271:191-202. doi: 10.1016/j.jns.2008.04.030.

[66] Dai H, Ciric B, Zhang GX, Rostami A. Interleukin-10 plays a crucial role in suppression of experimental autoimmune encephalomyelitis by Bowman-Birk inhibitor. J Neuroimmunol. 2012;245:1-7. doi: 10.1016/j.jneuroim.2012.01.005.

[67] Rynda-Apple A, Huarte E, Maddaloni M, Callis G, Skyberg JA, Pascual DW. Active immunization using a single dose immunotherapeutic abates established EAE via IL-10 and regulatory T cells. Eur J Immunol. 2011;41:313-323. doi: 10.1002/eji.201041104; 10.1002/eji.201041104.

[68] Gold R, Kappos L, Arnold DL, Bar-Or A, Giovannoni G, Selmaj K, Tornatore C, Sweetser MT, Yang M, Sheikh SI, Dawson KT; DEFINE Study Investigators.Placebo-

controlled phase 3 study of oral BG-12 for relapsing multiple sclerosis. N Engl J Med. 2012 Sep 20; 367[12]:1098-107.

[69] Fox RJ, Miller DH, Phillips JT, Hutchinson M, Havrdova E, Kita M, Yang M, Raghu-pathi K, Novas M, Sweetser MT, Viglietta V, Dawson KT; CONFIRM Study Investi-gators.Placebo-controlled phase 3 study of oral BG-12 or glatiramer in multiple sclerosis. N Engl J Med. 2012 Sep 20;367[12]:1087-97.

Chapter 5

Facilities for Obtaining Soybean Oil in Small Plants

Ednilton Tavares de Andrade,

Luciana Pinto Teixeira, Ivênio Moreira da Silva,

Roberto Guimarães Pereira,

Oscar Edwin Piamba Tulcan and

Danielle Oliveira de Andrade

Additional information is available at the end of the chapter

1. Introduction

In this chapter we will study the processes, procedures and types of equipment used in industrial processing of soybeans for obtaining vegetable oil and its adequatestorage in small plants.

Soybeans, despite its low oil content (18% to 22%) is the second most important oilseed crop of the world, after palm oil. In 2010, it represented 27.3% of vegetable oil total produced worldwide, compared with 33.7% palm oil (pulp and almond), 15.6% rapeseed oil and 8.7% sunflower oil, together are account for 85.3% of vegetable oil total produced globally [12]. The high protein content (37% to 40%) of soybean is the main feedstock in the manufacture of feed for domestic animals. Almost 70% of meal protein that makes up the animal feed comes from soybean [7].

The demand for vegetable oils will grow, mainly by increased consumption / capita in emerging countries. The average annual consumption of edible oil of a citizen of a developed country is about 50 liters, while the world average is about 20 liters / head / year. Another factor that will contribute to this increase is the use as biofuel (biodiesel and H-Bio), the new lever to consumption of vegetable oil.

The soybean oil currently holds the 2nd position in the world supply of oils and fats, according to Oilworld. In 1990, production of oil stood at around 16.1 million tonnes, followed by palm oil with 10.8 million tons. Other vegetable oils were the significant world production of rapeseed and sunflower, both with approximately 8 million tons, and cotton and peanuts, with approximately 4 million tons each. Although interchangeable, each one of these oils has specific characteristics that makes it more or less appropriate depending on its final use [12].

While the supply of vegetable oils is large, each of these oils have specific characteristics that make them more or less suitable for use as a biofuel [12]. The restriction on the use of soy for biodiesel is compared to the low oil content in their grains. The oil yield per hectare of soybean, considering an average oil content of 20% and within the grain yield per area of 400 to 800 kg in a crop that produces 2000 to 4000 kg / ha, respectively [21]. The yield of soybean is around 2.8 to 2.95 t / ha.

According to the USDA (U.S. Department of Agriculture United States) for the 2010/2011 harvest, the estimate of global soybean production was of 256.1 million tons (Table 1), down 1.46% compared to 259.89 million tons produced in 2009/2010. Likewise, the global consumption of 2010/2011 was estimated at 255.284.000 tons, an increase of 7.5% compared to 237.430.000 tons achieved in the previous crop. Still, world ending stocks of the product in 2010/2011 will be at 58.21 million tons, 3.26% below the world ending stocks of previous crop (2009/2010) of 60.17 million tons.

Country	Harvested Area (Million Hectares)	Production (1000 MT)
United States	29.800	83.172
Brazil	25.000	68.500
Argentina	18.600	46.500
China	7.650	13.500
India	10.270	11.000
Paraguay	2.600	5.000
Canada	1.542	4.246
Russia	1.180	1.749
Ukraine	1.100	2.200
Uruguay	1.000	1.700
Bolivia	900	1.580
World Total	103.094	245.065

Table 1. Estimates of global soybean production for 2011/2012 harvest [30]

2. The Product: The Meal and Soybean Oil

The products of the processes are simplified form the crude oil and meal (cake) semi-defatted, which can be used for the preparation of animal feed. It is planned to be used for human consumption, since such use would imply the need for greater care and sophistication in hygiene and microbiological control. There is no waste in the process, which may cause environmental problems.

Compared with the traditional processes, the main disadvantage is the high content of residual oil in the semi-defatted meal (about 8%). Soybean oil is very unsaturated and may lead to rancidity bran under these conditions. In a small scale production, it is expected that the soy meal is consumed as it is produced, ie the time of storage would be small. Furthermore, the oil in the meal will replace the oil which is usually added in the preparation of feeds.

The second biggest problem is the allocation of oil produced. Because crude soybean oil, cannot be consumed as food without the refining process - the taste is very bad. Although oil extracted by extraction with low concentrations of phosphatides, which is equivalent to the degummed oil, its storage is not recommended for long periods. Hardly the big oil refiners will buy this product.

In the case of commodities, the market promotes an intense coordination of the system, controlling the prices of commodities. As soybean production-level family does not benefit from economies of scale, the tendency is to seek new ways to add value to the product, incorporating new features into grains. The differentiation of the grains can open prospects for more efficient production, processing or use, making common grains, marketed as mere commodities, specializing in products with high added value and commercial [11, 10].

3. Production Process

There are two production processes oils and fats. For materials with high oil contents (over 30%), it uses the pressing process. For raw materials with lower levels of oil, it uses the solvent extraction. In the extraction by pressing the residual oil content of the raw material is around 10%, while in this extraction solvent content can be reduced to less than 1%.

In industrial processes, typically raw materials rich in oil is pressed up to a residual oil content of about 20% and the remaining oil is extracted by solvent. Thus, the soybean oil (20% oil) is usually extracted only solvent as sunflower oil (45% oil) is partially removed by pressing and the remaining solvent.

The oil obtained in these processes, known as crude oil, generally undergoes a purification process (refining) before being consumed as food. The only exception is the commercial-scale oil (olive oil) olive oil that is consumed without refining (oil "virgin"), although other oils such as sesame, sunflower, peanut oils can be consumed raw. The soybean oil, cotton and canola are consumed only after refined.

The residues of extraction, pie, if the pressing, bran, in the case of solvent extraction, less than 20% are used for human consumption. They are generally used for the preparation of animal feed.

The oil extraction process can be divided into three phases. The first involves the pre-cleaniness, drying and storage of product to be processed. The second step concerns the preparation of the grains for the oil extraction, by facilitating the extraction processes, such as the loss of grain, conditioning or heating, lamination, and expander. Finally, the third stage involves the extraction itself, which may develop by pressing or solvent. Figure 1 details the phases and steps of the extraction process.

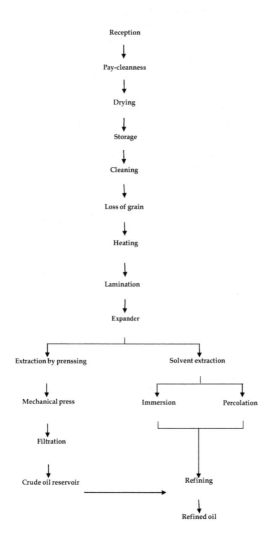

Figure 1. Flowchart for the production of soybean oil

3.1. Storage of raw material

Considering itself it importance of all previous the productive stages the harvest, since the election of the seeds to the cultural treatments, to one adjusted storage, the grains must pass for two important stages: the first one is the daily pay-cleanness, in which all the impurities must be removed, therefore it intervenes directly with the income of the process of oil extraction, useful life of the involved machines, beyond serving as inoculate of plagues and harmful microorganisms. The second stage is the process of drying that has as objective to guarantee the reduction of the moisture content of the product, of form to minimize the deterioration processes during the storage.

The stage of storage is essential for the maintenance of the quality of the raw material to be processed, influencing directly in the final product quality [23]. During the storage of the grains, that must be carried through in silos or specifically projected warehouses, the characteristics of temperature and moisture content of the product they must be monitored. Such characteristics are determinative in the minimization of the losses for deteriorations caused for microorganisms and attack of plagues, beyond being essential to prevent the occurrence of oxidative processes in the interior of the grains.

In what it says respect the quality of the oil during the storage, the variations in the moisture content of the product, as well as of the temperature can provide to enzymatic and oxidative reactions of the present oil in the interior of the grain, providing alterations in the characteristics and disposal of acid the fatty gifts. For the case of the soy grains, the oil if finds deposited in lipid bodies (the Spherosomes) distributed throughout its endosperm. Therefore, the control of such characteristics throughout all the previous stage the extraction of the oil is of basic importance for the guarantee of a quality by-product, as much in artisan scale as in industrial scale. In accordance with [23], low the quality of the crude oil influences in the increase of the losses and expenses with refining, providing a lesser income.

3.2. Preparation of the raw material

The extraction of the oil presents as by-products, beyond the involved lipid fraction with the rude oil, also proteins and carbohydrates that constitute the pie or bran, much used in the food industry, as much for animals as for human beings. Therefore, for the guarantee of the quality and not contamination of by-products, the value aggregation, and the integral exploitation of the processed product, must be carried through the preparation of the grain form to guarantee a maximum separation between the oil and the bran.

In what the oil production says respect proceeding from the soy grain, that presents about 20% of oil, two methods extraction can be used: the solvent extraction for and the extraction for pressing. Although some authors to characterize the extraction for pressing of the soy as not being economically interesting in reason of its low oil content, indicating the extraction for solvent, in this book will be presented the preparations of the necessary raw material for the two forms of extraction.

The purpose of the stage of preparation of the grain for the extraction of the oil involves to provide the increase of the susceptibility of disruption of lipid organelles, contained in endosperm, through thermal and mechanical treatments. For this, after the withdrawal of the mass of grains of the storing unit, with moisture content of approximately 10%, in wet base (b.w.), is indicated, again the separation of the impurities that still can be contained in the mass of grains.

After this stage, must be carried through the form grain in addition to lead the unfastening and the separation of the rinds, that are abrasive, and to favor the uniformity of the size of particles to be processed. The grain in addition can be carried through by means of a breaker called equipment of coil that consists of the disposal of two corrugated cylinders, parallel made use that turns with different speeds, while the mass of grains is lead between them. During this process, for the withdrawal of the remaining rind a pneumatic aspiration is carried through, jointly with bolters splitting.

In this process of in addition the grains the milling of the grain is not indicated, since it negative intervenes with the separation between the rind and the remaining structure of the grain, the cotyledon; beyond, for the case of the extraction for solvent, to make it difficult the separation of the solvent and the oil of the bran. Another important determinative characteristic in the success of this stage is the maintenance of the grains with moisture content of, approximately, 10%, in b.u., of form to prevent the embouchement of the machine for higher the moisture content cases, or the dust production for very low moisture contents.

After the grain in addition, must be carried through the baking of the processed mass of grains. This stage has as objective to provide to a fast increase of the moisture content of the mass of grains, producing a bigger plasticity to the mass, and minimizing the dust production. This process is presented as a facilitator to rupture of the Spherosomes cell walls, in order to facilitate the leakage oil.

In accordance with [18], the increasing of the moisture of the flakes, the cell wall disruption and subsequent increase in the permeability of cell membranes, facilitates the exit of the oil, reducing its viscosity and its surface tension, which allow agglomeration of the oil droplets and its subsequent extraction. The baking can be carried through by warm vertical cookers through chambers with warm vapor, or horizontal rotating drums for a warm tubular beam the vapor.

After this preparation of the raw material, mainly for the case of the extraction with solvent, is indicated the accomplishment of plus others two stages: the lamination and the expansion. These two stages have the objective in the distance to provide the minimization of between solvent and the oil, favoring the extraction process.

The lamination if bases on the flake attainment from pieces of grains, with the objective to minimize the internal resistance in favor of the extraction of the oil. For the confection of flakes, the grains broken and with raised temperature more are lead, in the rolling mill, enter a pair of made use smooth cylinders horizontally that 0,3 mm jam pieces of the grains of soy in blades of 0,2, forming flakes [23]. In this process, the flake production with homogeneous thickness directly is related with the efficiency of extraction of the oil and, mainly, with the quality and pureness of the produced bran, since this characteristic influences in the interaction between the oil and the solvent during the extraction.

The following stage is called of expansion. In this stage, the flakes are, again, humidified with warm water vapor and for attrition throughout screws without end that lead the material until a perforated plate. In reason of the difference of pressure before and after the ticket for this plate, the warm and humidified flake suffers the process from expansion. In accordance with [23], this expansion occurs in reason of the starch presence in the grain.

The ticket of the mass of grains for the expander, or expander - extruder, propitiates greater porosity to the mass and permeability to solvent, favoring the contact between solvent and the expanded cell, guaranteeing a more efficient percolating between the oil the solvent. According [23], the use of the expander implies in the increase of about 40% in the capacity of extraction for solvent. After the expansion, the material must be cooled until the temperature of extraction of the oil, and depending on the conditions of the processing, the mass must pass for the form drying to guarantee a bigger efficiency in the extraction process.

3.3. The extraction

The extraction of the oil for being carried through two different methods: the extraction mechanics through the pressing, or by means of the chemical extraction for solvent. In situations special, of form if to get the maximum efficiency of extraction it can be used the two methods sequentially. To follow, the two forms of oil extraction will be presented proceeding from oil seeds, as it is the case of the soy.

3.3.1. Extraction for pressing

The method of extraction for pressing consists of the withdrawal of the oil by means of the application of a external pressure on the mass of grains, through the pressing mechanics. As main involved advantages with the use of the press mechanics for the oil extraction low cost of installation is its, not the use of solvent, and not the necessity of posterior refining of the oil, what it implies in the reduction of the processing cost and, consequently, of the gotten oil, favoring the use of the same one for small producers.

Currently, the press more common mechanics is the continuous press of screw, also call of expeller, that hopper of feeding is composed for one, that leads the material to be pressed by means of a screw without end of step interrupted for steel rings, made use parallel; to the end of the set, they find if a cone tip that regulates the speed of exit of the pie and the chamber pressure on the mass.

The pressing process if develops from the introduction of the mass of grains in hopper that it feeds the screw without end, compressing it against steel rings, providing the elimination of the oil for the orifices. The extraction speed depends directly on the imposed pressure, that initially must understand of 300 the 400 kg cm^{-2}, but throughout the process due to the gradual accumulation of mass in the interior of the press, the pressure can be superior the 1,000 kg cm^{-2} [23].

During the process, the mass of grains equally is pressed, preventing the resorption of the oil for other parcels of the mass. After the ticket for the press, the rude oil must be filtered with the purpose to separate the solid residues proceeding from the remaining pie.

3.3.2. Extraction for solvent

The oil extraction for solvent can be used as only method of extraction, or same as comple-ment to the extraction mechanics, in the daily pay-extraction form. This method of extrac-tion if bases on the absorption of the solvent for the lipid cells, where in its interior it has the dissolution of the oil, that, later, for leaching, he is loaded for the exterior of the cell. This process, in adjusted conditions, approximately removes 99% of the oil contained in the mass of grains [23].

However, for the guarantee of the efficiency of the process, it is essential the adjusted prepa-ration of the grain and the choice of to be used extractor, of form to guarantee the maximum contact of the solvent with the cellular wall. Of this form, how much bigger it is the amount of cells breached throughout the preparation of the mass of grains, faster it is the extraction process, since the solvent will only go to dissolve the free oil, not needing to carry for diffu-sion the dissolved oil to the external region to the cell.

In accordance with [23], the transport of lipid throughout the cellular membrane occurs in function of the variation of its permeability (initially impermeable to the lipid) in function of the difference of the internal and external osmotic pressures to the cells. The increase of in-tracellular pressure in virtue of the action of the solvent it provides the expansion of the membrane and, consequently, the dilatation of the pores of the cellular membrane, allowing the ticket of the solvent oil solution and for the extracellular region, which had to the gradi-ent of existing concentration [23, 26].

The extraction process occurs in higher temperatures, seen its influence in the viscosity of the solvent oil mixture and, and in the solubilization of the oil in the solvent. The extraction speeds from beginning to end of the process of solvent extraction progresses differently. Ini-tially, when the oil of better quality is extracted, the process if develops of fast form, due to the biggest gradient of concentration, however, throughout the process this speed diminish-es, and the extracted oil to the end presents minor quality, in reason, mainly, of the presence of other cellular composites that provide losses throughout the refining.

From this process, the oil extraction for solvent can be developed of two forms, for immer-sion, where the mass of grains is kept immersed in the solvent for a definitive period of time, or by percolating, which the mass of grains is made use in layers to guarantee opti-mum contact of the solvent that it passes freely between the grains and it is renewed when it has saturation. To the end of the extraction, as much the solvent mixed to the oil, as the sol-vent gift next to the remaining bran can be recouped and be reintroduced in the process. For the grains to solvent extraction with low concentrations of oil is indicated by immersing the system [2].

For the extraction of the oil, frequent the commercial hexane is used as solvent. Proceeding from the refining of the oil, the hexane presents determinative characteristics for its good performance as solvent, as: it is a composition to apolar total being miscible in oil at the same time that it is immiscible in water, presents low latent heat of boiling, and it does not react with the constituent material of the equipment used for the extraction. Although its fa-vorable characteristics physicist-chemistries, this solvent requires special care with the security in the industrial plant and with its manuscript, seen its high inflammability, explosion and toxicity [23].

4. Refining

The refining if characterizes as the set of operations carried through after the process of extraction for the removal of residues gifts in the crude oil that can affect the color, stability, aroma, and flavor, beyond its physical characteristics. These residues are preceding from drag mechanics, and/or solubilization of other substances in the oil or solvent the occurrences during the extraction process.

The refining takes place in two stages: the first step is the physical removal of substances, while the second involves the refining processes through neutralization, clarification, and deodorization. In accordance with [31], initially the oil passes for a stage of physical separation, where, in a tank, the separation for gravity of insoluble substances is carried through. After this stage, the oil passes for the degumming that consists of the removal of phospholipids, sugars, resins, and breaks up of soluble proteins in water.

In accordance with [23], the presence of phospholipids in the soy oil favors the occurrence of losses, in reason of the formation of depositions with presence of about 35% of oil in the deep one the tanks of deposition, with aspect similar to a gum. In this in case that, the phospholipids can be presented of two forms: hydratable, being liable of withdrawal with addition of water and centrifugation, and not hydratable that needs the addition of acid citric phosphoric or for becomes it hydratable for its posterior withdrawal.

Still of according to [31], after the degumming process, must be proceeded with the neutralization, that the withdrawal of fatty acid involves, pigments, remaining phospholipids of the degumming, and soluble sulfur composites in water. In the neutralization process it has the caustic soda water addition, with fatty acid the purpose to reduce the text of free, to clot the phosphatides and the gums, to degrade part of the dyes gifts, and to load the insoluble substance for the clotted material. Of this process it has the production of you leave organic sodium or soaps, what it results in the necessity of one another stage, the *laundering*.

The laundering is a process that results in the withdrawal of the soap produced during the neutralization. It is based on the *distilled* water addition the temperature of 85-95°C (about 10-20% of the volume of oil) for the elimination of the soda water and the foam of the oil, however case the separation between the oil and the water is difficult indicates it dilution in the aluminum sulfate water. The laundering process can more than proceed a time until the total exemption from soap from the oil. To finish the process laundering, in reason of the water addition, the drying must be proceeded from the oil.

After the drying, the oil is directed to the *clarification (bleaching)* that objective the pigment elimination, residual products of the oxidation, metals weighed, and soaps, of form to guarantee the improvement of the flavor, odor and oxidative stability of the oil. Bleaching occurs from the adsorption of specific adsorbents substances for the elimination of substances that confer coloration, sulfur, soaps and metals to the soy oil. The adsorbents substances can be some types of silicates, diatoms lands, clays acid activated, silica and active coal [23].

In accordance with [14], throughout the refining, the clarification produces an oil with bigger susceptibility to the oxidation, being indicated for the storage the use of absent nitrogen stream bed of oxygen.

After the clarification, the oil follows for its last stage of refining, the deodorization. The deodorization of characterizes for the substance elimination formed during the storage and processing of the grain and other natural substances of the oil that can provide to awkward flavor and odor. For the withdrawal of these substances to use it distillation with chain of vapor to the vacuum, of form to guarantee the protection of the oil to the effect of the atmospheric oxidation, prevention of hydrolysis for the vapor, and the reduction of the necessary amount of vapor for the process.

For some types of oils, as it is the case of the palm oil and peanut, the physical refining is proceeded after the accomplishment from the deodorization. The physical refining consists of the elimination for evaporation of fatty acid the free gifts in the oil, providing bigger income of the fine oil. For the case of the soy oil, the physical refining is not indicated, in reason of its low acidity after to the end of the process and the difficulty in the degumming process.

5. Quality

Every product has a number of characteristic attributes. It's called quality, whose existence will define the success or failure in their marketing. This quality is mainly observed by two fundamental aspects: the first relates to the consumer who seeks desirable characteristics, whether from an economic standpoint, nutritional, aesthetic, etc. The second aspect refers to the legality, where the product goes through a series laboratory analysis and is classified into pre-established standards and its final quality is attested.

According [13], quality can be defined as a set of characteristics that will directly impact on the acceptability of the product. In this context, the subjectivity of the sensory analysis of each person are determinants for to preset of the quality, that can involve the appearance, texture, flavor and aroma.

The color, size, shape, integrity and consistency are factors directly involved in the appearance of the product, already the relateds with the physical senses of touch, and mouth are determined on the according to texture; the flavor factors involve the taste, the aroma, and correlate with the olfactory sensitivity. Besides these, you can also to relate the after taste that occurs according to a secondary analysis of the product [13].

Associated with these organoleptics factors for the acceptance of the product are also the nutritional value, the presence of toxic substances and the final price. All these characteristics are to related to the type of raw material and production method. In the case of vegetable oils, with is the case of the soybean oil, these factors are directly related to the conditions of storage and processing of the beans, as has been seen above, as well as the characteristics, extraction and treating the produced oil, and form of storage and shipping.

Overall, the changes and loss of quality of the agricultural products are related with the growth and activity of microorganisms, the action of enzymes, chemical reactions, attack by insects and rodents, and physical changes caused by mechanical agents. According [17], the major causes of food deterioration are, respectively, the attack of microorganisms and oxidative processes.

The vegetable oils have fewer characteristics in your reactive molecule, unlike proteins and carbohydrates presents in the grains. The reactivity of the oil is concentrated, mostly, in the hydrolysis that produces free fatty acids according to the moisture content present, lipolytic enzymes and temperature, and the oxidation reaction of lipid compound that is a function of the concentration of oxygen in the medium.

The hydrolysis process provides the breakdown of triglycerides and therefore the increase of free fatty acids, that influences the acidity of the vegetable oil. High levels of acidity in vegetable oils leads to high losses during the refining stage [23]. However, in general, for to monitor the deterioration of the vegetable oils is accomplished due to the oxidative reactions occurred in the triglyceride molecule.

Oils	Fatty acids composition (Wt.-%)								
	12:00	14:00	16:00	18:00	18:01	18:02	18:03	22:01	IV
Canola	0	0	4.5	1.5	58.5	25	9	1.5	118
Corn	0	0	11	2	36	50	1		121
Sunflower	0		6	4.5	32	57.5			126
Soybean	0		8	4	28	53	7		130

Table 2. Vegetable oils fatty acid composition

According [24], all vegetable oils consist primarily of triglycerides. The triglycerides have a three-carbon backbone with a long hydrocarbon chain attached to each of the carbons. These chains are attached through an oxygen atom and a carbonyl carbon, which is a carbon atom that is double-bonded to second oxygen. The differences between oils from different sources relate to the length of the fatty acid chains attached to the backbone and the number of carbon–carbon double bonds on the chain. Most fatty acid chains from plant based oils are 18 carbons long with between zero and three double bonds. Fatty acid chains without double bonds are said to be saturated and those with double bonds are unsaturated (Misra and Murthy, 2010).

In general, vegetable oils are made especially for fatty acids with chains between 12 and 24 carbons: Lauric (C12:0), Myristic (C14:0), Palmitic (C16:0); Palmitoleic (C16:1) Stearic (C18:0), Oleic (C18:1), Linoleic (C18:2); Linolenic (C18:3); Arachidic (C20:0); Gadoleic (C20:1); Behenic (C22:0), Erucic (C22:1); Lignoceric (C24:0). The proportions of the fatty acid composition can be determined by gas chromatography method. The weight composition of fatty acids found in soybean oil was: C16:0=11.6%, C16:1=0.1%, C18:0=32%, C18:1=20.4%; C18:2=57.1% and C18:3=5% [24].

Average values of the chemical composition of vegetable oils and values of iodine index are presented in the Table 2 [15] and physical properties of soybean oil in Table 3. The iodine Index or Number is one characteristic of the oil that measures its index of unsaturation, when evaluating the amount of iodine necessary to saturate the double links of the molecule.

PROPERTY	SOYBEAN OIL
Density kg/L (20 °C)	0.92037
Viscosity mm²/s (40 °C)	30.787
Flash Point °C	332
Cloud Point °C	-2
Pour Point °C	-14
Copper strip corrosion	1b

Table 3. Physical properties of soybean oil [24]

Both the hydrolysis reaction as the oxidation reaction can also be influenced by temperature, light, presence of unsaturated fatty acids, moisture content, and product type. As a result of these reactions is the development of different flavors and odors that compromise the quality and acceptability of the product. For these changes organoleptics give the name of rancidity. Addition, for the production of biodiesel, the presence of free fatty acids in vegetable oils affect the process of separation of soaps, and influence on the reactions of esterification, compromising the efficiency of the process.

Thus, in that it involves, specifically, the oxidation of lipids, the main causative factors are the composition of fats, the presence of oxygen, temperature, and luminosity. The composition of fats influence on the presence of unsaturation in the molecule, as the unsaturation increases the susceptibility of oxygen absorption, that the more present in the environment, more available for the reactions is presented. Both the temperature and luminosity influence, proportionally, in the rates of reactions. For the specific case of light, its presence influences on accelerating the development of rancidity in fats.

According [8] and [27], the quality of the oil can be followed depending on the index of acid determined by the presence of free fatty acids; index of saponification which demonstrates the presence of oils or fats high proportion fatty acids, color; foaming, viscosity, density, and index of peroxide that is determined by the presence of iodine. According to [24], the Table 3 shows the properties of the soybean oil.

It is shown in Table 4 that vegetable oils present corrosion within the pattern established for diesel oil (Standard corrosion = 1) according to ANP.

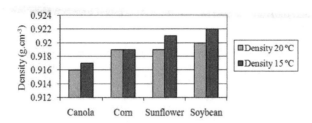

Figure 2. Density in temperatures of 15 and 20 ºC

Vegetable Oil	Corrosivity
Canola	1a
Sunflower	1a
Corn	1a
Soybean	1b

Table 4. Corrosivity of vegetable oils at temperature of 100 ºC

Looking at Table 5, 6 and 7, canola oil has the lowest density of 0.917 g.cm^{-3} at a temperature of 15 °C and 0.916 g.cm^{-3} at a temperature of 20 °C and soybean oil has the highest density, 0.922 g.cm^{-3} at a temperature of 15 °C and sunflower and corn oil showed a higher density at a temperature of 20 °C, 0.919 g.cm^{-3}. According to the specifications of the ABNT standard, diesel has a density of around 0.865 g.cm^{-3} (Figure 2).

In Table 3 the mean values and standard deviation of the viscosity of vegetable oils are shown.

Vegetable Oil	Kinematic Viscosity (mm^2.s^{-1})	Standard deviation
Canola	35.5278	0.0081
Sunflower	31.7275	0.0726
Corn	33.7713	0.0409
Soybean	31.6107	0.0093

Table 5. Kinematic viscosity values of vegetable oils obtained experimentally

The test results of kinematic viscosity of vegetable oils are shown in Figure 3 below:

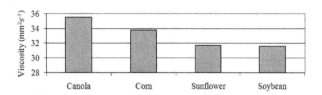

Figure 3. Kinematic viscosity of vegetable oils

Table 6 show the temperatures measured in the tests of cloud.

Vegetable Oil	Temperatures of Cloud Point	Temperatures of Flow Point
Canola	-1 °C	-20 °C
Sunflower	1 °C	-18 °C
Corn	3 °C	-12.5 °C
Soybean	-2 °C	-18.5 °C

Table 6. Temperatures of cloud points and flow points of vegetable oils

Vegetable oils have a density slightly larger than that of diesel fuel and below water, they are products of easy handling and processing. Their viscosities are far from being similar to diesel fuel, but this allows their use in cases of thermo-chemical conversion without major problems, and are easy to carry through pipelines without large deformation of tension and energy.

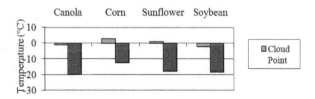

Figure 4. Results obtained experimentally for cloud point and flow point of vegetable oils

6. Control the Production Management of Soybean Oil

The acquisition process of vegetable oils occurs through of relatively simple procedures, starting with the attainment of oilseeds, proceeding to mechanical extraction, filtering and marketing of oil. This understanding contributes to the study of the entire production chain.

Generally the oil extracted is sent to the food industry, or the production of biofuels, and in some cases is used as feedstock in other industrial processes. Soybean oil has its applications in the first two cases cited. [16] agribusiness of the soybean oil has a high processing capacity, with oil production exceeding 50 million liters per year.

Vegetable oil	Density 15°C	Standard deviation	Density 20°C	Standard deviation	Corrosion	Kinematic Viscosity (mm².s⁻¹)	Standard deviation	Cloud point	Flow point
Canola	0.917	0.0005	0.916	0.0005	1a	35.5278	0.0081	-1 °C	-20 °C
Sunflower	0.921	0.0010	0.919	0.0005	1a	31.7275	0.0726	1 °C	-18 °C
Corn	0.919	0.0005	0.919	0.0005	1a	33.7713	0.0409	3 °C	-12.5 °C
Soybean	0.922	0.0000	0.920	0.0006	1b	31.6107	0.0093	-2 °C	-18.5 °C

Table 7. Physical characterization of vegetable oils [29, 1]

It is known specific attributes which must be met after defining the destination of oil extracted, not all characteristics of a product can be easily evaluated. In the case of agri-food products there is the existence of personal or subjective parameters that hinder your judgment. Thus, the oil-producing units paying attention to product quality, this will help to stay in the market and seeing new opportunities.

[3] reported that the quality of a product is resulting from the joining of items:

a. Quality of product design: that is the result of the activities that translate the knowledge of market needs and opportunities in information technology for the production.

b. Quality of design process: is results from the efficient translation of the design specifications of the product in process at various levels, such as process flow diagram, layout, design tools and equipment, project work etc..

c. Quality of conformation: is the result of the efficiency level of actual production.

d. Quality of services associated with the product: it is the result of the quality level of the stages of distribution and marketing, beyond quality of after-sales services.

[16] reported that the limit of the concept of quality has advanced in segments related to the management of the productive sector. It is clear that the companies' survival now depends on the quality of products or services it offered. Several agents have sought to monitor quality through product recalls, allowing verification of consignments dispatched by identifying the weaknesses of the production complex, such as inefficiency in logistics and handling conditions and storage of the product.

The production of vegetable oil in small units can be designed more efficiently by understanding the concept of agro-industrial chain, because according to [20] in recent years the so-called industrialization of agriculture has led to increasing dependence on agriculture to industry. Thus small farmers can participate in a competitive market where strategic alliances allow to produce and market their products. Given this perception, competitiveness is to migrate the level of individual firms to competition between agribusiness chains. Thus, because it is an integrated system there is a need to coordinate the supply chain, and the search for quality is now undertaking the various agents.

According [3] the coordination of quality is a set of activities planned and controlled by a coordinator agent, in order to improve quality management in the chain and ensure product quality through a process of transaction information, helping to improve the matching of customers and to reduce costs and losses at all stages of the chain.

This concept of quality coordination, "to plan", "controlling" and "improve" the quality management, "the process of information transaction" and "coordinating agent" are:

- Planning: Planning activities in order to create a process capable of producing products that meet consumers;

- Control: Control processes and activities with the objective of evaluating the performance of real quality and act if there is a diversion;

- Improve: set of activities that aims to improve the quality of processes and products;

- Transition process of information: the acquisition, management and distribution of information throughout the production chain;

- Coordinating agent: a key to the coordination of quality aims to make information related to product quality and quality management are identified, communicated and controlled throughout the chain.

The coordinator agent can be a company, a group of companies, an association of representatives of the segments of the agribusiness chain or a third party contracted to perform the tasks for the agent coordinated.

The coordination process quality in agro-industrial chain of soybean oil, produced in small units, can be strengthened by adopting measures to encourage the involvement of its agents. Usually the resources of small producers to cover production are limited, requiring therefore of a reliable outlook for the sale of its production. At this point, there is the need of productive sector strategies are aligned with unusual strategies for the development of the productive chain.

The literature dedicated to the study of agribusiness systems suggests practices to be implemented from the producer to the customer. Measures such as discounted prices, flexibility in time and remuneration for services, contribute to the strengthening of alliances among its members.

The strategies that comprise the agribusiness chain agents, shall consider agribusiness products are subject to special features such as seasonality of production, the need for special conditions for transport and storage and other care as they are perishable. The following paragraphs deal with aspects related to the quality of agro-industrial products, particularly vegetable oils.

6.1. Methodologies applied to quality management

Methodologies will be presented that can be applied to quality management in the production of vegetable oils in small units.

6.1.1. Application of ISO 9000

The International Organization for Standadization is responsible for standardization in global character, is the ISO 9000 group linked the quality management systems, which was created searching to standardize the rules for products and industries.

In addition to ISO 9000, regarding the standards focused on vocabulary and fundamentals, there is the importance of ISO 9001 that dealing with Quality Management Systems and Guidelines for SGQs auditing.

A process model of ISO 9001 may be seen as an implementation of the PDCA cycle, according [19], in which:

- PLan: It means setting objectives and processes necessary to deliver products or services in accordance with the requirements and policies of the organization;
- Do: It means the implementation of procedures;
- Check: It means to monitor and measure processes and products by comparing them to the policies, objectives and product requirements and report the results;
- Action: It means to take actions to continually improve process performance.

The most common steps for the implementation of ISO 9001 are:

- Conviction of top management
- Choice of coordinating implementation
- Assessment of current situation
- Preparation of schedule
- Training leveling
- Establishment of working groups
- Specific training
- Process Mapping
- Development / deployment documentation
- Training of internal auditors
- Performing internal audits
- Corrective Actions
- Training end

- Pre-certification audit, and

- Setting and maintaining the quality management system ISO 9001

6.2. Hazard Analysis and Critical Control Points (HACCP)

According [4] the HACCP is based in the analysis method of the mode and effects and fault causes or failure, mode and effect analysis (FMEA), developed by Kaoru Ishikawa in the Japanese industry. The HACCP allows to identify and assess hazards associated with different stages of the food chain, and define the necessary means for its control. The HACCP should be considered as a quality system, a rational practice, organized and systematic, indicated to provide the necessary confidence that a agri-food product will meet the health and safety requirements expected by the consumer.

Studies based on the concept of prevention, developed by [5], concluded that the HACCP represents an advance in food safety, when the adoption of preventive measures promoted the effective design of food safety and the processes in which, a priori, was analyzed quality (microbiological, physico-chemical and sensory) of the products already processed.

6.3. Traceability

At first glance, it is believed to be unnecessary to use tools that require high technological standard in small units of oil extraction. However, products offered in different ways may present a promising alternative for small farmers. Traceability aims to identify the origin of the product from the farm to the consumer. This principle is indifferent to processes in which the product has undergone. Thus, it is expected to reach the final with a quality product, and with a known source.

To [9] an efficient screening process should consist of the following elements:

Standards and / or the quality that aim to protect / secure;

Procedures allowed, prohibited, tolerated and required;

Grace periods or transition established as provided in the rules;

Requirement that producers are provided with proof of purchase, sales, everything that allows inspectors to check compliance of standards by the operator (owner of the process);

Periodic visits to the default setting;

Visits "surprise" the establishment.

According to [19] in soybean production, traceability is an essential tool for identification and separation of genetically modified organisms that are restricted in certain countries or markets. Thus, traceability systems are necessary for the management and development of the agribusiness chain, ensuring quality food offered.

6.4. Good Manufacturing Practices

Such standards can be applied in the production of oil intended for human consumption. Procedures are designed to get quality products. The Good Manufacturing Practices (GMP), according to Tomich et al. (2005), are a set of standards used in products, processes, services and buildings, for the promotion and certification of quality and food safety.

According to [6] requires that each establishment has its Manual of Good Manufacturing Practices, which details the conditions of hygiene and sanitary food handling procedures, cleaning equipment, utensils, facilities and buildings of establishments, in addition to establishment of minimum health requirements of buildings, facilities, equipment and tools, control of water supply, health and hygiene of food handlers, the integrated control of pests and vectors, and control and quality assurance of final products.

The following legislation provides for the establishment of Good Manufacturing Practices (GMP), in Brazil, whereas other related items: Standards of Identity and Quality, sanitary inspection, Standard Operating Procedures and Checklist of GMP.

- Ordinance no. MS 1.428/1993
- Ordinance no. 368/1997 of the MAP
- Ordinance no. 326/1997 of the MS / SVS
- Resolution RDC / ANVISA no. 275/2002

6.5. Cleaner Production: CP

The methodology of CP is the application of an economic strategy, environmental and technical, integrated processes and products in order to increase the efficiency of use of raw materials, water and energy, not the generation, waste minimization and recycling with environmental and economic benefits to the productive processes.

To have success in the development of cleaner production, are required to exercise responsible environmental management and technology assessment, the initial effort is fundamental in changing attitudes, including a subjective evaluation of the entire process.

All factors imply aggregate values and services, with the primary concern of consuming less material and generate less pollution.

[25] studied the management of agro-industrial residues. As a result it was proposed a roadmap for Environmental Management System (EMS). Operational actions were suggested as: Mass Balance identifying and quantifying infrastructure resources; Anticipation and monitoring the adoption of measures to prevent accidents or damage to the ecological nature of the life cycle analysis of products, Cleaner Production (CP) and Reverse Logistics, which is to collect, package, transport and dispose of waste that were generated in the activities of obtaining Raw Material and managerial actions suggesting adoption of sustainable strategies and policies, Environmental Management System as a mechanism to monitor the administrative and managerial performance the organization, Environmental Audit and Environmental Education.

7. From Control to Management

Currently, the concern with quality extends beyond the aspects mentioned above, no longer a mere bureaucratic requirement of regulatory agencies and inspection, but a fundamental strategy and essential to ensure competitiveness. The quality is replaced by a much broader approach, involving all levels of the organization and process.

8. Quality attributes

The main attributes that describe the desirable characteristics in grain for the quality of the bran are high in protein concentration, profile and level of amino acids, especially lysine, and high energy. To control the meal must consider factors such as moisture content, oil content, protein content, the urea activity and the rate of protein digestibility. And in the case of crude oil, emphasizes the determination of fatty acids and their state of rancidity.

These attributes directly influence the quality and safety of the product and the forms of controls are given in the text. It should be laid down the acceptable quality level (EQS) according to the intended use of the product. Obviously, the level of residual bran oil obtained by pressing has maximum and minimum other than the one extracted by solvent. The determination of the EQS should consider the characteristics of the process to optimize your control.

This item will be established quality standards obtained for soybean oil for immediate and latent analysis.

9. Process Control

As animal feed, the control of raw materials and products may be carried out with simpler and less frequently. The analyzes have suggested the purpose of controlling the quality of raw materials, products and the process yield. The quality and characteristics of the raw material has great influence on product quality and yield. Soybeans should be free of mold, broken grains, greens and other defects.

During the extrusion process, there is an internal friction of grain against the internal elements of the extruder rising to temperature and pressure with the complete inactivation of the activities: ureatics, antitrypsin and hemoglutinant (anti-nutritional factors). Exposure of the grain at high temperature and pressure for an extremely short time (20-40 seconds) favors the obtaining a quality product without compromising the nutritional quality of soybean. Therefore, factors such as temperature and pressure during the process should be monitored and recorded continuously. In annex presents tables to assist in process control. Below we suggest some analysis should be performed periodically.

10. Installations and Equipments

The following relate to the main points that should be taken into consideration when choosing a site to be deployed to agribusiness:

- the potential for obtaining the raw material in the region should be higher than the projected demand of the plant and enable future expansion in production;

- water supply reliable and good quality (drinking water);

- providing sufficient electrical power without interruption;

- availability of skilled labor, including technical personnel;

- no contaminants of any kind on the outskirts of agribusiness;

- road infrastructure in working condition and easily accessible;

- availability of sufficient area to implement the agricultural industry and its future expansion.

All new property before it began, it must seek approval of their facilities by the regional body of the Labor Ministry or agency responsible, and this after doing a preview, issue a certificate of approval for facilities. This procedure is adopted in order to ensure that the new establishment activities free of accident hazards and / or occupational diseases, which is why the establishment does not meet regulations will be subject to the impediment of its operation until the standard is met.

Figure 5. Seed cooker

Figure 6. Crusher seeds

Figure 7. Press small

Figure 8. Filter with collect the oil bucket

Figure 9. Small plant for the production of biodiesel

Figure 10. Settling tank to the glycerin

A unit of oil extraction on the farm can be mounted in a simplified form with the elements described below. Figure 5 shows a seed cooker being installed, its use may be waived in work with seeds whose oil extraction is done cold, as in the case of sunflower seeds. The crusher shown in Figure 6 can be used to prepare larger seed. Equipment such as the press shown in Figure 7 can perform pressing in an average yield of 40 kg per hour and is suitable for small applications. The filtration must be done efficiently preventing the oil is lost, thus a device for filtering, with collect bucket oil is of great importance in the process (Figure 8). A small plant, for production of oil and biodiesel, you can compose the oil processing unit (Figure 9). Glycerin resulting product from the transesterification of vegetable oils may be separated by centrifugation or decantation. In the second case, the composition of glycerin biodiesel more tanks can be conducted as shown in Figure 10, which should preferably have conical structure inside facilitating the decanting of the glycerine.

11. Conclusion

In conclusion we have with the expansion of soy agribusiness, there is a need to add value to the product. And in case of small farmers or family farmers in the production of oil will allow your own farm profit by adding value, besides the possibility of using this product on the farm.

The values of cloud points, close to zero degrees Celsius, and flow, near minus fifteen degrees, makes manipulation of these oils possible in tropical climates, where temperatures barely reach those levels, do not show stoppage problems, flow and clogging the lines of the process involved with the use of oil.

Tests for corrosion highlights low levels of attack in copper sheets, and oil is a product that does not generate large losses of material in the areas of the process which they have contact with, on the other hand the formation of clogging points and false surfaces as a result of polymerization of oil when working at higher temperatures can be more dangerous.

The physical and chemical characteristics evaluated in this work present a great relation with the composition of the tested oils, where there is a relation between the density, viscosity and corrosion index with the iodine index of the fluid. On the other hand, the cloud and flow points are related with the presence of saturated and unsaturated fatty acids. In respect to the saturation of the oil, it can be favorable in problems such as corrosion, of low temperatures or in the case where lower viscosity values are needed.

Author details

Ednilton Tavares de Andrade[1], Luciana Pinto Teixeira[3], Ivênio Moreira da Silva[3], Roberto Guimarães Pereira[2], Oscar Edwin Piamba Tulcan[4] and Danielle Oliveira de Andrade[3]

1 Federal Fluminense University/TER/PGEB/PGMEC, Brazil

2 Federal Fluminense University/TEM/PGMEC/PGEB/MSG, Brazil

3 Federal Fluminense University/TER/PGMEC, Brazil

4 National University of Colombia, Bogota, Colombia

References

[1] Andrade, D. O., Tulcan, O. E. P., Andrade, E. T., & Pereira, R. G. (2008). Determination of the physical characterization of vegetable oil. *International Conference of Agricultural Engineering*, 4.

[2] Bockisch, M. (1993). Fats and oils handbook. *Champaign, AOCS*, 345-718.

[3] Borrás, Miguel. Ângelo, & Toledo, José. Carlos. (2006). A Coordenação de Cadeias Agroindustriais: Garantindo a Qualidade e Competitividade no Agronegócio. *Zuin, Luiz Fernando Soares; Queiroz, Timóteo Ramos. Agronegócios: Gestão e Inovação. São Paulo: Saraiva*, 464, cap. 2.

[4] Borrás, Miguel. Ângelo, & Toledo, José. Carlos. (2006). Qualidade dos Produtos Agroalimentares: A Importância da Gestão da Qualidade no Agronegócio. *Zuin, Luiz Fernando Soares; Queiroz, Timóteo Ramos. Agronegócios: Gestão e Inovação. São Paulo: Saraiva*, 464, cap. 7.

[5] Brum, Jaime. Victor Ferreira. (2004). Análise de Perigos e Pontos Críticos de Controle em Indústria de Laticínios de Curitiba- PR.

[6] Carrazza, Luis. Roberto, Maciel, Luis. Gustavo, Borges, Moacir. Chaves, Cordeiro, Augusto. César Rodrigues., & Ávila, João. Carlos Cruz. (2011). Caderno de Normas Fiscais, Sanitárias e Ambientais Regularização de Agroindústrias Comunitárias de Produtos de Uso Sustentável da Biodiversidade. *SPN- Instituto Sociedade, População e Natureza(ISPN).Brasília- DF*, 43.

[7] Dall', Agnol. A., Lazarotto, J. J., & Hirakuri, M. H. (2010). Desenvolvimento, Mercado e Rentabilidade da Soja Brasileira. *Circular Técnico 74. Embrapa Soja*, April, 18.

[8] Dobarganes, M. C. (2000). Frying fats: quality control. In: IUPAC workshop on fats, oils and oilseed analysis. *Rio de Janeiro. EMBRAPA. Book of conferences*, 29-45.

[9] Dulley, Richard. Domingues, & Toledo, Alessandra. A. Gayoso Franco. de. (2003). Rastreabilidade dos Produtos Agrícolas. *Informações econômicas, SP*, 33(3), mar.

[10] Embrapa Soja. (2012b). Uso industrial. http://www.cnpso.embrapa.br/index.php?op_page=27&cod_pai=31, Accessed: 26 March.

[11] Embrapa Soja. (2012a). Uso. http://www.cnpso.embrapa.br/index.php?op_page=25&cod_pai=29, Accessed: 26 March.

[12] FAO (Food and Agricultural Organization of the United Nations). (2012). http://faostat.fao.org/site/339/default.aspx, Accessed: 26 March.

[13] Gava, A. J., Silva, C. A. B., & Frias, J. B. G. (2008). Tecnologia de alimentos: princípios e aplicações. *São Paulo: Nobel.*

[14] Hui, Y. H. (1996). Baley's industrial oil and fats products. 5. Ed. *Nova York, Wiley,* 4.

[15] Knothe, G., Dunn, R., & Bagby, M. O. (2003). Biodiesel: The Use of Vegetable Oils and Their Derivatives as Alternative Diesel Fuels. *Fuels and Chemicals from Biomass. Washington, D.C.: American Chemical Society.*

[16] Lima, Suzana. Maria Valle., & Castro, Antonio. Maria Gomes. de. (2010). A Agroindústria de Óleo Vegetal para a Produção de Biodiesel. *Castro, Antonio Maria Gomes de; LIMA, Suzana Maria Valle; SILVA, João Flávio Veloso. Complexo Agroindutrial de Biodiesel no Brasil: Competitividade das Cadeias Produtivas de Matérias-Primas. Brasília, DF: Embrapa Agroenergia,* cap. 6.

[17] Lindley, M. G. (1998). The impact of food processing on antioxidants in vegetable oil, fruits and vegetables. *Trend in Food Science and Technology,* 9.

[18] Mandarino, José. Marcos Gontijo., & Roessing, Antônio. Carlos. (2001). Tecnologia para produção do óleo de soja: descrição das etapas, equipamentos, produtos e subprodutos. *Londrina: Embrapa Soja,* 40.

[19] Martins, Roberto. Antônio. (2009). Gestão da Qualidade Agroindustrial. *BATALHA, Mário Otávio. Gestão Agroindustrial. GEPAI. 3ª Ed. São Paulo: Atlas,* cap. 8.

[20] Mendes, Judas. Tadeu Grassi., & Padilha, Junior. João Batista. (2007). Agronegócio: Uma abordagem Econômica. *São Paulo: Pearson Prentice Hall.*

[21] Moraes, R. M. A. (2007). Potencial da soja na produção de biodiesel. http://www.cisoja.com.br/index.php?p=artigo&idA=1, Accessed: 26 March. 2012.

[22] Moraes, R. M. A. (2007). Potencial da soja na produção de biodiesel.. http://www.cisoja.com.br/index.php?p=artigo&idA=1, Accessed: 26 March 2012.

[23] Oetterer, M., Regitano-D'Arce, M. A. B., & Spoto, M. H. F. (2006). Fundamentos de Ciência e Tecnologia de Alimentos. *Barueri, SP: Manole,* 612.

[24] Pereira, R. G. ., Tulcan, O. P. ., Lameira, V. J. ., Espirito, Santo., Filho, D. M. ., & Andrade, E. T. (2011). Use of Soybean Oil in Energy Generation. *Dora Krezhova. (Org.). Recent Trends for Enhancing the Diversity and Quality of Soybean Products. Recent Trends for Enhancing the Diversity and Quality of Soybean Products. Rijeka: InTech,* 01, 301-320.

[25] Schenini, Pedro. Carlos. (2011). Gerenciamento de Resíduos da Agroindustria. *II Simpósio Internacional sobre Gerenciamento de Resíduos Agropecuários e Agroindustriais- II SIGERA.15 a 17 de março de- Foz do Iguaçu, PR, Volume I- Palestras.*

[26] Schneider, F. H. (1980). Zur extraktiven lipid-freisetzung aus pflanzlichen zellen. *Fette Seifen Antstrichmittel, Hamburgo,* 82(1), 16-23.

[27] Shahidi, F., & Wanasundra, U. N. (1998). Methods for measuring oxidative rancidity in fats and oils. *Akoh, C & Min, D. O. Foods lipids- chemistry, nutricion and biotechnology. Nova York, Marcel Dekker*.

[28] Shahidi, F., & Wanasundara, U. N. (1998). Omega-3 fatty acid concentrates: Nutritional aspects and production technologies. *Food Sci. Technol*, 9, 230-240.

[29] Tulcan, O. E. P., Andrade, D. O., Andrade, E. T., & Pereira, R. G. (2008). Analisys of physical characteristics of vegetable oil. *International Conference of Agricultural Engineering*, 4.

[30] USDA- United States Department of Agriculture. (2012). http://www.fas.usda.gov/psdonline/psdQuery.aspx, Accessed: 26 March.

[31] Young, V. (1980). Processing of oils and fats. *Hamilton, R. J. & Bhati, A. (eds). Fats and oils: chemistry and technology. Londres, Appl. Sci. Publ*, 135-165.

Chapter 6

Effect of Dietary Plant Lipids on Conjugated Linoleic Acid (CLA) Concentrations in Beef and Lamb Meats

Pilar Teresa Garcia and Jorge J. Casal

Additional information is available at the end of the chapter

1. Introduction

Beef and lamb, are a food category with positive and negative nutritional attributes. Rumi-nant meats are major sources for many bioactive compounds including iron, zinc and B vita-mins. However they are associated with nutrients and nutritional profiles that are considered negative including high levels of saturated fatty acids (SFA) and cholesterol. It is well know that the low PUFA/SFA and high n-6/n-3 ratio of meats contribute to the imbal-ance in the in the fatty acid intake of today consumers [1]. Consumers are becoming more aware of the relationships between diet and health and this has increased consumer interest in the nutritional value of foods. Nutritionist advisers recommended a higher intake of poly-unsaturated fatty acids (PUFA), especially n-3 PUFA at the expense of n-6 PUFA.

The nutritional beef and lamb profile could be further improved by addition of potentially health promoting nutrients. There are many references of improved fatty acid composition in grass fed beef. Besides the beneficial effects of n-3 fatty acids on human health one fatty acid that has drawn significant attention for its potential health benefits in the last two deca-des is conjugated linoleic acid (CLA). Conjugated linoleic acids (CLA) are implicated as anti-carcinogenic, anti-atherosclerosis, and anti-inflammatory agents in a variety of experimental model systems. It has been shown that in ruminants grazing have potential beneficial effects on PUFA/SFA and n-6/n-3 ratios, increasing the PUFA and CLA content and decreasing the SFA concentration of beef [2].

The total CLA content of beef varies from 0.17 to 1.35% of fat [3]. This wide range is related to the type feed, breed differences, and management strategies used to raise cattle [3, 4]. Grazing beef steers on pasture or increasing the amounts of forage (grass or legumes hay) in the diet has been shown to increase the CLA content in the fat of cattle. Also, supplementing high-grain diets of beef cattle with oils (e.g., soybean oil, linseed oil, sunflower oil) may increase the CLA content of beef [3, 5].

There has been an increased interest in the substitution of animal fat sources with vegetable oils in animal nutrition. Vegetable oils have been attributed with reducing the level of saturation in monogastric animal tissues due to their unsaturated fatty acid concentration when compared with animal fat. In ruminants, dietary lipids were undergo two important transformations in the rumen. The initial transformation is the hydrolysis of the ester bond by microbial lipases. This initial step is a pre requisite for the second transformation, the biohydrogenation of unsaturated fatty acids [6, 7].

Several factors influence the CLA content of beef as breed, sex, seasonal variation, type of muscle, production practices but diet plays the most important role. Dietary CLA from beef can be increased by manipulation of animal diets. CLA concentration in beef can be influenced by dietary containing oils or oilseeds high in PUFA, usually linoleic or linolenic fatty acids.These dietary practices can increase CLA concentrations up to 3 fold [5, 8]. Moreover, trans-11 18:1 (vaccenic acid,VA) is the precursor of cis-9,trans-11 18:2 (rumenic acid, RA) is the major CLA isomer in animal and humans and, therefore, it might be considered as a fatty acid with beneficial properties.

Soybean oil is one of the few plant sources providing ample amounts of both essential fatty acids 18:2 n-6 and 18:3 n-3. The fatty acid content of soy foods is often unrecognized by health professionals, perhaps because there is so much focus on soy proteins. Soybeans are used in cattle, poultry and pigs diets and could be a more important source of 18:3 n-3 for animal nutrition and also increase 18:3 n-3 and its fatty acids metabolites in meats. Genomics, specifically marker assisted plant breeding combined with recombinant DNA technology, provided powerful means for modifying the composition of oilseeds to improve their nutritional value and provide the functional properties required for various food oils [9].

Thus, the manipulation of the fatty acid composition in ruminant meat to reduce SFA content and the n-6/n-3 ratio whilst, simultaneously increasing the PUFA and CLA contents, is the major importance in meat research. The supplementation of ruminant diets with PUFA rich lipids is the most effective approach to decrease saturated FA and promote the enrichment of CLA and n-3 PUFA.

2. CLA structure, biosynthesis and potential beneficial effects on human health

The CLA acronym refers to a group of positional and geometric isomers of linoleic acid, in which the double bands are conjugated. At least twenty four different CLA isomers have been reported as occurring naturally in food, especially from ruminant origin [10]. Isomerisation and incomplete hydrogenation of PUFA in the rumen produce several of octadecenoic, octadecadienoic and octadecatrienoic isomeric fatty acids [11] and, at least some of them, have powerful biological properties. The formation of conjugated dienes in the rumen during biohydrogenation of lipids in feed was observed previously, however, the anticarcinogenic effect of beef extracts was first observed and later identified [12, 13, 14].

The dominant CLA in ruminant meats is the cis-9, trans-11 isomer (RA) which has being identified as possessing a range of health promoting biological properties including antitumoral and anticarcinogenic activities [15]. The rumenic acid is mostly produced in tissues by delta 9 desaturation of trans-11 18:1, (VA) and by ruminal biohydrogenation of dietary PUFA. The higher deposition of CLA in the neutral lipid fraction, 88% of total CLA relatively to phospholipid fraction, has been reported [16].The majority of the main natural isomer cis-9,trans-11 CLA does not originate directly from the rumen. Instead, only small amounts of CLA escape the rumen and trans-18:1 isomers are the main biohydrogenation intermediates available. El absorbed trans-11 18:1 is desaturated in the tissues by Δ9-desaturase to form RA [17]. Stearoyl-CoA (SCD) is a rate-limiting enzyme responsible for the conversion of SFA into monounsaturated fatty acids (MUFA). This enzyme, located in the endoplasmic reticulum, inserts a double band between carbons 9 and 10 into SFA and affects the fatty acid composition of membrane phospholipids, triglycerides and cholesterol esters [18]. SCD is also a key enzyme in the endogenous production of the cis-9,trans-11 isomer of conjugated linoleic acid (CLA). Trans octadecenoates (trans 18:1) are the major intermediates formed during rumen biohydrogenation of C18 PUFA. High trans-10 18:1 have been observed in tissues of concentrated-fed ruminants, whereas vaccenic acid is consistently associated with forage feeding [11, 19]. Evidence is accumulating that different trans 18:1 isomers have differential effects on plasma LDL cholesterol. Trans-9 and trans-10 18:1 are more powerful in increasing plasma LDL cholesterol than trans-11 18:1 [20]. Comparison of antiproliferative activities of different CLA isomers present in beef on a set of human tumour cells demonstrates that all CLA isomers possess antiproliferative properties. It appears that important to determine the variations of the distribution of CLA isomers in beef since these proportions could influence the biological properties of bioformed CLA [21].

3. Factors influencing CLA concentrations on beef lipids

Amounts of CLA in beef vary mainly with feeding conditions, nature and quality of forages, proportions between forage and concentrate, oil-seed supplementations, but also with intrinsic factors such as breed and sex and age of animals [22].

3.1. Breed, sex and age (Table 1)

Breed or genotype and production system are determinant factors of the fatty acid composition of the ruminant meats. Breed affects the fat content of meat and fat content itself is a factor determining fatty acid composition. Genetic variability relates to differences between breeds or lines, variation due to the crossing of breeds and variation between animals within breeds reported that it can be difficult to assess the real contribution of genetics to variation in the CLA content.

	CLA	Reference
Breed		
LD Limousin	2.24 g/100g	[29]
LD Angus	1.96 g/100g	[29]
LD Angus	0.51 b% FAME	[25]
LD Charolais x AA	0.57 a% FAME	[25]
LD Holando x AA	0.58a% FAME	[25]
LD Nguni grass	0.34% FA	[28]
LD Bonsmara grass	0.31% FA	[28]
LD Angus grass	0.33% FA	[28]
LD Holstein grass	0.84 % FA	[24]
LD Simmental grass	0.87% FA	[24]
LD Holstein concentrate	0.75% FA	[24]
LD Simmental concentrate	0.72% FA	[24]
SM Pasture and Silage Steers Longhorn	6.75a mg/100g	[23]
SM Pasture and Silage Steers Charolais	3.29b mg/100g	[23]
SM Pasture and Silage Steers Hereford	2.93b mg/100g	[23]
SM Pasture and Silage Steers B. Gallowey	5.09a mg/100g	[23]
SM Pasture and Silage Steers Beef Shorton	4.01ab mg/100g	[23]
Sub Pasture and Silage Steers Longhorn	1210a mg/100g	[23]
Sub Pasture and Silage Steers Charolais	651b mg/100g	[23]
Sub Pasture and Silage Steers Hereford	584b mg/100g	[23]
Sub Pasture and Silage Steers B. Gallowey	796 b mg/100g	[23]
Sub Pasture and Silage Steers Beef Shorton	808b mg/100g	[23]
Mertolenga PDO beef	0.39ab g/100g FA	[27]
Mertolenga PDO veal	0.46a g/100g FA	[27]
Vitela Tradicional do Montado PGI veal	0.35b g/100g FA	[27]
LT & LL Veal Limousin	1.09% FAME	[26]
LT & LL Veal Tudanka x Charolais	1.00% FAME	[26]
Sex and age		
LL bulls 14 month	0.37 % FA	[31]
LL bulls 18 month	0.39% FA	[31]
LL heifers 14 month	0.44% FA	[31]
LL heifers 18 month	0.41 % FA	[31]
L lumborum steers	0.20 % FA	[32]
L.lumborum bulls	0.21 % FA	[32]

Table 1. Conjugated linoleic acid (CLA) concentrations on beef according to breed, sex and age .a b Indicates a significant differences (at least p < 0.05) between breed, sex or age reported within each respective study. Abbreviations LD: Longissimus dorsi ; SM: Semimembranosus ; LL Longissimus lumborum; LT Longissimus thoracis; Sub: Subcutaneous fat.

Significant between-breed differences in CLA content were observed in both muscle and subcutaneous adipose tissue of five breeds of cattle with the highest values in Longhorn and with the lowest in Hereford [23]. German Holstein bulls accumulated a higher amount of CLA compared with German Simmental bulls [24]. CLA percentages were affected by breed with the low values for Angus beef compared with Charolais x Angus and Holstein Argentine steers [25]. The content of trans -10 C18:1 isomer tended to be higher in Limousin compared to Tudanca meat when expressed as mg/100g of meat, and the difference was only significant when expressed in terms of relative percent. The higher level of trans-10 C18:1 was consistent with the greater consumption of concentrate by Limousin calves [26]. Within a similar production system the age/weight, gender and crossbreeding practices have minor effects on muscle FA composition but Mertolenga-PDO veal has higher total CLA contents that PDO beef and PGI veal [27]. On the contrary the cis-9, trans-11 CLA levels among steers of Nguni, Bonsmara and Angus breeds raised on natural pasture were similar [28]. Similar results were found comparing the CLA content of Limousin and Aberdeen Angus beef [29].

Sex and age differences in muscle FA contents are often be explained by the degree of fatness and associated changes in the triacylglycerol/phospholipid ratio [30]. Sex-dependent differences in the FA composition of muscle an adipose tissue from cattle slaughtered at different ages were demonstrated [31]. Concentration CLA in meat beef not affected by castration [32].

3.2. Type muscle and anatomical location (Table 2)

Little work has been conducted to assess the effects of slaughter season and muscle type on meat CLA profile. The type of muscles strongly influenced proportions of total CLA and of all CLA isomers classes in intramuscular fatty acids (Table 2). CLA is mainly associated to the triacylglycerol fraction which is linked to the fat content of tissues [21]. VA and CLA percentages were lower in lean muscle than subcutaneous fat or marbling [33]. The CLA content of steaks differs depending on the location of the fat, CLA level was almost doubled in outer subcutaneous fat compared to lean muscle [34]. There was significant differences in the concentration of CLA among depot sites through-out a bovine carcass. The brisket contained a higher concentration of cis-9, trans-11 CLA but no significant differences in the concentrations of trans-10, cis-12 CLA among the locations [35].

3.3. Season and pasture type (Table 2)

No differences between dietary grass silage and red clover silage were detected on CLA content of LD muscle of dairy cull cows [36]. Total CLA content was lower ($p < 0.05$) in intensively produced beef than in Carnalentejana-PDO meat, which did not show significant differences ($p<0.05$) when the slaughter season was compared. Furthermore *Longissimus thoracis* (LT) muscle had a higher ($p<0.001$) total CLA content relative to *Longissimus dorsi* (LD) muscle. In addition no significant differences ($p<0.05$) regarding specific CLA content were observed when slaughter season, production system and muscle type were analyzed [37]. Significant interactions between the slaughter season and muscle type were obtained for several fatty acid and CLA isomers and for total lipid and CLA. Mirandesa –PDO veal showed seasonal differences in the levels of CLA isomers but the CLA content was affected by much more influence by the muscle type [38]. The variation of CLA milk fat content during pasture season might be related to the alfa-linolenic/linoleic acid ratio in the pasture. The ratio in the average pasture sample decreased from 4.36 in May to 1.97 in August, and subsequently it increased to 3.14 in September, thus close to that at the beginning of pasture season. Thus the seasonal variation of the ratio in pasture were directly proportional to the corresponding content of CLA in ewe milk fat [39].

	CLA	Reference
Muscle and adipose tissue location		
Steak muscle	0.30b % FA	[33]
Steak marbling	0.50a % FA	[33]
Outer subcutaneous fat	0.50a % FA	[33]
Inner subcutaneous fat	0.50a% FA	[33]
Seam	0.40ab % FA	[33]
Adipose tissue brisket	0.70a g/100g FA	[35]
Adipose tissue chuck	0.62ab g/100g FA	[35]
Adipose tissue flank	0.56b g/100g FA	[35]
Adipose tissue loin	0.53b g/100g FA	[35]
Adipose tissue plate	0.57b g/100g FA	[35]
Rib	0.52b g//100g FA	[35]
Round	0.63ab g/100g FA	[35]
Sirloin	0.57b g/100g FA	[35]
LT concentrate	4.45 mg/g fat	[37]
ST concentrate	3.88 mg/g fat	[37]
Season		
LT Autumn	5.07 mg/g fat	[37]
LT Spring	4.92 mg/g fat	[37]]
ST Autumn	3.82 mg/g fat	[37]
ST Spring	5.06 mg/g fat	[37]
L L Spring	0.30 a g/100g FA	[34]
L L Autumn	0.31a g/100g FA	[34]
ST Spring	0.23 b g/100g FA	[34]
ST Autumn	0.19b g/100g FA	[34]]
Pasture type		
LD Tall fescue	0.28%	[91]
LD Alfalfa	0.37%	[91]
LD Red clover	0.30%	[91]
LD cull cows grass silage	0.22 % TFA	[36]
LD cull cows red clover silage	0.17 % TFA	[36]

Table 2. Conjugated linoleic acid (CLA) concentrations on beef according to muscle and adipose tissue location, season and grass composition a b Indicates a significant differences (at least $p<0.05$) between anatomical location, season or pasture type reported within each respective study. Abbreviations LD: Longissimus dorsi ; SM: Semimembranosus ; LL Longissimus lumborum; LT Longissimus thoracis

3.4. Grass vs. concentrate (Tables 3 & 5)

A direct linear relation between grass percentage in cattle diet and meat CLA content has been described by [2] although the mechanism remains controversial. They suggested that grass in the diet enhances the growth of ruminal bacterium *Butyrivibrio fibrisolvens* which convert 18:2 n-6 into cis-9, trans-11 CLA isomer through the action of a linoleic acid isomerase. Others [40] proposed that the increased content of CLA in animals fed forage- based diets is associated with an increase in trans-11 18:1, which is the substrate of stearoyl-CoA desaturase in tissues. It is generally accepted that the concentrations of CLA can be increased in beef by increasing the forage to concentrate ratio, and by feeding fresh grass instead of grass silage [4, 22] (Table 3). Beef contains both of the bioactive CLA isomers, namely, cis-9, trans-11 and trans-10, cis-12. Many reports demonstrated that cis-9, trans 11 CLA is a major fatty acid in tissue and little or no trans-10, cis 12 CLA was detected [5,41]. High trans-10 18:1 have been observed in tissues of concentrated-fed ruminants, whereas vaccenic acid is consistently associated with forage feeding [11, 19]. Significantly higher contents of trans-18:1 were found in animals fed on concentrate diets relative to the pasture diet. This is mainly due to the trans-6, trans-8, trans-9 and trans-10 isomers, since the trans-11 and trans-12 18:1 remains unaffected by the dietary treatments. The feeding systems, pasture only, pasture feeding followed by 2 or 4 months of finishing on concentrate, and concentrate only, had a major impact on the concentration of CLA isomers from bull LD muscles. Beef fat from pasture-fed animals had a higher nutritional quality relative to that from concentrate-fed bulls and the feeding regimen had a major impact on the CLA isomeric distribution of beef affecting 10 of 14 CLA isomers. The CLA isomeric profile showed a clear predominance of the cis-9, trans-11 isomer for all diets [42]. The grass silage diets increased the proportions of trans-11 18:1 and cis-9, trans-11 18:2. Feeding a high forage diet may therefore have increased the rate of appearance of trans-11 18:1 in the rumen, provid- ing more substrate for the endogenous production and deposition of CLA in bovine tissues [43]. This hypothesis is consistent with an increase of trans-11 18:1 concentration with no effect on cis-9, trans-11 in duodenal content of Hereford steers fed increasing levels of grass hay [44]. The relative flow of PUFA through the major biohydrogenation pathways, trans-10 or trans-11, 18:1, can be judged by the 11t-/10t- 18:1 ratio with a higher ratio denoting an improvement in its healthfulness to its human consumers [45]. Backfat composition was compared in steers fed either a control (barley grain based) diet or diets containing increasing levels of corn or wheat derived dried distillers'grains with solubles (DDGS). Back fat from control and wheat de- rived DDGS fed steers had lower levels of trans-18:1 and a higher 11trans/10 trans 18:1 ratio compared to back fat from corn derived DDGS fed steers [45]. The explanation might be found in ruminal biohydrogenation pathway of LA and ALA. Most of the cis-9, trans-11 CLA iso- mer present in tissues derive from endogenous desaturation of trans-11, 18:1, which origi- nates during biohydrogenation of 18:2n-6 and 18:3n-3. The CLA concentrations in three different muscles of pasture- or feedlot-finished cattle were greater from pasture-finished than from cattle feedlot-finished [46]. The absolute cis-9, trans-11 CLA was about twice as high in Asturi- ana de la montaña (AV) and Asturiana del Valle (AV) animals than in other AV genotypes, probably due to the much higher fat content of the AM and AV animals [47]. This effect was also found in other studied were cis-9, trans-11 content variation was influenced by the total lipid content, and hence with variation in the neutral lipid fraction [48]. A linear correlation between VA and cis-9, trans-11 CLA was observed in several studies [8] and in other studied no significant correlation was found [49]. Breed or genotype effects could act by enhancing or inhibiting the Δ9-desaturase activity. The major isomers in beef fed a high barley diet is trans-10, 18:1 rather than trans-11, 18:1. In feedlot finished beef fed a diet containing 73% barley was found 2.13% of trans-10 18:1 and only 0.77% of trans-11 18:1 in subcutaneous fat [19, 50]. Feeding ruminants diets with high levels of barley (low fiber, high starch) reduces rumen pH, alters the bacterial flora and causes a shift in the biohydrogenation pathway towards producing trans-10 18:1 instead of trans-11 18:1 [51]. Subcutaneous fat is quite sensitive to changes in diet and rumen function. This is due to adipose tissue having a high proportion of neutral lipids which accumulate greater levels of PUFA biohydrogenation products relative to polar lipids [24]. In addition, subcutaneous fat is easily accessible, inexpensive and levels of trans-18:1 have been reported to be linearly related to those found in muscle [52]. Vaccenic acid made up the greatest concentration of total trans fats in grass-fed beef, whereas CLA accounted for approximately 15% of the total trans fats [53].

	CLA	Reference
LD Grazing	10.8a mg/g fat	[2]
LD Concentrate-fed	3.7b mg/g fat	[2]
LD Grazing	5.3a mg/g fat	[90]
LD Concentrate-fed	2.5 b mg/g fat	[90]
LD Pasture	0.72 % FAME	[25]
LD Pasture +0.7% corn	0.61 % FAME	[25]
LD Pasture+1.0 %corn	0.58% FAME	[25]
LD Feedlot	0.31 % FAME	[25]
LD Grass silage (GS)	3.62% FA	[43]
LD GS +Low concentrate	2.50% FA	[43]
LD GS+ High concentrate	2.72% FA	[43]
LT Semi-intensive 12 month	0.49a %	[92]
LT Semi-intensive 14 month	0.49a %	[92]
LT Intensive 12 month	0.25b %	[92]
LT Intensive 14 month	0.29b %	[92]
Ground control	0.50b g/100g	[53]
Ground grass	0.94a g/100g	[53]
Steaks control	0.38b g/100g	[53]
Steaks grass	0.66a g/100g/	[53]
Control	0.82 % FA	[45]
Back fat 20% DDGS corn	0.88 % FA	[45]
Back fat 20% DDGS wheat	0.88 % FA	[45]
Back fat 40%DDGS corn	0.97 % FA	[45]
Back fat 40% DDGS wheat	0.81 % FA	[45]
LT concentrate	4.45 mg/g fat	[37]
ST concentrate	3.88 mg/g fat	[37]

Table 3. Conjugated linoleic acid (CLA) concentrations on beef under dietary grass or concentrate a b Indicates a significant differences (at least $p<0.05$) between dietary grass or concentrate reported within each respective study. Abbreviations LD: Longissimus dorsi ; Longissimus thoracis ; ST: Semitendinosus

3.5. Oil supplementation (Tables 4 & 5)

The most common method of enhancing the CLA and VA content of ruminant meat and dairy products is to provide the animal with additional dietary unsaturated fatty acids, usually from plants oils such as soybean oil (SBO), for use as substrates for ruminal biohydrogenation [4]. Steers fed a corn-based diet supplemented with SBO may enhance TVA without impacting CLA, while reducing the MUFA content of lean beef [54]. Both oilseed and free oils affect CLA content in a similar manner. Free plant oils with high PUFA concentrations are normally not included in ruminant diets as high levels of dietary fat disturb the rumen environment and inhibit microbial activity. The main sources of supplementary fatty acids in ruminant rations are plant oils and oilseeds, fish oils, marine algae and fat supplements. Since dietary inclusion of fatty acids must be restricted to avoid impairment of rumen function, the capacity to manipulate the fatty acid composition by use of ruminally available fatty acids is limited [55]. Many researchers have found higher CLA content in muscle lipids by supplementing with different oils. However, some studies reported no significant differences in CLA content due to oil supplementations. The differences in responses to plant oils were probably due to variations in stage of growth of cattle, levels of oil supplementation, levels of oil in total ration and amount of linoleic acid in oils. Researchers have successfully increased CLA content by supplementation of different oils [4,48,56]. Others [3] supplementing with 4% SBO to diets did not affect the CLA. Similar to [41] who reported that feeding 5% SBO no affected CLA but increased trans10-cis-12 CLA. The addition of different vegetable oils to the bulls diet (soybean or linseed, either protected or not protected from rumen digestion) increased the CLA content, with an average CLA value of 0.72 %. The increase of CLA was also due to the addition of oils presenting large quantities of its precursor LA in diets with unprotected soybean and linseed oils [57]. Diets containing silage and concentrate or sugarcane and sunflower seeds fed Canchim- breed animals, produce an improvement in CLA levels (0.73g/100g vs. 0.34g/100) [58]. Rapeseed oil and whole rapeseed do not seem to have positive effects. Of the three studied none showed increased CLA concentrations in the LDi after supplementation with 6% rapeseed oil [41]. Soybean oil (SBO) has been used as a source de LA throughout the finishing period to promote greater CLA accretion in lean tissues with equivocal results [56, 41]. and where CLA accretion was increased with SBO addition, growth performance was reduced [56]. Fed steers with 5% of soybean oil in a finishing experiment for 102 days had no effects in meat cis-9, trans, 11 CLA [41]. In a study with steers, supplementation of 4% soybean oil to a finishing diet based on concentrate and forage (80:20) resulted in a depression of the CLA deposition in muscle tissues (2.5 vs. 3.1 mg/g FAME) compared to the same diet without soybean oil On the other hand, comparing 4% with 8% added soybean oil in a 60:40 concentrate : forage diet showed a numerical increase of the CLA content with the higher soybean supplementation (2.8/3.1 mg/g FAME) [59]. The inclusion of sunflower oil in the diets (80% barley, 20% barley silage) of finishing cattle at 0%, 3%, or 6% increased the CLA content of the beef by 75% when cattle were fed 6% sunflower oil [4]. Although supplementation with oil or oil seeds increased CLA content in muscle, the inclusion of linoleic acid –rich oil or oilseeds such as safflower or sunflower, in the diet of ruminants appears to be the most effective [60]. Supplementation of cattle with a blend of oils rich in n-3 PUFA and linoleic acid results in a synergistic accumulation of ruminal and tissue concentrations of TVA [61]. VA is the substrate for Δ9 –desaturase- catalyzed de novo tissue tissue synthesis of cis-9 trans-11 isomer of CLA. However, despite increases in its substrate, muscle tissue concentrations of cis-9, trans-11 CLA have not increased by using this strategy [62]. Inclusion of extruded linseed in the diet of Limousin and Charolais cattle, increase CLA [63]. The importance of the contribution of TVA to total CLA intake is further reinforced by a French study [64] in which a huge 233% increase of VA was shown, along with 117% increase of RA, which was caused by adding extruded linseeds into the animal fodder. Several authors reported that diets containing proportionally high levels of linolenic acid, such as fresh grass, grass silage, and concentrates containing linseed, resulted in increased deposition of the cis-9, trans-11 CLA isomer in muscle [65]. The biohydrogenation by rumen microorganism does not include the cis-9, trans-11 CLA isomer as an intermediate. The trans-11 18:1 is the common intermediate during the biohydrogenation of dietary linoleic acid and linolenic acid to stearic acid [6]. Since only a relatively small percentage the cis-9, trans-11 CLA isomer, formed in the rumen, is available for deposition on the muscles, the major source of this isomer in muscle results from the endogenous synthesis involving Δ 9 desaturase and vaccenic acid [17].. Hereford steers cannulated in the proximal duodenum were used to evaluated the effects of forage and sunflower oil level on ruminal biohydrogenation and conjugated linoleic acid. Flow of trans-10 18:1 decreased linearly as dietary forage level increased whereas trans -11 18:1 flow to the duodenum increased linearly with increased dietary forage. Dietary forage or sunflower oil levels did not alter the outflow of cis-9, trans-11 CLA [44]. Linseed supplementation was an efficient way to increase CLA proportion in beef (+22% to 36%) but was highly modulated by the nature of the basal diet, and by intrinsic factors as breed, age/sex, type of muscle, since these ones could modulate CLA proportions in beef from 24% to 47% [21]. Soybean oil, which is rich in linoleic acid, has been found in several studied [66,67,] to be more efficient than linseed oil, which is rich in linolenic acid, in increasing the CLA content of milk. In beef cattle the addition of 3% and 6% sunflower oil to a barley based finishing diet results in increased CLA content in LD muscle: 2.0 vs 2.6 vs. 3.5 mg/g lipid for control, 3%,, and 6% sunflower oil, respectively. A more substantial increase in the CLA concentration was found when sunflower oil was added to both the growing and finishing diet of beef cattle.[68,69]. 4.3, 6,3 and 9.1 mg CLA / g FAME in LD muscle lipids of heifers, were found, after supplementing the feed with 0, 55, and 110 g sunflower oil per kg of the diet for 142 days before slaughter [48]. Supplementation of a high forage fattening diet with either soybean oil or extruded full fat soybeans at a level of 33g added oil per kg of diet DM resulted in a 280-410 % increase in the concentration of CLA in the intramuscular and subcutaneous lipid depots of fattening Friesian bull calves The content of VA in both lipid depots were also increased about three-fold by this oil supplementation [70].

	CLA	Reference
Concentrate IMF fat	3.4 b mg/g fat	[70]
Soybean oil IMF fat	13.0 a mg/g fat	[70]
Extruded soybean IMF fat	15.4 a mg/g fat	[70]

	CLA	Reference
Concentrate Sub fat	5.2 c mg/g fat	[70]
Soybean oil Sub fat	20.3 b mg/g fat	[70]
Extruded soybean Sub fat	26.6 a mg/g fat	[70]
LD concentrate / silage	0.41d % FA	[93]
LD Grass	0.70c % FA	[93]
LD grass +sunflower oil	1.34a % FA	[93]
LD grass +linseed oil	0.93b% FA	[93]
LD Wagyu Control	0.27 b % FA	[68]
LD Wagyu 6% sunflower oil	1.29a % FA	[68]
LD Limousin Control	0.28b % FA	[68]
LD Limousin 6% sunflower oil	1.19a % FA	[68]
LM grass	0.73c % FA	[48]
LM grass+ sunflower oil	1.78a % FA	[48]
LM grass+linseed oil	1.26b % FA	[48]
LM Corn oil 0%	0.68b % FA	[94]
LM Corn oil 0.75%	0.85a % FA	[94]
LM Corn oil 1.5%	0.81ab % FA	[94]
LT Control	0.33% FA	[95]
LT Control + Vit E	0.36 % FA	[95]
LT Control	0.34 % FA	[95]
LT Control+ flaxseed	0.34 % FA	[95]
LD Control	0.35 c mg/100g FA	[57]
LD Soybean oil	0.94a mg/100g FA	[57]
LD Linseed oil	0.80a mg/100g FA	[57]
LD Protected linseed oil	0.55b mg/100g FA	[57]
LM grass NL	0.78c g/100g FA	[48]
LM grass+sunflower oil NL	1.90a g/100g FA	[48]
LM grass+linseed oil NL	1,35b g/100g FA	[48]
LM grass PL	0.32c g/100g FA	[48]
LM grass+sunflower oil PL	0.71a g/100g FA	[48]
LM grass+linseed oil PL	0.51b g/100g FA	[48]

Table 4. Conjugated linoleic acid (CLA) concentrations on beef under dietary oils supplementation. a b Indicates a significant differences (at least p<0.05) between dietary oil supplementations reported within each respective study. Abbreviations LD: Longissimus dorsi ; SM: Semimembranosus ; LT Longissimus thoracis; Sub: Subcutaneous fat; NL: neutral lipids; PL: Phospholipids.

Diet P vs. C	Trans- 11 18:1	Trans- 10 18:1	Reference
C	0.92	1.21b	[42]
P+4month C	1.10	0.81b	[42]
P+2month C	1.15	0.98b	[42]
P	1.35	0.20a	[42]
Ground control	1.14	2.69	[53]
Ground grass	4.14	0.75	[53]
Steaks control	0.51	3.60	[53]
Steaks grass	2.95	0.60	[53]
Grass silage (GS)	2.03a	Na	[43]
GS +Low C	1.37b	Na	[43]
GS+ High C	1.15b	Na	[43]
Control	0.65	2.02	[45]
20% DDGS corn	0.78	2.37	[45]
20%DDGS wheat	0.74	1.60	[45]
40% DDGS corn	0.92	3.16	[45]
40%DDGS wheat	0.69	1.33	[45]
LT et LL Tudanca x Charolais	2.68	0.36b	[26]
LT et LL Limousin	2.24	1.01 a	[26]

Table 5. Trans-11 and trans 10 C18:1 isomer proportions on beef under different conditions. a b Indicates a significant differences (at least $p<0.05$) between trans-10 C18:1 and trans-11 C18:1 reported within each respective study. "na" indicates that the value was not reported in the original study. Abbreviations LD: Longissimus dorsi ; LL Longissimus lumborum LT: Longissimus thoracis.

4. Factors influencing CLA concentrations on lamb lipids

In lamb production, more than other species, each country or region has its own specific weight/age and type of carcass criteria, depending on the culture and the customs of the people. Many factors including breed, gender, age/body weight, fatness, depot site, environmental condition, diet and rearing management influence lamb fat deposition and composition. Further studied are needed to understand how animal circadian rhythms, diurnal rumination patterns and daily changes in herbage chemical composition could affect lamb fatty composition [71].

4.1. Production system (Tables 6 & 8)

No differences were detected in the muscle CLA/ trans-11 18:1 index of herbage or concentrate –fed lambs but the supplementation of tanino produced strong effects on the accumulation of fatty acids which are involved in the biohydrogenation pathway [72]. During two years (Y1 and Y2) lambs were under four diets. Only silage both pre and post weaning (SS), only silage until weaning, silage plus concentrate thereafter (SC), silage plus concentrate both pre and post weaning (CC) and silage plus concentrate before weaning, only silage after (CS). Treatment differences for trans-11 18:1 were presented only in Y1, with muscle from the lamb fed silage before weaning having the highest levels. The same groups has the highest levels of cis-9, trans 11 CLA in Y1. Similar in Y2 the group SS has the highest CLA level, while the CC group has the lowest [73]. The feeding strategy around parturition influence the CLA and VA content of lamb meat. Pre-partum grazing, regardless of post-partum feeding, can improve the fatty acid composition, increasing the CLA content in lamb meat [74]. The meat of lambs slaughtered at Christmas has a higher CLA content than those reared in winter (slaughtered at Easter) as a result of the traditional feeding system which provided that lambs born and reared in autumn receive milk from ewes permanently pastured while those reared in winter are suckled by ewes permanently stall-fed [75]. The grazing on *T.subterraneum* as monoculture, associated with *L. multiflorum* in the proportion T/L=66/33 incremented cis-9,trans-,11 CLA of *L. dorsi* muscle of lambs [76]. The meat fatty acid profile was affected by the grazing management: compared to a morning-grazing or to a whole day-grazing management. Allowing lambs to gaze in the afternoon resulted in a meat fatty acid profile richer in CLA. In particular, in the 4hPM meat there is a greater proportion of those fatty acids arising from ruminal biohydrogenation, among them the CLA [71].

	CLA mg/g fat	Reference
LD Grass pellets	1.29a % FA	[40]
LD Concentrate diet	1.02a % FA	[40]
LD Concentrate diet *Ad libitum*	0.74b % FA	[40]
Muscle Concentrate+concentrate CC	0.46b g/100g lipids	[73]
Muscle Silage+concentrate SC	0.61a g/100g lipids	[73]
Muscle Concentrate+silage CS	0.45b g/100g lipids	[73]
Muscle Silage+silage SS	0.65a g/100g lipids	[73]
LT Pre-partum hay	1.42b % FA	[74]
LT Pre-partum grazing	1.66a % FA	[74]
LT Post-partum hay	1.35b % FA	[74]
LT Post –partum grazing	1.73a % FA	[74]
LD Grassed 9 am to 5 pm	1.85b g/100g FAME	[71]
LD Grassed 9 am to 1 pm	1.45b g/100g FAME	[71]
LD Grassed 1 pm to 5 pm	2.39a g/100g FAME	[71]
LL Sucking lamb Autumn	1.10% IM Fat	[75]
LL Sucking lamb Winter	0.56 % IM Fat	[75]

	CLA mg/g fat	Reference
LD Grazing subterraneous clover	0.46a % FA	[76]
LD Grazing Italian rye grass	0.26 b % FA	[76]
Pasture LD	0.90b % FAME	[96]
Pasture Leg muscles	1.27a % FAME	[96]
Pasture LD total lipids	0.90 % total FAME	[96]
Pasture LD Triacylglycerols	0.62 %total FAME	[96]
Pasture LD Phospholipids	0.11 % total FAME	[96]

Table 6. Conjugated linoleic acid (CLA) concentrations on lamb meat according to concentrate, pasture, muscle type and season. a b Indicates a significant differences (at least p<0.05) between values reported within each respective study. Abbreviations LD: Longissimus dorsi ; ST: Semitendinosus ; LL Longissimus lumborum; ; LT: Longissimus toracsis; SM: Semimembranosus.

4.2. Oil supplementation (Tables 7& 8).

Several strategies have been tested in recent years to improve CLA isomers in meat of inten-sively–reared lambs, keep indoors and fed high-concentrate diets rich linoleic acid and poor in linolenic. Incorporating linseed rich, in linolenic acid, the proportion of trans-11, 18:1 and cis-9, trans-11 18:2 were higher in the muscle and in the adipose tissues of linseed –fed lambs than in control lambs [77]. This increased is in contrast to results of [78] but in agree-ment with [79]. Discrepancias between these studies may due to differences in the level of intake the linoleic and linolenic acids or the different level of Δ9- desaturase inhibition as it has been shown that Δ9 desaturase is inhibited by PUFA with increasing inhibition as the degree of fatty acid unsaturation increases. Fed lambs from weaning to slaughter with diets that contained 5% supplemental from high oleic acid safflower or normal safflower in-creased the meat cis-9,trans,11 CLA compared with the control group [80] In lambs inclu-sion of 8% of soybean oil to a lucerne hay-based diet resulted in an intramuscular (M. *Longissimus thoracis*) CLA content of 23.7 compared with 5.5 mg/g FAME in the control group [81]. Feeding soybean and linseed oils to lambs pre and post weaning did not increase CLA content of muscle, whereas post weaning oil supplementation minimally increased CLA concentration in subcutaneous fat [82]. Conflicting results have been reported on alter-ing FA content of meat supplementing ruminant diets with lipid sources high in linoleic and linolenic acids. Some research suggests supplementing CLA, linoleic or linolenic acids in high concentrate fed to lambs can increase CLA content in muscle [83], whereas supplemen-tation of linoleic in finishing diets fed to cattle had no effects on CLA in adipose or muscle tissue [3,41]. Feeding lipid sources rich in linoleic and linolenic increases the cis-9,trans-11 18:2 content of ruminant meats [21,81, 83,84]. However, feeding linseed oil, rich in linolenic acid, seems to be less effective in the increases of cis-9, trans-11 18:2 in muscle than sunflow-er oil, rich in linoleic acid [45,48]. Seems to be that a blend of sunflower and linseed oils may be a good approach to obtain an enrichment in CLA in lamb meat. Maximun CLA concen-trations (42.9 mg/100 g fresh lamb tissue) was observed with 100% of sunflower, decreasing linearly at 78% by sunflower oil with linseed oil replacement [86]. A consistent significant increase in CLA content in lamb tissues was observed with dietary supplementation with

6% of safflower oil. The CLA concentration in several lamb tissues was increased by more than 200% [39]. These results indicated that supplementation of lamb feedlot diets with a source of LA was a successful method of increasing CLA content of tissues. Merino Branco ram lambs initially fed with concentrate showed a lower proportions of cis-9,trans-11 18:2 CLA (0.98% vs. 1.38% of total fatty acids) than lambs initially fed with Lucerne. Initial diet did not compromise the response to the CLA promoting diet (dehydrated lucerne plus 10% soybean oil) and the proportion of cis-9,trans-11 C18:2 CLA in intramuscular fat increased with the duration of time on the CLA-promoting diet (1.02% vs. 1.34% of total fatty acids) [87]. Supplementation of oilseed with different levels of oleic (rapeseed), linoleic (sunflower and safflower seeds), and linolenic acid (linseed) on trans-11 18:1 and CLA isomers on ewe different tissues showed that the percentage of trans-11 C18:1 averaged around 4.56 % of total fatty acids for all supplements and tissues [88]. Increasing dietary forage and soybean oil did not change the sheep mixed ruminal microbes concentration of vaccenic acid but increased rumenic acid [89].

	CLA mg/g fat	Reference
LD Sunflower oil	2.13a mg/100 g muscle	[86]
LD Sunflower oil+ 33% linseed oil	2.06 a mg/100 g muscle	[86]
LD Sunflower oil + 66% linseed oil	1.84b mg/100g muscle	[86]
LD Linseed oil	1.56 c mg/100g muscle	[86]
Leg control	1.78a mg/g fat	[97]
Leg CLA	1,50a mg/g fat	[97]
Leg Safflower oil	4.41b mg/g fat	[97]
Adipose tissue Control	2.77 b mg/g fat	[97]
Adipose tissue CLA	2.60 b mg/g fat	[97]
Adipose tissue Safflower oil	7.33 a mg/g fat	[97]
LD control	3955 b ppm in muscle	[98]
LD Control + 5% sunflower oil	8491a ppm in muscle	[98]
Fat control	4947b ppm in fat	[98]
Fat control +5% sunflower oil	11313a ppm in fat	[98]
LT Control	0.75c % of FA	[87]
LT Control+Lucerne+10% soybean oil	1.21b % of FA	[87]
LT Lucerne	1.28 b% of FA	[87]
LT Lucerne+Lucerne+10% soyben oil	1.47 a% of FA	[87]
Muscle Control no fat	0.05 b mg/g muscle	[77]
Muscle control+wheat+linseed	0.11 a mg/g muscle	[77]
Muscle control+corn+linseed	0.12 a mg/g muscle	[77]
LL Control	0.60b g/100g FAME	[88]
LL Linseed	0.72b g/100g FAME	[88]
LL Rapeseed	0.70b g/100g FAME	[88]
LL Safflower seed	0.96a g7100g FAME	[88]
LL Sunflower seed	0.98a g/100g FAME	[88]

Table 7. Conjugated linoleic acid (CLA) concentrations on lamb meat according to dietary oil. a b Indicates a significant differences (at least p<0.05) between valus reported within each respective study. Abbreviations LD: Longissimus dorsi ; LL Longissimus lumborum; LT Longissimus thoracis

	Trans- 11 18:1	Trans- 10 18:1	Reference
LD Grazing subterraneous clover	4.22	Na	[76]
LD Grazing Italian rye grass	3.65	Na	[76]
LD Grass 9 am to 5 pm	1.55a	Na	[71]
LD Grass 9 am to 1 pm	1.06b	Na	[71]
LD Grass 1 pm to 5 pm	1.60a	Na	[71]
LD grass pellets	2.25a	0.38b	[40]
LD concentrate	1.39b	1.54a	[40]
LD concentrate ad lib	0.85b	1.73a	[40]
LD concentrate	79.4	Na	[72]
LD Herbage	31.4	Na	[72]
Silage-silage	1.54a	Na	[73]
Silage-concentrate	1.45a	Na	[73]
Concentrate-concentrate	1.08b	Na	[73]
Concentrate-silage	1.14b	Na	[73]
LM Pre Control	0.25	5.13b	[82]
LM Pre Control +oil	0.29	6.02a	[82]
LM Post Control	0.25	4.30b	[82]
LM Post Control+ oil	0.29	6.81a	[82]
Sub Pre Control	0.29	9.85b	[82]
Sub Pre Control +oil	0.31	8.25b	[82]
Sub Post Control	0.28	7.09b	[82]
Sub Post Control+ oil	0.33	11.01a	[82]

Table 8. Conjugated linoleic acid (CLA) isomer proportionson lamb under different condicions. a b Indicates a significant differences (at least $p<0.05$) between values reported within each respective study. "na" indicates that the value was not reported in the original study. Abbreviations LD: Longissimus dorsi ; LM: Semimembranosus; Sub: Subcutaneous

5. Conclusions

Several factors influence the CLA content of ruminant meats as breed, sex, seasonal varia-tion, type of muscle, production practices but diet plays the most important role. CLA con-centration in beef and lamb can be influenced by dietary containing oils or oilseeds high in PUFA, usually linoleic or linolenic fatty acids. The supplementation of ruminant diets with PUFA rich lipids is the most effective approach to decrease saturated FA and promote the enrichment of CLA and n-3 PUFA. The differences in responses to plant oils were probably due to variations in stage of growth of animals, levels of oil supplementation, levels of oil in total ration and amount of linoleic acid in oils. Thus, the manipulation of the fatty acid com-position in ruminant meat to reduce SFA content and the n-6/n-3 ratio whilst, simultaneous-ly increasing the PUFA and CLA contents, is the major importance in meat research.

Author details

Pilar Teresa Garcia[1*] and Jorge J. Casal[2]

*Address all correspondence to: pgarcia@cnia.inta.gov.ar

1 Area Bioquimica y Nutricion. Instituto Tecnologia de Alimentos. Centro de Investigacion en Agroindustria. Instituto Nacional Tecnologia Agropecuaria. INTA Castelar, CC77(B1708WAB) Moron, Pcia Buenos Aires, Argentina

2 Universidad de Moron. Facultad de Ciencias Agroalimentarias. Cabido 137. Moron , Pcia Buenos Aires, Argentina

References

[1] Russo, G. L. (2009). Dietary n-6 and n-3 polyunsaturated fatty acids: From biochemical to chemical implications in cardiovascular prevention. Biochemical Pharmacolog , 77, 937-946.

[2] French, P., Stanton, C., Lawless, F., O`, Riordan. E.g, Monahan, F. J., Caffrey, P. J., & Moloney, A. P. (2000). Fatty acid composition, including conjugated linoleic acid, of intramuscular fat from steers offered grazed grass, grass silage or concentrate-based diets. *Journal Animal Science*, 78, 2849-2855.

[3] Dhiman, T. R., Nam, S. H., & Ure, A. L. (2005). Factors affecting conjugated linoleic acid content in milk and meat. *Crit. Rev Food Sci. Nutr.*, 45(6), 463-482.

[4] Mir, P. S., Mc Allister, T. A., Scott, S., Aalhus, J., Baron, V., Mc Cartney, D., et al. (2004). Conjugated linoleic acid-enriched beef production. *American Journal of Clinical Nutrition*, 79, 1207S-1211S.

[5] Madron, M. S., Peterson, D. G., Dwyer, D. A., Corl, B. A., Baumgard, L. H., Beermann, D. H., & Bauman,d, E. (2002). Effect of extruded full-fat soybeans on conjugated linoleic acid content of intramuscular, intermuscular, and subcutaneous fat in beef steers. *Journal Animal Science*, 80, 1135-1143.

[6] Harfoot, C. G., & Hazelwood, G. P. (1997). Lipid metabolism in the rumen. *In The Rumen Microbial Ecosystem, 2nd ed. P.N. Hobson, ed. Elservier Science Publishing Co., Inc., New York, NY.*

[7] Jenkins, T. C., Wallace, R. J., Moate, P. J., & Mosley, E. E. (2008). Board-invited review: Recent advances in biohydrogenation of unsaturated fatty acids within the rumenmicrobial ecosystem. *Journal Animal Science*, 86, 397-412.

[8] Enser, M., Scollan, N. D., Choi, N. J., Kurt, E., Hallett, K., & Wood, J. D. (1999). Effect of dietary lipid on the content of conjugated linolenic acid (CLA) in beef muscle. *Animal Science*, 69, 149-156.

[9] Jimenez, J. J., Bernal, J. L., Nozal, M. J., Toribio, L., & Bernal, J. (2009). Profile and relative concentrations of fatty acids in corn and soybean seed from transgenic and isogenic crops. *Journal of Chromatography*, 1216, 7288-7295.

[10] Sehat, N., Rickert, R., Mossoba, M. M., Kramer, J. K., Yurawez, M. P., Roach, J. A., et al. (1999). Improved separation of conjugated fatty acids methyl esters by silver ion-high performance liquid chromatography. *Lipids*, 34, 407-413.

[11] Bessa, R. J. B., Santos-Silva, J., Ribeiro, J. M. R., & Portugal, A. V. . (2000). Reticulo-rumen biohydrogenation and the enrichment of ruminant edible products with linoleic acid conjugated isomers. *Livestock Production Science*, 63, 201-211.

[12] Pariza, M. W., Ashoor, S. H., Chu, F. S., & Lund, D. B. (1979). Effects of temperature and time on mutagen formation in pan-fried hamburger. *Cancer Letters*, 17, 63-69.

[13] Pariza, M. W., Park, Y., & Cook, M. E. (2001). The biologically active isomers of conjugated linoleic acid. *Progress in Lipid Research*, 40, 283-298.

[14] Ha, Y. L., Grimm, N. K., & Pariza, M. W. (1987). Anticarcinogens from fried ground beef: heat- altered derivatives of linoleic acid. *Carcinogenesis* , 8, 1881-1887.

[15] Belury, M. A. (2002). Dietary conjugated linoleic acid in health: physiological effects and mechanisms of action. *Annual Review of Nutrition*, 22, 505-531.

[16] Wood, J. D., Enser, M., Fisher, A. V., Nute, G. R., Sheard, P. R., Richardson, R., et al. (2008). Fat deposition, fatty acid composition, and meat quality: A review. *Meat Science*, 78, 343-358.

[17] Griinari, J. M., & Bauman, D. E. (1999). Biosynthesis of conjugated linoleic acid and its incorporation in meat and milk of ruminants. In Advances in Conjugated Linoleic Acid Research. M.P. Yur awcez, Mossoba, M.M.,, 1

[18] Ntambi, J. M., & Miyazaki, M. (2004). Regulation of stearoyl-CoA desaturases and role in metabolism. *Progress Lipid Research*, 43, 91-104.

[19] Dugan, M. E., Kramer, R., J. K., G., Robertson, W. M., Meadus, W. J., Aldai, N., & Rolland, D. C. . (2007). Comparing subcutaneous adipose tissue in beef and muskus-with emphasis on trans 18:1 and conjugated linoleic acids. *Lipids*, 42, 509-518.

[20] Willet, W. C. (2005). The scientific basis for TFA regulation- is it sufficient? *In Proceedings of the first international symposium on "Trans Fatty Acids and Health", Rungstedgaard, Denmark*, 11-13 September, 24.

[21] De la Torre, A., Gruffat, D., Durand, D., Micol, D., Peyron, A., Scilowski, V., & Bauchard, D. (2006). Factor influencing proportion and composition of CLA in beef. *Meat Science*, 73, 258-268.

[22] Raes, K., De Smet, D., & Demeyer, D. (2004). Effect of dietary fatty acids on incorpo-ration of long chain polyunsaturated fatty acids and conjugated linoleic acid in lamb,beef and pork meat: A review. *Animal Feed Science and Technology*, 113, 199-221.

[23] Dance, I. J. E., Matthews, K. R., & Doran, G. (2009). Effect of breed on fatty acid com-position and stearoyl-CoA desaturase protein expression in the Semimembranosus muscle and subcutaneous adipose tissue of cattle. *Livestock Science*, 125, 291-297.

[24] Nuernberg, K., Dannenbrger, D., Nuernberg, G., Ender, K., Voigt, J., Scollan, N. D., Wood, J. D., Nute, G. R., & Richardson, R. I. (2005). Effect of grass-based and a con-centrate feeding system on meat quality characteristics and fatty acid composition of longissimus muscle in different cattle breeds. *Livestock Production Science,,* 94, 137-147.

[25] Garcia, P. T., Pensel, N. A., Sancho, A. M., Latimori, N. J., Kloster, A. M., Amigone, M. A., & Casal, J. J. (2008). Beef lipids in relation to animal breed and nutrition in Ar-gentina. *Meat Science*, 79, 500-508.

[26] Aldai, N., Lavin, P., Kramer, J. K. G., Jaroso, R., & Mantecon, A. R. . (2012). Beef effect on quality veal production in mountain areas: emphasis on meatfatty acid composi-tion. *Meat Science, (in press)*.

[27] Monteiro, A. C. G., Fontes, M. A., Bessa, R. J. B., Prates, J. A. M., & Lemos, J. P. C. (2012). Intramuscular lipids of Mertolenga PDO beef, Mertolenga PDO veal and "Vi-tela Tradicional do Montado" PGI veal. *Food Chemistry*, 132, 1486-1494.

[28] Muchenje, v., Hugo, A., Dzama, K., Chimonyo, M., Strydom, P. E., & Raats, J. G. (2009). Cholesterol levels and fatty acid profiles of beef from three cattle breeds raised on natural pasture. *Journal of Food Composition and Analysis*, 22, 354-358.

[29] Ward, R. E., Wood, B., Otter, N., & Doran, O. (2010). Relationship between the ex-pression of key lipogenic enzymes, fatty acid composition, and intramuscular fat content of Limousin and Aberdeen Angus cattle. *Livestock Science*, 127, 20-29.

[30] De Smet, S., Raes, K., & Demeyer, D. (2004). Meat fatty composition as affected by fatness and genetic factors: A review. *Anim. Res.*, 53, 81-98.

[31] Barton, L., Bures, D., Kott, T., & Rehak, D. (2011). Effects of sex and age on bovine muscle and adipose tissue fatty acid composition and stearoyl-CoA desaturase mRNA expression. *Meat Science*, 89, 444-450.

[32] Monteiro, A. C. G., Santos-Silva, J., Bessa, R. J. B., Navas, D. R., & Lemos, J. P. C. (2005). Fatty acid composition of intramuscular fat of bulls and steers. *Livestock Pro-duction Science*, 99, 13-19.

[33] Jiang, T., Busboom, J. R., Nelson, M. L., O`, Fallon. J., Ringkob, T. P., Joos, D., & Piper, K. (2010). Effectof sampling fat location and cooking on fatty acid composition of beef steaks. *Meat Science*, 84, 86-92.

[34] Pestana, J. M., Costa, A. S. H., Alves, S. P., Martins, S. V., & Alfaia, C. M. (2012). Seasonal changes and muscle type effect on the nutritional quality of intramuscular fat in Mirandesa-PDO veal. *Meat Science*, 90, 819-827.

[35] Turk, S. N., & Smith, S. B. (2009). Carcass fatty acid mapping. *Meat Science*, 81, 658-663.

[36] Lee, M. R. F., Evans, P. R., Nute, G. R., Richardson, R. I., & Scollan, N. D. (2009). A comparison between red clover silage and grass silage feeding on fatty acid composition, meat stability and sensory quality of the M. Longissimus muscle of dairy cull cows. *Meat Science*, 81, 738-744.

[37] Alfaia, C. M. M., Ribeiro, V. S. S., Lourenco, M. R. A., Quaresma, M. A. G., Martins, S. I. V., Portugal, A. P. V., et al. (2006). Fatty acid composition, conjugated linoleic acid isomers and cholesterol in beef from crossbred bullocks intensively produced and from Alentejana purebred bullocks reared according to Carnalentejana-PDO specifications. *Meat Science*, 72, 425-436.

[38] Alfaia, C. P. M., Castro, M. L. F., Martins, S. I. V., Portugal, A. P. V., Alves, S. P. A., Fontes, C. M. G. A., et al. (2007). Influence of slaughter season and muscle type on fatty acid composition, conjugated linoleic acid isomeric distribution and nutritional quality of intramuscular in Arouquesa-PDO veal. *Meat Science*, 76, 787-795.

[39] Mel'uchova, B., Blasko, J., Kubinec, R., & Gorova, R. (2008). Seasonal variations in fatty acid composition of pasture forage plant and CLA content in ewe milk fat. *Small Ruminant Research*, 78, 56-65.

[40] Daniel, Z. C. T. R., Wynn, R. J., Salter, A. M., & Buttery, P. J. (2004). Differing effects of forage and concentrate diets on the oleic acid and the conjugated linoleic acid content of sheep tissues. *Journal of Animal Science*, 82, 747-758.

[41] Beaulieu, A. D., Drackley, J. K., & Merchen, N. R. (2002). Concentrations of conjugated linoleic acid (cis-9, trans-11-octadecadienoic acid) are not increased in tissue lipids of cattle fed a high concentrate diet supplement with soybean oil. *Journal Animal Science*, 80, 847-861.

[42] Alfaia, C. P. M., Alves, S. P., Martins, S. I. V., Costa, A. S. H., Fontes, C. M. G. A., Lemos, J. P. C., Bessa, R. J. B., & Prates, J. A. M. (2009). Effect of the feeding system on intramuscular fatty acids and conjugated linoleic acid isomers of beef cattle, with emphasis on their nutritional value and discriminatory ability. *Food Chemistry*, 114, 939-946.

[43] Faucitano, L., Chouinard, P. Y., Fortin, J. J., Mandell, I. B., Lafreniere, C., Girard, C. L., & Berthiaume, R. (2008). Comparison of alternative beef production systems based on forrage finishing or grain-forage diets with or without growth promotans: 2. Meat quality, fatty acid composition, and overall palatability. *Journal Animal Science*, 86, 1678-1689.

[44] Sackman, J. R., Duckett, S. K., Gillis, M. H., Realini, C. E., Parks, A. H., & Eggeslton, R. B. (2003). Effect of forage and sunflower oil levels on ruminal biohydrogenation of fatty acids and conjugated linoleic acid formation in beef steers fed finishing diets. *Journal Animal Science*, 81, 3174-3181.

[45] Aldai, N., Dugan, M. E. R., Juarez, M., Martinez, A., & Koldo, O. (2010). Double-musculin carácter influences the trans-18:1 conjugated linoleic acid profile in concentrate-fed yearling bulls. *Meat Science*, 85, 59-65.

[46] Rule, R. D., Mac, Neil. M. D., & Short, R. E. (1997). Influence of sire growth potential, time on feed, and growing-finishing strategy on cholesterol and fatty acids of the ground carcass and Longissimus muscle of beef steers. *Journal of Animal Science*, 75, 1525-1533.

[47] Aldai, N., Murray, B. E., Olivan, M., Martinez, A., Troy, D. J., Osoro, K., & Najera, A. I. . (2005). The influence of breed and mh-genotype on carcass conformation, meat physic-chemical characteristics, and the fatty acid profile of muscle from yearling bulls. *Meat Science*, VER.

[48] Noci, F., French, P., Monahan, F., & Napier, J. A. . (2007). The.J.& Moloney, A.P.The fatty acid composition of muscle fat and subcutaneous adipose tissue of grazing heifers supplemented with plant oil-enriched concentrates. *Journal Animal Science*, 85, 1062-1073.

[49] Raes, K., & Smet, S. D. (2001). Effect of double-muscling in begian blue young bulls on the intramuscular fatty acid composition with emphasis on conjugated linoleic acid and polyunsaturated fatty acids. *Animal Science*, 73, 253-260.

[50] Bauman, D. E., Baumgard, L. H., Corl, B. A., & Griinari, J. M. (1999). Biosynthesis of conjugated linoleic acids in ruminants. *Proceedings American. Society Animal Science.*, VER.

[51] Dugan, M. E. R., Rollan, D. C., Aalhus, J. L., Aldai, N., & Kramer, J. K. G. (2008). Subcutaneous fat composition of youthful and mature Canadian beef. *Canadian Journal of Animal Science*, 88, 591-599.

[52] Basarab, J. A., Mir, P. S., Aalhus, J. L., Shah, M. A., Baron, B. S., Okine, E. K., & Robertson, W. M. (2007). Effect of sunflower seed supplementation on the fatty acid composition of muscle and adipose tissue of pasture-fed and feedlot finished beef. *Canadian Journal Animal Science*, 87, 71-86.

[53] Leheska, J. M., Thompson, L. D., Howe, J. C., Hentges, E., Boyce, J., & Brooks, J. C. (2008). Effects of conventional and grass-feeding systems on the nutrient composition of beef. *Journal Animal Science*, 86, 3575-3585.

[54] Ludden, P. A., Kucuk, O., Rule, D. C., & Hess, B. W. (2009). Growth and carcass fatty acid composition of beef steers fed soybean oil for increasing duration before slaughter. *Meat Science*, 82, 185-192.

[55] Scollan, N. D. (2005). Effect of a grass-based and a concentrate feeding system on meat quality characteristics and fatty acids composition of longissimus muscle in different cattle breeds. *Livestock Production Science*, 94, 137-147.

[56] Engle, T. E., Spears, J. W., Fellner, V., & Odle, J. (2000). Effects of dietary soybean oil and dietary copper on ruminal and tissue lipid metabolism in finishing steers. *Journal Animal Science*, 78, 2713-2721.

[57] Oliveira, E. A., Sampaio, A. A. M., Henrique, W., Pivaro, T. M., Rosa, B. L., Fernandes, A. R. M., & Andrade, A. T. (2012). Quality traits and lipid composition of meat from Nellore young bulls fed with different oils protected or unprotected from rumen degradation. *Meat Science*, 90, 28-35.

[58] Fernandez, A. R. M., Sampaio, A. A. M., Henrique, W., Oliveira, E. A., Oliveira, R. V., & Leonet, F. R. . (2009). Composicao em acidos e qualidade da carne de tourinhos Nelore e Canchim alimentados com dietas a base de cana-de-azucar edos niveis de concentrado. *Revista Brasileña de Zootecnia*, 38, 328-337.

[59] Griswold, K. E., Apgar, G. A., Robinson, R. A., Jacobson, B. N., Johnson, D., & Woody, H. D. (2003). Effectiveness of short-term feeding strategies for altering conjugated linoleic acid content of beef. *Journal. Animal Science*, 61, 1862-1871.

[60] Cassut, M. M., Scheeder, M. R., Ossowski, D. A., Sutter, F., Sliwinski, B., Danilo, A. A., et al. (2000). Comparative evaluation of rumen protected fat, coconut oil and various oilseeds supplemented to fattening bulls. *Archiv der Tierernahrung*, 23, 25-44.

[61] Abu, Ghagazaleh. A. A., Schingoethe, D. G., Hippen, A. R., Kalscheur, K. F., & Whitlock, L. A. . (2002). Fatty acid profile of milk and rumen digesta from cows fed fish oil, extruded soybean or their blend. *Journal Dairy Science.*, 85, 2266-2276.

[62] Kenny, D. A., Kelly, J. P., Monahan, F. J., & Moloney, A. P. (2007). Effect of dietary fish and soya oil on muscle fatty acid concentrations and oxidative lipid stability in beef cattle. *Journal Animal Science*, 85(1), 911.

[63] Barton, L., Marounek, M., Kudrna, V., Bures, D., & Zahradkova, R. (2007). Growth performance and fatty acid profiles of intramuscular and subcutaneous fat from Limousin and Charolais heifers fed extruded linseed. *Meat Science*, 76, 517-523.

[64] Weill, P., Schmitt, B. C. G., Chesneau, G., Daniel, N., Safraou, F., & Legrand, P. (2002). Effects of introducing linseed oil in livestock diet on blood fatty acid composition of consumers of animal products. *Annals of Nutrition and Metabolism*, 46, 182-191.

[65] Dannenberger, D., Nuernberg, K., Nuernberg, G., Scollan, N., Steinhart, H. y., & Ender, K. (2005). Effect of pasture vs. concentrate diet on CLA isomers distribution in different tissue lipids of beef cattle. *Lipids*, 40, 589-598.

[66] Dhiman, T. R., Satter, L. D., Pariza, M. W., Galli, M. P., Albright, K., & Tolosa, M. X. . (2000). Conjugated linoleic acid (CLA) content of milk from cows offered diets rich in linoleic and linolenic acid. *Journal Dairy Science*, 83, 1016-1027.

[67] Chouinard, P. Y., Corneau, L., Butler, W. R., Chilliard, Y., Drackley, J. K., & Bauman, D. E. . (2001). Effect of dietary lipid source on conjugated linolei acid concentrations in milk fat. Journal Dairy Science , 84, 680-690.

[68] Mir, P. S., Mir, Z., Kuber, P. S., Gasking, C. T., Martin, E. L., Dodson, M., Elias, V., Calles, J. A., Johnson, K. A., Busboom, J. R., Wood, A. J., Pittenger, G. L., & Reeves, J. J. . (2002). Growth, carcass characteristics, muscle conjugated linoleic acid (CLA) content, and response to intravenous glucose challenge in high percentage Wagyu, Wagyu x Limousin, and Limousin steers fed sunflower oil-containing diets. *Journal Animal Science*, 80, 2996-3004.

[69] Mir, P. S., Mc Allister, T. A., Zaman, S., Jones, S. D. M., He, M. L., Aalhus, J. L., Jeremiah, L. E., Goonewardene, L. A., Weselake, R. J., & Mir, Z. (2003). Effect of dietary sunflower oil and vitamin E on beef cattle performance, carcass characteristics and meat quality. *Canadian Journal Animal Science*, 83, 53-66.

[70] Aharoni, Y., Orlow, A., Brosh, A., Granit, R., & Kanner, J. (2005). Effects of soybean supplementation of high forage fattening diet on fatty acid profiles in lipid depots of fattening bull calves, and their levels of blood vitamin E. *Animal Feed Science and Technology*, 119, 191-202.

[71] Vasta, V., Pagano, R. I., Luciano, g., Scerra, M., Caparra, P., Foti, F., et al. (2012). Effect of morning vs. afternoon on intramuscular fatty acid composition in lamb. *Meat Science*, 90, 93-98.

[72] Vasta, V., Priolo, A., Scerra, M., Hallet, K. G., Wood, J. D., & Doran, O. (2009). Δ9 desaturase protein expression and fatty acid composition of longissimus dorsi muscle in lamb fed green herbage or concentrate with or without added tannins. *Meat Science*, 82, 357-364.

[73] Bernes, G., Turner, T., & Pickova, J. (2012). Sheep fed only silage supplement with concentrates. 2. Effects on lamb performance and fatty acid profile of ewe milk and lamb meat. *Small Ruminant Research*, 102, 114-124.

[74] Joy, M., Ripoll, G., Molino, F., Dervishi, E., & Alvarez, Rodriguez. J. (2012). Influence of the type of forage supplied to ewes in pre- and post-partum periods on the meat fatty acid of suckling lambs. *Meat Science*, 90, 775-782.

[75] Mazzone, G., Giammarco, M., Vignola, G., Sardi, L., & Lambertini, L. (2010). Effects of rearing season on carcass and meat quality of suckling Apennine light lambs. *Meat Science*, 86, 474-478.

[76] Chiofalo, B., Simonella, S., Di Grigoli, A., Liotta, L., Frenda, A. S., Lo, Presti. V., Bonanno, A., & Chiofalo, V. (2010). Chemical and acidic compostion of Longissimus dorsi of Comisana lambs fed with Trifolium subterraneum and Lolium multiflorum. *Small Ruminant Research*, 88, 89-96.

[77] Berthelot, V., Bas, P., & Schmidely, P. (2010). Utilization of extruded linseed to modi-fy fatty acid composition of intensively-reared lamb meat: Effect of associated cereals (wheat vs. corn) and linoleic acid content of the diet. *Meat Science*, 84, 114-124.

[78] Bas, P., Berthelot, V., Pottier, E., & Normand, J. (2007). Effect of linseed on fatty acid composition of muscles and adipose tissues of lamb with emphasis on trans fatty acids. Meat Science , 77, 678-688.

[79] Wachira, A. M., Sinclair, L. A., Wilkinson, R. G., Enser, M., Wood, J. D., & Fisher, A. V. (2002). Effects of dietary fat source and breed on the carcass composition, n-3 poly-unsaturated fatty acid and conjugated linoleic acid content of sheep and adipose tis-sue. *Journal of Nutrition*, 88, 697-709.

[80] Bolte, M. R., Hess, B. W., Means, W. J., Moss, G. E., & Rule, D. C. (2002). Feeding lambs high-oleate or high-linoleate safflower seeds differentially influences carcass fatty acid composition. *Journal Animal Science*, 80, 609-616.

[81] Santos-Silva, J., Bessa, R. J. B., & Mendes, I. A. (2003). The effect of supplementation with expanded sunflower seed on carcass and meat quality of lambs raised on pas-ture. *Meat Science*, 65, 1301-1308.

[82] Radunz, A. E., Wickersham, L. A., Loerch, S. C., Fluharty, F. L., Reynolds, C. K., & Zerby, H. N. (2009). Effects of dietary polyunsaturated fatty acid composition in muscle and subcutaneous adipose tissue of lamb. *Journal Animal Science*, 87, 4082-4091.

[83] Bessa, R. J. B., Portugal, V. P., Mendes, I. A., & Santos-Silva, J. (2005). Effect of lipid supplementation on growth performance, carcass and meat quality and fatty acid composition of intramuscular lipids of lambs fed dehydrated Lucerne or concentrate. *Livestock Production Science*, 96, 185-194.

[84] Bessa, R. B., Alves, S. P., Figueredo, R., Texeira, A., Rodriguez, A., Janeiro, A., et al. (2006). Discrimination of production system and origin of animal products using chemical markers. In Animal Products from the Mediterranean area. EAAP Publica-tion The Netherland: Wageningen Academic Publishers.(119), 231-240.

[85] Bessa, R. J. B., Alves, S. P., Jeronimo, E., Alfaia, C. M., Prates, J. A. M., & Santos-Silva, J. (2007). Effect of lipid supplements on ruminal biohydrogenation intermediates and muscle fatty acids in lambs. *European Journal of Lipid Science Technology*, 109, 868-878.

[86] Jeronimo, E., Alves, S. P., Prates, J. A. M., Santos-Silva, J., & Bessa, R. J. B. (2009). Ef-fect of dietary replacement of sunflower oil with linseed oil on intramuscular fatty acids of lamb meat. *Meat Science*, 84, 499-505.

[87] Bessa, R. J. B., Lourenco, M., Portugal, P. V., & Santos-Silva, J. (2008). Effects of previ-ous diet and duration of soybean oil supplementation on light lamb carcass composi-tion, meat quality and fatty acid composition. Meat Science , 80, 1100-1105.

[88] Peng, Y. S., Brown, M. A., Wu, J. P., & Liu, Z. (2010). Different oilseed supplements alter fatty acid composition of different adipose tissues of adult ewes. *Meat Science*, 85, 542-549.

[89] Kucut, O., Hess, B. W., & Rule, D. C. (2008). Fatty acid composition of mixed ruminal microbes isolated from sheep supplemented with soybean oil. *Research in Veterinary Science*, 84, 213-224.

[90] Realini, C. E., Duckett, S. K., Brito, G. W., Rizza, M., & De Mattos, D. (2004). Effect of pasture vs. concentrate feeding with or without antioxidants on carcass characteristics, fatty acid composition, and quality of Uruguayan beef. *Meat Science*, 66, 567-577.

[91] Dierking, R. M., Kallenbach, R. L., & Grun, I. U. (2010). Effect of forage species on fatty acid content and performance of pasture-finished steers. *Meat Science*, 85, 597-605.

[92] Humada, M. L., Serrano, E., Sañudo, C., Rolland, D. C., & Dugan, M. E. R. (2012). Production system and slaughter age effects on intramuscular fatty acids from Young Tudanca bulls. *Meat Science*, 90, 678-685.

[93] Sarries, M. V., Murray, B. E., Moloney, A. P., Troy, D., & Berian, M. J. (2009). The effect of cooking on the fatty acid composition of Longissimus dorsi muscle from beef steers fed rations designed to increase the concentration of conjugated linoleic acid in tissue. *Meat Science*, 81, 307-312.

[94] Pavan, E., & Duckett, S. K. (2007). Corn oil supplementation to steers grazing endophyte-free tall fescue. II Effects on Longissimus muscle and subcutaneous adipose fatty acid composition and stearoyl-CoA desaturase activity and expression. *Journal Animal Science* , 85, 1731-1740.

[95] Juarez, M.., Dugan, M. E. R., Aalhus, J. L., Aldai, N., Basarab, J. A., Baron, B. S., & Mc Allister, T. A. (2011). Effects of vitamin E and flaxseed on rumen-derived fatty acid intermediates in beef intramuscular fat. *Meat Science*, 88, 434-440.

[96] Garcia, P. T., Casal, J. J., Fianuchi, S., Magaldi, J. J., Rodriguez, F. J., & Nancucheo, J. A. (2008). Conjugated linolenic acid (CLA) and polyunsaturated fatty acids in muscle lipids of lamb from the Patagonian area of Argentina. *Meat Science*, 79, 541-548.

[97] Mir, Z., Rushfeldt, M. I., Paterson, L. J., & Weselake, P. (2000). Effects of dietary supplementation with CLA or linoleic rich oil in the lamb content of lamb tissues. *Small Ruminant Research*, 36, 25-31.

[98] Kott, R. W., Hatfield, P. G., Bergman, J. W., Flyn, C. R., Van Wagoner, H., & Boles, J. A. (2003). Feedlot performance, carcass composition, and muscle and fat CLA concentrations of lambs fed diets supplemented with safflower seeds. *Meat Science*, 49, 11-17.

Chapter 7

Soybean Oil Derivatives
for Fuel and Chemical Feedstocks

Joanna McFarlane

Additional information is available at the end of the chapter

1. Introduction

Plant-based sources of hydrocarbons are being considered as alternatives to petrochemicals because of the need to conserve petroleum resources for reasons of national security and climate change [1]. Changes in fuel formulations to include ethanol from corn sugar and methyl esters from agricultural products are examples of this policy in the United States and elsewhere as biofuels from efficiently grown and processed biomass are claimed to be carbon neutral. In the United States, the mandate to include biofuels has been implemented as the Renewable Fuels Standards (RFS1 and RFS2) [2] with biobased diesel fuel as one of the categories. The production of biodiesel in the United States has varied considerably over the last few years, but was 241×10^6 gallons in the first quarter of 2012, a high number but one that still only represents 2% of the total volume of diesel fuel produced for heating and vehicles [3]. Most of the biodiesel comes from soybean oil, more than double the contribution of the other major feedstocks combined: canola oil, yellow grease, and tallow.

Replacements for commodity chemicals are also being considered, as this value stream represents much of the profit for the oil industry and one that would be affected by shortages in oil or other fossil fuels. While the discovery of large amounts of natural gas associated with oil shale deposits have reduced this as an immediate concern for instance the estimated recoverable reserves in the Western US have now reached 800×10^9 bbls [4] -research into bio-based feedstock materials continues for the expected long-term benefit. In particular, this chapter reviews a literature on the conversion of bio-based extracts to hydrocarbons for fuels and for building block commodity chemicals, with a focus on soybean derived products.

2. Fuels

Although commercially produced, more economical conversion of methyl esters from soy-bean triglycerides is an active area of research to make the product more cost competitive in comparison with standard petrochemical diesel [5]. The processes of esterification and trans-esterification to produce methyl esters that can be burned directly in compression -gnition engines has been reviewed elsewhere [6, 7].The fatty acid chains on the lipid molecule that constitutes soybean oil, also called a triacylglycerol or TAG, are split from the glycerol back-bone and esterified with an alcohol, generally methanol, in the presence of a homogeneous base or acid catalyst, Reaction 1.

Commercial processing of biodiesel through homogeneous catalysis suffers from high feed-stock costs and batch processing that requires long residence times to achieve good conver-sion. Hence, ongoing research continues to explore methods how to best use low-quality feedstocks, and to reduce the reagent requirements, energy usage, processing time, and com-plexity [8]. Figure 1 shows results from simulation of a continuous process to make biodiesel, varying temperature (a) and methanol content (b) to determine conditions for the optimal pro-duction of high quality grade biodiesel. As the process is limited by kinetics and mass trans-fer, the effect of mixing has also been investigated by considering the available volume fraction of reagents (c) [9], defined as the molar ratio of reagents in the reaction zone versus the overall reagent volume in the vessel. The available volume can be changed by increasing the contact zone between the immiscible reagents where the reactions take place. The interfacial surface area is dependent on the intensity of mixing in the multiphase system. Interfacial area can also be increased by reducing the size of the dispersed phase droplets, such as by bub-bling reagent methanol into the oil through a frit. Novel approaches to biodiesel production continue to be explored, particularly for lower grade and waste feedstocks, such as the direct extraction of fatty acid chains through use of a solvent such as an ionic liquid to pretreat esteri-fication to the methyl ester [10]. Other work has examined methanol-based transesterification of waste cooking oil under quite mild conditions (110°C in 2 h) in contact with tungsten oxide solid acid catalysts, giving yields of fatty acid methyl esters (FAME) that are close to the Amer-ican Society of Testing and Materials (ASTM) standard for biodiesel [11]. The authors of that study, Komintarachat and Chuepeng, reported several advantages of working with a WO_x/Al_2O_3 catalyst. Prior separation of free fatty acids, in their sample of waste cooking oil report-ed as 15%, was not necessary to achieve high yields in a one step process. In addition, they found the catalyst has desirable properties for scale-up, being low cost, reusable, and less reac-tive than traditional homogeneous catalysts.

The choice of acid or base homogeneous catalysis depends on the concentration of free fatty acids (FFA) in the triglyceride feedstock. Virgin soybean oil has a low FFA content, <4%, and so can be converted to biodiesel by transesterification without an acid-catalyzed esterifica-tion pretreatment. However, oil that has been degraded by heat, such as waste oil, requires a two-step conversion. FFA produced during heating have to be esterified, otherwise they be-come saponified during transesterification. New processes are being developed to simplify the conversion of waste oil, such as the use of a supported heteropolyacid catalyst that si-multaneously promotes both the esterification and transesterification processes [13].

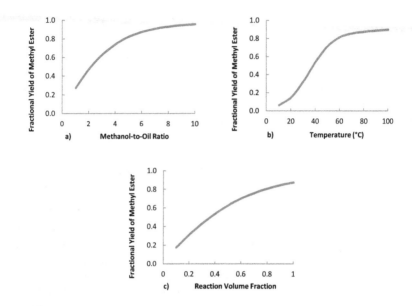

Figure 1. Results of parametric studies on methyl ester production in a continuous reactor showing (a) asymptotic approach to a maximum yield with methanol-to-oil molar ratio and (b) with reactor temperature. The dependence of yield on volume fraction simulated the effect of mixing in the reactor. These calculations were programmed in Mat-Lab® [12], for nominal reaction conditions of 50°C, 7:1 methanol-to-oil molar ratio, and an effective reaction volume of 60% [9].

$$C_3H_5(CO_2R^1)(CO_2R^2)(CO_2R^3) + CH_3OH \rightarrow CH_3O_2R^1 + C_3H_5(OH)(CO_2R^2)(CO_2R^3)$$
$$C_3H_5(OH)(CO_2R^2)(CO_2R^3) + CH_3OH \rightarrow CH_3O_2R^2 + C_3H_5(OH)_2(CO_2R^3) \qquad (1)$$
$$C_3H_5(OH)_2(CO_2R^3) + CH_3OH \rightarrow CH_3O_2R^3 + C_3H_8O_3$$

Triglyceride + 3 Methanoln3 Methyl Esters + Glycerine

The chemical conversion to FAME produces a low viscosity, high-cetane number fuel that can be mixed directly with conventional diesel. The physical properties of diesel and biodiesel, or FAME, are compared in Table 1 [14]. The properties in the table are given for the liquid phase at 25°C and for vapor phase at 527°C, corresponding to pre-ignition conditions in a compression ignition engine. Although similar in carbon chain length and cetane number, biodiesel differs from diesel significantly in its vapor pressure, liquid viscosity, and vapor diffusion coefficient. The properties of the biodiesel depend on the length and unsaturation of the fatty acid chains, Table 1. Where the data are lacking for the soybean-derived methyl esters, the vapor phase properties for biodiesel have been replaced by those of methyl oleate.

Physical Property	Diesel*	Biodiesel*
Density [kg·m^{-3}]	762*	884*
vapor pressure [Pa]	2.22*	6.10x10^{-5}-1.18x10^{-3}*
surface tension [J·m^{-2}]	2.68x10^{-2}*	2.49x10^{-2}*
liquid viscosity [Pa·s]	3.14 x10^{-3}*	9.10 x10^{-3}*
liquid thermal conductivity [J·m^{-1}·s^{-1}·K^{-1}]	0.144*	0.100*
latent heat [J·kg^{-1}]	3.60x10^{5}*	3.38x10^{5}**'
liquid specific heat [J·kg^{-1}·K^{-1}]	2.27x10^{3}*	2.01x10^{3}*
vapor specific heat [kJ·mol^{-1}·K^{-1}]	0.643#	0.848#
vapor diffusion coefficient [m^{2}·s^{-1}]	8.50x10^{-6}#	9.44x10^{-7}#
vapor viscosity [Pa·s]	1.00 x10^{-5}#	1.21 x10^{-5}**#
vapor thermal conductivity [J·m^{-1}·s^{-1}·K^{-1}]	4.35x10^{-2}#	2.60x10^{-2}#

* Values at 25°C for liquids

Values at 527°C for vapors

** Values for methyl oleate as a representative biodiesel component.

Table 1. Diesel (C$_{14}$H$_{30}$) versus biodiesel properties at 25 or 527°C

Although successfully blended up to 20 volume% for commercial and military use [16], methyl ester content in vehicle fuel is limited by a number of factors, including the perform-ance in cold weather, the effect of oxygen content on engine components (particularly in the case of older engines), shelf-life and thermal stability [17], and higher NO$_x$ emissions from engines that are not tuned to handle the higher temperature conditions of methyl ester com-bustion [18]. Results from simulations presented in Figures 2 and 3 show on a microscopic scale how the combustion of biodiesel can differ from diesel (represented as n-heptane in the engine simulations) in terms of temperature and emissions [15]. The development of en-gines that can accommodate biodiesel have focused on the effects of physical properties on spray parameters and droplet formation that will greatly affect the ignition conditions and combustion characteristics, Figures 2 and 3, and enhanced by early oxidation in the low tem-perature heat release phase of combustion, Figure 4. Figure 4 shows the progression of the combustion of 20% biodiesel as a function of crank angle position, with 360° corresponding to top-dead-center. The key radicals in the low temperature heat release portion of the cycle include OH• and HO$_2$• , but OH• dominates after the main ignition event

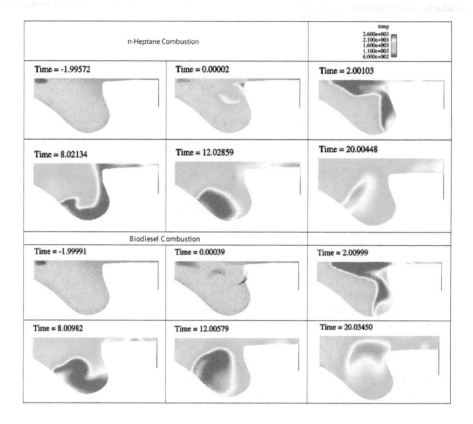

Figure 2. Predicted differences in the temperature in n-heptane and biodiesel combustion after injection into the cylinder. The n-heptane, representing diesel fuel, shows combustion occurring very rapidly after injection, 2×10^{-8}s. The biodiesel, while slower to vaporize and ignite, shows a higher temperature at 12 ms after injection. The simulations were done assuming a cycle of 2000 Rev/min. The temperature key in the upper right is given in degrees K. Reprinted with permission from SAE paper 2008-01-1378 Copyright © 2008 SAE International [15].

These factors have led to interest in synthesizing a hydrocarbon fuel starting with methyl esters, a so-called "green diesel" that will maintain the high cetane number of biodiesel, but will achieve better performance in an automobile: through enhanced mixing, injection, and combustion; reduced downstream issues such as NO_x emissions; and better upstream handling associated with fuel manufacture and distribution. Bunting and colleagues have reviewed the development of fungible and compatible biofuels [20]. That report considers a wide variety of products, from ethanol to pyrolysis oils as well as soy-derived biodiesel. Concerns that arise when developing alternative fuels include refining, blending, and distribution, regulatory barriers, verification of performance, and changes in operating practices throughout the distribution system. In this chapter, we focus on the chemistry of the fuel.

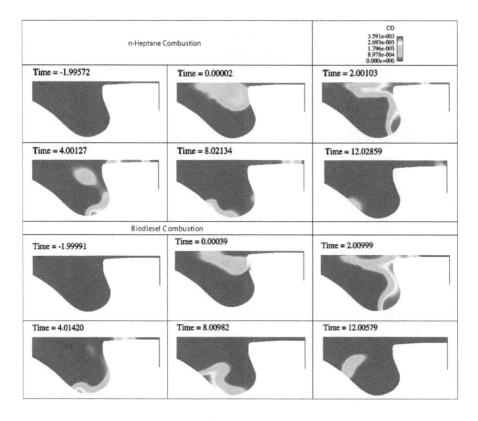

Figure 3. Predicted differences in the CO concentration in n-heptane and biodiesel combustion after injection into the cylinder. As with the temperature profiles, the production of CO from biodiesel lagged that from n-heptane during the event. However, the final concentration of CO was higher for biodiesel than for n-heptane. The CO concentration key in the upper right is given in mole fraction. Reprinted with permission from SAE paper 2008-01-1378 Copyright © 2008 SAE International [15].

Unsaturated methyl esters have more affinity for water and contaminants than does hexadecane, a typical component of diesel fuel. Water affinity is often expressed in the form of the octanol-water distribution coefficient or K_{ow}, with lower values of K_{ow} indicating more hydrophilic compounds, K_{ow}=moles(octanol)/moles(water). K_{ow} values for a few select components of fossil-based diesel [21] and long chain methyl esters typical of those derived from soy oil [22] are presented in Table 2. Water can be problematic in fuel distribution systems which are made of low carbon or low grade stainless steel, but water can be gravity separated when the fuel is held in storage vessels. Separation is less likely to occur with oxygenated fuels, particularly those that have degraded to shorter chain components through autooxidation. Because of the issues with materials compatibility, potential contamination of pipelines by residues, and high viscosity at low temperatures, biodiesel must be added to standard diesel fuel at a terminal loading facility, where the fuel is mixed and then loaded onto trucks for distribution. However, mixing at a distribution terminal affords less quality control than at a refinery, with the latter having the ability for online testing of properties and composition, followed by adjustment to meet ASTM specifications if necessary [23].

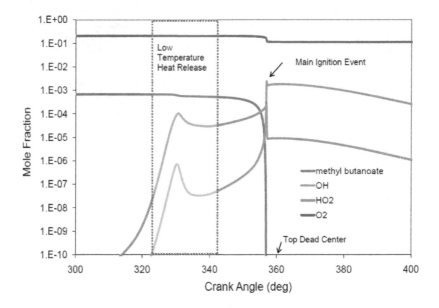

Figure 4. Combustion of a 20% blend of methyl butanoate (simulating biodiesel) and n-heptane (simulating petroleum diesel) showing the importance of reactive species during different times of the cycle [19].

Biodiesel derived from soybean oil comprises long fatty acid chains, C16-C18, with a high cetane number, and a high degree of unsaturation, 84-87%, for better cold flow properties (reduced viscosity) relative to other plant-based methyl esters. However, the unsaturation can also lead to issues with shelf life and thermal stability [28] in comparison with more hydrogenated oils such as palm oil. The process of oxidation and its effects on the properties of biodiesel has been studied using chemical and thermal analysis by Tan and colleagues [29], and reviewed by Mushbrush [17] and Knothe [30]. Oxidation of the double bonds can occur through an autocatalytic mechanism simplistically depicted below, Reaction (2), initiated by hydrogen abstraction from an unsaturated carbon atom. The greater the unsaturation, the more stable the allylic radical, $R_n\bullet$, thus a precursor mono-unsaturated FFA has greater stability that the doubly- and triply-unsaturated chains. Once formed, the radical can combine with O_2, allowing formation of a secondary intermediate in the chain, a reactive hydroperoxide [31, 32], Reaction (3). Light increases the rate of decomposition, because photosensitization allows a direct reaction between the O_2 and the carbon-carbon double bond offering another pathway to oxidation. The hydroperoxides, once formed, can convert to a variety of products, cyclized five and six membered rings, malonaldehyde or $C_3O_2H_4$, hydroxy and epoxy esters, and allylic hydroxyl- and ketone compounds, among other oxygenated derivatives. Cleavage reactions form reactive radicals, continuing the process to produce volatile compounds such as carbonyls, alcohols, esters, and short chain hydrocarbons, Reactions (4a and b), some of which react to form furans, aldehydes, ketones, lactones, alkynes, and aromatics [33]. Because of an associated increase in viscosity and acid number, these oxidation products are generally undesirable in a combustion engine [34].

	Log K_{ow}	water solubility (mg/L)at 25°C
Biodiesel Components		
methylpalmitate, C16:0 (10-12%) (hexadecanoic acid, methyl ester)	7.38	4.00×10^{-3}
methyl stearate, C18:0 (3-4%) (octadecanoic acid, methyl ester)	8.35	3.01×10^{-3}
methyloleate, C18:1 (23-25%) (octadecenoic acid, methyl ester)	7.45	3.68×10^{-3}
methyllinolate, C18:2 (53-56%) (octadecadienoic acid, methyl ester)	6.82	2.10×10^{-2}
methyllinolenate, C18:3 (6-8%) (octadecatrienoic acid, methylester)	6.29	9.18×10^{-2}
Diesel #2 Components		
Monoaromatics and small cyclic compounds (10.0%)	2-5 Hexylbenzene is 5.52	1.02
cycloparaffins (34.0%)	3-5 Cyclohexane is 3.44	55 [27]
Naphthalenes and PAH (14.7%)	3-5 Naphthalene is 3.3	30
n- and i-paraffins (41.3%)	3.3-7.06 for short chain HC [21] 8.2 for $C_{16}H_{34}$	9×10^{-4}

Table 2. Hydrophobicity expressed as octanol-water partition coefficients [24, 25] for organic derivatives of petroleum and biodiesel [26]

$$R_n\text{-}H \rightarrow R_n\cdot + H\cdot$$
$$R_n\cdot + O_2 \rightarrow R_nOO \tag{2}$$
$$R_nOO\cdot + R_{n'}\text{-}H \rightarrow R_nOO\text{-}H + R_{n'}$$

$$R_n\text{-}C\text{=}C\text{-}R_{n'} + 2O_2 \rightarrow R_n\text{-}C\left(OOH\right)\text{-}C\text{=}C\text{-}R_{n'\text{-}1} + R_{n\text{-}1}\text{-}C\text{=}C\text{-}C\left(OOH\right)\text{-}R_{n'} \tag{3}$$

$$R_n\text{-}CH\text{=}CH\text{-}C\left(O\cdot\right)H\text{-}R_{n'} \rightarrow R_{n'}\text{-}CHO + R_n\text{-}CH\text{=}CH\cdot \text{ (a)}$$
$$R_n\text{-}CH\text{=}CH\text{-}C\left(O\cdot\right)H\text{-}R_{n'} \rightarrow R_n\text{-}CH\text{=}CH\text{-}CHO + R_{n'}\cdot \text{ (b)} \tag{4}$$

Besides light and heat, oxidative stability is also greatly influenced by the choice of storage tank materials and the presence of minor components or contaminants in the mixture. Hence, deterioration can be slowed by the use of additives in the fuel. Commonly used antioxidants include phenol derived compounds such as tert-butyl hydroquinone (TBHQ) [35], butylatedhydroxytoluene (BHT), butylatedhydroxanisole (BHA), and propyl gallate. Effective additive concentrations of 1000 mg kg^{-1} (1000 ppm) do not appear to affect combustion or physical properties [36], but are sufficient to increase the induction period for autooxidation by binding with the peroxy radicals as shown created in Reaction (2) [37]. Phosphorylated antioxidants, including phosphites, phosphonites and phosphines [38], either hinder hydrogen atom extraction or promote the decomposition of hydroperoxides [39], Reaction (5). They are often used in combination with the phenolic antioxidants for additional efficacy. Organosulfites can also be used to stabilize methyl esters, as they react with hydroperoxides to form sulfates [40].These compounds would be less than desirable as fuel additives; however, as the non-radical decomposition is catalyzed by the presence of acid, and ultimate products include SO_x and acids H_2SO_3 and H_2SO_4.

$$R_nOO\cdot + A\text{-}H \leftrightarrow R_nOO\text{-}H + A\cdot \tag{5}$$

3. Chemical conversions

For soybean oil and soybean-derived feedstocks to be used as drop-in replacements for petroleum derived products, deoxygenation processing has to be undertaken. Depending on the desired products, this can involve a number a steps, listed below. Not all of the steps are needed for each product. In general, the desire is to shift increase the carbon-to-oxygen ratio to be closer to that of petroleum, and reduce the carbon to hydrogen ratio, as depicted in Figure 5, plotted with data collected by Choudhary [41]. The various catalysts used to achieve the deoxygenation of triglycerides has been reviewed in a number of publications, for instance by Morgan [42].

i.   Hydrogenation (saturation of double bonds)[43]

ii.   Thermal cracking – heating in an inert atmosphere without addition of H_2

iii.   Acid- or base-catalyzed cracking over metal oxides or zeolites [44]

iv.   Hydrodeoxygenation (HDO removal of O as H_2O) – usually at higher H_2 pressure and lower temperatures than the cracking processes. The catalyst has transition metal + heteroatom (S or N) like $NiMo/Al_2O_3$ or $CoMo/Al_2O_3$. Non-sulfided forms have also been studied to a lesser extent, Ni/Al_2O_3 or Ni/SiO_2.

v.   Decarboxylation (remove O as CO_2) – more H_2 efficient than HDO, unless product CO_2 becomes methanized. Often uses supported platinum catalysts in a batch reaction, for example: Pt/Al_2O_3 or Pd/C, 270-360°C, 17-40 bar H_2.

vi.   Decarbonylation or removal of oxygen as CO – same catalysts as decarboxylation

vii.   Removal of other heteroatoms (S, N, P, metals) – especially from used cooking oil or if sulfur and nitrogen compounds have been added in earlier processes to maintain the catalyst activity.

viii.   Various side reactions including: hydrocracking, water-gas shift, methanization, cyclization, and aromatization

ix.   Isomerization [45], often deliberately designed to produce better fuel characteristics, such as cold flow behavior. For example, linear paraffins, n=16-18, freeze at 18-28°C, while iso-paraffins of the same carbon number freeze at -11 to 3°C

x.   Co-processing of soybean oil with diesel fuels in an oil refinery by fluidized catalytic cracking (FCC) – achieves isomerization as well as separations. May get inhibition of deoxygenation because of S groups in the diesel fuel components and vice versa.

4. Oil feedstocksfor hydrocarbon fuels

Deoxygenation drives the overall process for converting soybean oil into hydrocarbons. Processing requirements will be similar for both fuels and chemical feedstocks if the product from deoxygenation can be introduced as a feed in a petrochemical refinery where it will undergo further reactions and separation. If the goal is to only make hydrocarbon fuels; however, separations may not be as important after deoxygenation as they would be for isolating particular building block chemicals. Conversion in a smaller scale independent biorefinery may be feasible, solving the issue of the distributed production of soybeans. The drawback to this scheme is that if the fuels are to be introduced directly into the distribution system, issues related to quality control of the product may become siginificant. In this case, processing will have to account for the variability of bio-based feedstocks, even within a particular crop. Hence, for fuel production at a distributed processing facility, the goals would include achieving sufficient deoxygenation to allow incorporation upstream of the distribution point, enabling pipeline transportation, and providing reliability and quality control.

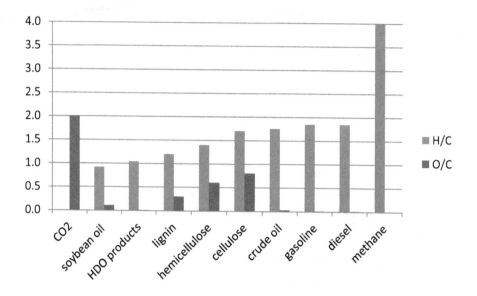

Figure 5. Progression in H/C mole ratio and O/C mole ratio for different sources of organic materials

Thermal cracking has the advantage of not requiring hydrogen for processing. Uncata-lyzed thermal cracking has been used to convert soybean oil to hydrocarbons, holding the oil at 430-440°C under vacuum for over one hour [46]. Not including the free fatty acid byproducts, more than half of the products comprised linear alkanes (51%). Other signifi-cant hydrocarbon products were identified by gas chromatography – mass spectroscopy as cycloalkanes (11%), alkenes (20%), cycloalkenes (5%), aromatics (8%), and polyaromatic hydrocarbons or PAH (5%). The authors contend that the rings, both saturated and aro-matic, came from cyclization of the fatty acid chains rather than from a Diels Alder addi-tion. The latter is usually considered the mechanism for ring formation from olefins, but in the case of the soybean oil conversion, the precursor dienes were not observed. The FFA also became decarboxylated, releasing hydrogen for saturation of double bonds, and producing byproducts CO_2 and CO. The significant fraction of PAH could be problematic for direct combustion of the resulting fuel, as these compounds can survive conditions through the combustion pathway in the engine and be emitted into the atmosphere. How-ever, pretreatment, to reduce the acid number and separation of the FFA, would allow this feedstock to be transported to a refinery for further processing. In thermal cracking, the most important variables governing product distribution include the temperature fol-lowed by the residence time [46-48]. Other literature has described investigations of cata-lysts that have shown promise for thermal cracking, including supported Ni, Pt or Pd on carbon. In particular, Morgan and coworkers have discovered a Ni/C catalyst and demon-strated a 92% conversion of soybean-derived triglyceride at 400°C, with a 70% yield in liq-uid form [42].

Biogas oil, which is a mixture of normal and isoparaffins having boiling points close to that of diesel, may be made from selective hydrotreating of natural triglycerides. Although some research on thermal cracking has suggested that introduction of hydrogen is not necessary [47], the HDO process allows the conversion to be carried out at lower temperatures, with fewer issues related to byproduct char and gas formation. Hydrodeoxygenation has also received much attention in the literature because this process shows promise to operate with less hydrogen than needed for hydrogenation [41]. HDO, in fact, involves a series of hydogenolysis and hydrogenation steps and is analogous to hydrodesulfurization of petroleum. Catalysts have already been developed for hydroprocessing of heavy oils. HDO investigations have been carried out on a number of seed oils, including soybean oil. For instance, rapeseed oil was deoxygenated at 260-280°C under 3.5 MPa H_2. The rate of reaction ranged over 0.25-4 h^{-1} when tested with a number of different catalysts, in order of performance: Ni-Mo/Al$_2$O$_3$> Mo/Al$_2$O$_3$>Ni/Al$_2$O$_3$ [49]. HDO of sunflower oil was performed in DMSO, with a NiMo/Al$_2$O$_3$/F catalyst, in a bench-scale continuous operation at high pressure. The oil hydrocracking entailed combined processes, including olefinic double bond saturation, oxygen removal, and isomerization [50], to give a conversion of 90%. HDO of sunflower oil (310-360°C, 2.0MPa) on a Pd/SAPO-31 catalyst gave excellent conversion to C17 and C18 straight chain and branched alkanes. However, the catalyst became fouled after a few hours [51]. Mixtures of sunflower oil and gas oil have been hydroprocessed over a sulfided catalyst, NiO(3%)–MoO$_3$(12%)–γ-Al$_2$O$_3$ incorporating 0, 15 or 30 wt.% zeolite beta (BEA). The reaction took place at conditions of 330°C, 60 bar, at a weight hourly space velocity (WHSV) of 2 h^{-1}, giving 100% conversion into hydrocarbons. The distribution arising from cracking giving the relative fractions of liquids/gases/and char was not discussed in the paper [52].

Hydrocracking has also been scaled up from the laboratory bench. A larger scale test of hydrocracking of fresh sunflower oil and used cooking oil was carried out by Bezergianni and colleagues [6]. The cracking process was carried out over a number of days until steady state was reached, and then an analysis was performed on the products. A presulfided commercial HDO catalyst was used, with sulfur in the form of dimethyldisulfide and nitrogen as tetra-butyl amine added to maintain the activity. The liquid hourly space velocity (LHSV) was 1.5 h^{-1} and the H_2-to-liquid ratio was 1098 Nm^3/m^3 (at 14 MPa). The difference in performance between the conversion of used and new oil was very small, with high yields of product in the diesel fuel boiling point range (70-80%). Less cracking to small molecules was observed at lower temps, i.e., 350°C, than 390°C, which is desirable for fuel manufacture.

The effect of sulfur on hydrodeoxygenation is of interest because it relates to the performance in an oil refinery with hydrodesulfurization (HDS) as well as HDO processing. Experiments over Pt/H-Y, Pt/H-ZSM-5, and sulfidedNiMo/γ-Al$_2$O$_3$ have been carried out in a batch reactor over a temperature range of 300-400°C and initial hydrogen pressures from 5 to 11 MPa. The reaction time was limited to 3 h [53]. Investigation of the performance of CoMo/Al$_2$O$_3$ at different sulfur levels (1% to < 10 mg kg^{-1}) and its effect on the HDO of sunflower oil were done under the following conditions: 300-380°C, 20-80 bar, 1-3 h^{-1}, and H_2/oil volume 200-800 Nm^3/m^3 [54]. Up to 75% of the target product C18-paraffins were made at the highest temps and lowest LHSV. At higher H_2 pressures, more hydrocracking occurred, forming light gases such as propane. Adding presulfided catalysts got better yields (5-8%) under less severe conditions. But the addition of sulfur can produce H_2S, which needs to be removed from product and recycled. Sulfur can react to form mercaptans, which are corrosive, adding cost and complexity to the process. H_2S has been found not to prevent catalyst deactivation as was anticipated. The effect of sulfur has also been investigated for the HDO of aliphatic ester model compounds [55].

Hydrogenation and deoxygenation to n-paraffins followed by isomerization is expensive and complicated, but now is performed on an industrial scale [41]. A Finnish company, Neste Oil, has built and operated three NexBTL plants around the world, in Singapore, Rotterdam, and Porvoo in 2009, to convert 190,000 t/a of C12-C16 triglycerides, fatty acid esters, and fatty acids to green diesel. The conversion involves hydrotreatment followed by isomerization to produce green diesel and is described in a Neste patent [56]. The Neste Singapore plant is rated at 800,000 t/a of palm oil, used oil and waste animal fat. A plant in Rotterdam started production in mid 2011 that uses a variety of feedstocks.

Another company, UOP/ENI S.p.A., has a process to produce green diesel that involves a number of catalytic steps to achieve deoxygenation and conversion to branched hydrocarbons. The process, based on hydrodeoxygenation, produces fuel that can be blended directly with petroleum, or added to an input stream in an oil refinery [56]. Emerald Biofuels plans to build a 85×10^6 gallon production facility based on the UOP technology (licensed by Honeywell) at a Dow Chemical site in Plaquemine, LA. Dynamic Fuels is already in production (75×10^6 gal) and Diamond Green Diesel (137×10^6 gal) also has a plant under construction at the mouth of the Mississippi, to take advantage of the proximity to shipping and petroleum refineries. The existing capacity, along with the operating Neste plants, currently produces 600×10^6 gal/a [57].

ExxonMobile is building a hydrotreating plant in Singapore to deliver bio-derived low sulfur diesel of up to 16×10^6 L/d, and has similar plants planned for Baytown and Baton Rouge, LA, in the USA [58].Other planned green biodiesel projects include sites at Norco LA (Darling International, Diamond Green Diesel, LLC, and Valero Energy Corp) to produce 137×10^6 gal/a from waste oil, and animal fats. KiOR plans construction of a plant to produce refinery intermediates in Columbus MS. Joule Unlimited will be constructing a biofuels demonstration plant (75×10^6 gal/a green diesel) in New Mexico [57].

Although hydrodeoxygenation is fairly mature, with industrial-scale production, technical problems remain that would benefit from further research. Minimization of hydrogen used in the conversion of biomass must happen to make the process economically viable [59]. A large fraction of the products from hydrodeoxygenation are linear paraffinic hydrocarbons, which tend to form waxes that can cause cold flow problems. Research into isomerization reactions and selection of catalysts to promote branched alkanes would be beneficial. In the case of soybean oil, however, thermal cracking has produced a preponderance of aromatic compounds, suggesting that mixing of fractions produced through different pathways may give rise to a fungible fuel. A third area of interest is the effect of acylglycerides and HDO products on catalysts that are used in oil refining, especially if the bio derivatives are to be introduced into the feed stream along with petroleum. In particular, there is a concern that the acylglycerides may affect the performance of catalysts such as CoMo, used in hydrode-sulfurization [41].

The activities undertaken by industry show that the underlying drivers for biofuel production low sulfur requirements in diesel, low net CO_2 emission during production and use, and potentially disruptive oil supply disruptions are pushing major companies to make investments in this area. Although the feedstock streams for biorefineries are not specific to soybean oil, the engineering efforts contribute to the general knowledge of producing green diesel from a variety of sources. Yet, the industry is sensitive to changes in feedstock and oil prices, and smaller initiatives have lost traction during changes in the market, particularly during the last few years. Currently, green diesel remains a niche player in the larger petro-leum refining industry.

5. Oil feedstocksfor materials

Soybean plants have been used to fabricate a variety of materials and products. Soybean de-rived materials have been used in the development of bio-based fibers and yarns [60]. In particular soybean protein fiber has been identified having potential uses [61] in the manu-facturing of fabrics. Materials production may not require the degree of chemical conversion and breakdown of the triglyceride that hydrocarbon production requires, but modification is still required to give the desired properties. For instance, soybean protein separated through precipitation after the oil has been removed from the seed requires further process-ing to crosslink the derived fibers and reduce brittleness and degradation [62, 63]. Soybean straw, available after harvesting of the beans, can be converted to technical fibers through alkali processing. The straw-derived fibers have a higher lignin content than cotton or linen, but after processing the cellulose content is comparable to these other agricultural sources, and could represent a bioresource estimated to be 55 million tons derived from 220 million tons of straw [64]. Soybean fiber left over from oil and protein extraction can be converted to ethanol through a two-step process: (i) pretreatment with aqueous ammonia to remove lig-nin (by 74% after 12 hours), and (ii) simultaneous saccharification and fermentation, giving 0.25g ethanol per gram of fiber [65]. The straw can also be converted to a bio-oil through fast pyrolysis [66].

6. Oil feedstocksfor chemicals

Soybean oil can be used to manufacture a number of different compounds including surfac-tants, fuel additives, detergents, polymers such as polyurethanes [67], and adhesives [68]. Polymer production from biomaterials has recently been reviewed by Lligadas and collea-gues [69]. Fatty acids can be converted to a polyurethane through a di-isocyanate intermedi-ate [70].Resin alternatives have been prepared from a number of different plant-derived materials, including soybean oil. To achieve the physical properties required for a thermo-

plastic polymer, that is having sufficient rigidity and tensile strength to be used in fabrication, soybean oil derived resins must be crosslinked via epoxidation (introduction of epoxy
groups into double bonds on the fatty acid chains) or mixed with petroleum-based materials. Good results obtained without incorporating conventional polymers used a combination
of epoxidized soybean oil and an anhydrided soyate. The linking process was catalyzed
with hexamethylenediamine, and gave a fiber with a tensile strength of >10 MPa [71]. Adhesives made of renewable polymers have been made from mixing of dimeric fatty acids and
diols with maleic-anhydrided soybean triglycerides. The gel is formed from cross linking of
the esters and extending the fatty acid chains within the structure [72]. Hybrid coatings have
been prepared from mixing blown soybean oil and sol-gel precursors (titanium and zirconium peroxides) to improve properties such as tensile strength, adhesion, flexibility, hardness
and impact resistance [73].

The conversion of plant-based acylglycerides to nitrogen containing compounds has been
reviewed by Biswas and colleagues [74].Properties of various products from palmitic acid
have been predicted based on a combinatorial approach, and then linked to an optimization
routine to select the product of choice based on predefined criteria (lubricity, critical micelle
concentration, or hydrophilic-lipophilic balance) [75]. The authors, Carmada and Sunderesa-
ni, wanted to refocus the synthesis paradigm. They developed a method to choose a chemical structure that would give thermophysical properties optimized to a particular
application. The structure then determined which synthetic method would be needed to
produce the desired end product.

7. Methyl ester feedstocks for chemicals

Methyl esters derived from soybean oil can also be used as starting materials for the production of hydrocarbons for fuels or chemical feedstocks. Various catalytic pathways from oxygenated precursor to hydrocarbons include: pyrolysis [76], deoxygenation and
hydrogenation [77, 78], and hydrotreatment [79]. The focus of many of these studies has
been production of fuels that are miscible or fungible with petroleum products, e.g., the
work published by the group of Daniel Resasco at the University of Oklahoma [80], for fuel
production rather than chemicals. In addition, much of the published literature focuses on
simpler chemical representatives of the methyl esters from soybean oil; but these results are
directly applicable to the production of chemical feedstocks, such as the synthesis of ethyl-
benzene that can be used for a variety of products: polymers, solvents, and reagents [77]. Although differences in the product distribution would be expected from TAG and single
methyl ester conversions, comparison studies carried out by Kubatova and colleagues
[46-48] on individual acylglycerides, as well as soybean mixtures, showed similar conversion chemistry.

Because it appears as if the products from these deoxygenation processes will require further processing to make fuels or chemical feedstocks, the FCC of triglycerides has been studied by groups such as Melero and colleagues [81]. In particular, the effect of soy-based

biomass on the FCC process is of interest. Melero investigated both the cracking of 100% soybean oil, and a mixture of 30% soybean oil with petroleum, the latter being a typical feedstock for an oil refinery. The feed injection was held at 70°C, but the FCC itself was carried out at 560°C, representing typical conditions in a refinery. The boiling point range for the soybean biodiesel was from 545.7 to 636.0°C in comparison with the diesel range from 200-330°C [82]. Because of the unsaturated chains in the soybean oil, the aromatic fraction in the product was enhanced relative to that of pure petroleum, but the PAH was reduced. Saturated fatty acid fragments gave rise to light alkanes, light oil, and diesel fractions. There could be some concern about enhanced corrosion in the cracker because of the presence of the FFA, but research suggests that the lifetime of these compounds is very short at these temperatures, and so they may not present a problem. Many of the FCC products are similar from acylglycerides and petroleum, although the composition distributions are different. In the case the biomass, of FCC products come from reactions in the cracker and in the case of petroleum, most hydrocarbons are present in the original feedstock. Issues such as gum formation or the effect of impurities, such as entrained alkali metals that could be present in biological materials, were not studied by Melero [81].

8. Conclusions

Although many chemical pathways have been demonstrated in the laboratory, the scale-up to handle large quantities of bio-derived material presents a number of challenges in comparison with petroleum refining. These range from additional transportation costs because of distributed feedstock production, to catalyst cost and regeneration. Seasonal variations in the cultivation and harvesting of soybeans and production of oil can result to chemical changes in the feedstock material and minor components. However, it appears as if the chemical modification processes are robust to minor changes in FFA distribution. Impurities and their impact on catalyst performance and lifetime may be significant and difficult to test outside of an industrial setting. Impurity and effects of minor components are highly dependent on unpredictable phenomena such as feedstock composition or process variability. Thus, these effects may not appear, much less be quantified, in a bench-scale operation using laboratory grade chemicals. Hence, operation of pilot and demonstration scale facilities will be very informative. The feasibility of the production of hydrocarbons from soybean triglycerides or methyl esters derived from these triglycerides is often dependent on the availability of low cost hydrogen. Other technical hurdles include the optimization of interfacial reactions and separations before soybean oil can make a significant contribution to the hydrocarbon economy. The question of whether feedstocks from soybean oil should be introduced into a stream in an oil refinery, or converted in a small scale refinery to fungible products depends on the final application and cost issues. However, once converted to hydrocarbons, separations to commodity chemicals or fuel should be analogous to handling conventional petroleum products.

Nomenclature and abbreviations

ASTM - American Society of Testing and Materials

BHA - butylatedhydroxyanisole

BHT - butylatedhydroxytoluene

FAME - Fatty Acid Methyl Esters

FCC - Fluidized Catalytic Cracking

FFA - Free Fatty Acid (monoglyceride)

HDO - Hydrodeoxygenation

LHSV - Liquid hourly space velocity

PAH - Polyaromatic Hydrocarbon

TAG- Triacylglyercol or triglyceride

TBHQ- Tert-butyl hydroquinone

WHSV - Weight hourly space velocity

Acknowledgements

Oak Ridge National Laboratory is managed by UT-Battelle, LLC, for the U.S. Department of Energy under contract DE-AC05-00OR22725.

Author details

Joanna McFarlane

Energy and Transportation Science Division, Oak Ridge National Laboratory, Oak Ridge, USA

References

[1] A. McIlroy and G. McRae, *Basic Research Needs for Clean and Efficient Combustion of 21st Century Transportation Fuels.* 2006.

[2] USEPA. *RFS2 – Federal Policy Drivers for Increased Biofuels Usage.* US Environmental Protection Agency, January 2009, accessed August 1, 2012;

[3] EIA. *Monthly Biodiesel Production Report*. May 2012, accessed August 1 2012;

[4] API. *US Oil Shale: Our Energy Resource, Our Energy Security, Our Choice*. 2011, accessed Aug 1, 2012; Available from: http://www.api.org/~/media/Files/Oil-and-Natural-Gas/Oil_Shale/Oil_Shale_Factsheet_1.ashx.

[5] D.J. Murphy, *Oil Crops as Potential Sources of Biofuels*, in *Technological Innovations in Major World Oil Crops*, S.K. Gupta, Editor. 2012, Springer: NY. p. 269-284.

[6] S. Bezergianni, S. Voutetakis, and A. Kalogianni, "Catalytic Hydrocracking of Fresh and Used Cooking Oil".*Industrial & Engineering Chemistry Research*, 2009. 48(18): p. 8402–8406.

[7] J. McFarlane, *Processing of Soybean Oil into Fuels*, in *Recent Trends for Enhancing the Diversity and Quality of Soybean Products*, D. Krezhova, Editor. 2011, InTech.

[8] J. McFarlane, C. Tsouris, J.F.J. Birdwell, Jr., D.L. Schuh, H.L. Jennings, A.M. Pahmer-Boitrago, and S.M. Terpstra, "Production of Biodiesel at Kinetic Limit Achieved in a Centrifugal Reactor/Separator".*Industrial and Engineering Chemistry Research* 2010. 49: p. 3160–3169.

[9] X.-W. Zhang and A. McWhirter, *Conserving Fossil Fuel Resources Using Reaction Simulations to Maximize Biodiesel Production*. 2012, Oak Ridge High School: Oak Ridge.

[10] M.S. Manic, V. Najdanovic-Visak, M.N. da Ponte, and Z.P. Visak, "Extraction of Free Fatty Acids from Soybean Oil Using Ionic Liquids or Poly(ethyleneglycol)S".*Aiche Journal*, 2011. 57(5): p. 1344-1355.

[11] C. Komintarachat and S. Chuepeng, "Solid Acid Catalyst for Biodiesel Production from Waste Used Cooking Oils".*Industrial & Engineering Chemistry Research*, 2009. 48(20): p. 9350-9353.

[12] MatLab. 2010, Mathworks: Natick, MA.

[13] A. Baig and F.T.T. Ng, "A Single-Step Solid Acid-Catalyzed Process for the Production of Biodiesel from High Free Fatty Acid Feedstocks".*Energy & Fuels*, 2010. 24: p. 4712-4720.

[14] V.K. Chakravarthy, J. McFarlane, C.S. Daw, Y. Ra, J.K. Griffin, and R. Reitz, "Physical Properties of Soy Bio-Diesel and N-Heptane: Implications for Use of Bio-Diesel in Diesel Engines".*SAE 2007 Transactions, Journal of Fuels and Lubricants*, 2008. 116: p. 885-895.

[15] Y. Ra, R.D. Reitz, J. McFarlane, and C.S. Daw, "Effects of Fuel Physical Properties on Diesel Engine Combustion Using Diesel and Bio-Diesel Fuels".*SAE International Journal of Fuels and Lubricants*, 2009. 1(1): p. 703-718.

[16] G.W. Mushrush, J.H. Wynne, H.D. Willauer, C.T. Lloyd, J.M. Hughes, and E.J. Beal, "Recycled Soybean Cooking Oils as Blending Stocks for Diesel Fuels".*Industrial & Engineering Chemistry Research*, 2004. 43(16): p. 4944-4946.

[17] G.W. Mushrush, J.H. Wynne, J.M. Hughes, E.J. Beal, and C.T. Lloyd, "Soybean-De-
      rived Fuel Liquids from Different Sources as Blending Stocks for Middle Distillate
      Ground Transportation Fuels".*Industrial & Engineering Chemistry Research*, 2003.
      42(11): p. 2387-2389.

[18] D.L. Purcell, B.T. McClure, J. McDonald, and H.N. Basu, "Transient Testing of Soy
      Methyl Ester Fuels in an Indirect Injection, Compression Ignition Engine".*Journal of
      the American Oil Chemists Society*, 1996. 73(3): p. 381-388.

[19] R. Ashen, K.C. Cushman, and J. McFarlane, *Chemical Kinetic Simulation of the Combus-
      tion of Bio-Based Fuels*. in press, Oak Ridge National Laboratory: Oak Ridge TN.

[20] B. Bunting, M. Bunce, T. Barone, and J. Storey, *Fungible and Compatible Biofuels: Litera-
      ture Search, Summary, and Recommendations*. 2010, Oak Ridge National Laboratory:
      Oak Ridge, TN.

[21] J. Risher, *Toxilogical Profile for Fuel Oils*. 1995, Sciences International, Inc: Research
      Triangle Institute

[22] T.A. Foglia, K.C. Jones, and J.G. Phillips, "Determination of Biodiesel and Triacylgly-
      cerols in Diesel Fuel by Liquid Chromatography".*Chromatographia*, 2005. 62(3-4): p.
      115-119.

[23] ASTM, *Specification for Diesel Fuel Oil, Biodiesel Blend (B6 to B20)*. 2008, American Soci-
      ety of Testing and Materials.

[24] D. Mackay, *Illustrated Handbook of Physical Chemical Properties and Environmental Fate
      for Organic Chemicals*. Vol. 5. 1997, Boca Raton, FL: Lewis Publishers.

[25] J. Sangster, "Octanol-Water Partition Coefficients". *Journal of Physical and Chemical
      Reference Data*, 1989. 18(3): p. 1111-1230.

[26] H.B. Krop, M.J.M. van Velzen, J.R. Parsons, and H.A.J. Govers, "N-Octanol-Water
      Partition Coefficients, Aqueous Solubilities and Henry's Law Constants of Fatty Acid
      Esters". *Chemosphere*, 1997. 34(1): p. 107-119.

[27] USEPA. *Chemicals in the Environment: OPPT Chemical Fact Sheets*. October 5, 2010;
      Available from: http://www.epa.gov/chemfact/.

[28] G.R. Stansell, V.M. Gray, and S.D. Sym, "Microalgal Fatty Acid Composition: Impli-
      cations for Biodiesel Quality". *Journal of Applied Phycology*, 2012. 24(4): p. 791-801.

[29] C.P. Tan and Y.B.C. Man, "Differential Scanning Calorimetric Analysis for Monitor-
      ing the Oxidation of Heated Oils".*Food Chemistry*, 1999. 67(2): p. 177-184.

[30] G. Knothe, "Some Aspects of Biodiesel Oxidative Stability".*Fuel Processing Technology*,
      2007. 88(7): p. 669-677.

[31] R.C. Simas, D. Barrrera-Arellano, M.N. Eberlin, R.R. Catharino, V. Souza, and R.M.
      Alberici, "Triacylglycerols Oxidation in Oils and Fats Monitored by Easy Ambient

Sonic-Spray Ionization Mass Spectrometry". *Journal of the American Oil Chemists Society*, 2012. 89: p. 1193-1200.

[32] T.D. Crowe and P.J. White, "Adaptation of the AOCS Official Method for Measuring Hydroperoxides from Small-Scale Oil Samples". *Journal of the American Oil Chemists Society*, 2001. 78(12): p. 1267-1269.

[33] E.N. Frankel, "Lipid Oxidation: Mechanisms, Products and Biological Significance". *Journal of the American Oil Chemists Society*, 1984. 61(12): p. 1908-1917.

[34] N. Canha, P. Felizardo, J.C. Menezes, and M.J.N. Correira, "Multivariate near Infrared Spectroscopy Models for Predicting the Oxidative Stability of Biodiesel: Effect of Antioxidants Addition".*Fuel*, 2012. 97: p. 352-357.

[35] T.S. Yang, Y.H. Chu, and T.T. Liu, "Effects of Storage Conditions on Oxidative Stability of Soybean Oil". *Journal of the Science of Food and Agriculture*, 2005. 85(9): p. 1587-1595.

[36] S. Schober and M. Mittelbach, "The Impact of Antioxidants on Biodiesel Oxidation Stability". *European Journal of Lipid Science and Technology*, 2004. 106(6): p. 382-389.

[37] G. Knothe, "A Technical Evaluation of Biodiesel from Vegetable Oils Vs. Algae. Will Algae-Derived Biodiesel Perform?". *Green Chemistry*, 2011. 13(11): p. 3048-3065.

[38] I. Kriston, A. Orban-Mester, G. Nagy, P. Staniek, E. Foldes, and B. Pukanszky, "Melt Stabilisation of Phillips Type Polyethylene, Part I: The Role of Phenolic and Phosphorous Antioxidants".*Polymer Degradation and Stability*, 2009. 94: p. 719-729.

[39] D. Lomonaco, F.J.N. Maia, C.S. Clemente, J.P.F. Mota, A.E.J. Costa, and S.E. Mazzetto, "Thermal Studies of New Biodiesel Antioxidants Synthesized from a Natural Occurring Phenolic Lipid".*Fuel*, 2012. 97: p. 552-559.

[40] A. Günther, T. König, W.D. Habicher, and K. Schwetlick, "Antioxidant Action of Organic Sulphites-I. Esters of Sulphurous Acid as Secondary Antioxidants ". *Polymer Degradation and Stability*, 1997. 55: p. 209-216.

[41] T.V. Choudhary and C.B. Phillips, "Renewable Fuels Via Catalytic Hydrodeoxygenation". *Applied Catalysis A: General*, 2011. 397(1-2): p. 1-12.

[42] T. Morgan, D. Grubb, E. Santillan-Jimenez, and M. Crocker, "Conversion of Triglycerides to Hydrocarbons over Supported Metal Catalysts". *Topics in Catalysis*, 2010. 53(11-12): p. 820-829.

[43] D. Jovanovic, R. Radovic, L. Mares, M. Stankovic, and B. Markovic, "Nickel Hydrogenation Catalyst for Tallow Hydrogenation and for the Selective Hydrogenation of Sunflower Seed Oil and Soybean Oil". *Catalysis Today*, 1998. 43(1-2): p. 21-28.

[44] C. Perego and A. Bosetti, "Biomass to Fuels: The Role of Zeolite and Mesoporous Materials". *Microporous and Mesoporous Materials*, 2011. 144(1-3): p. 28-39.

[45] V. Calemma, S. Peratello, and C. Perego, "Hydroisomerization and Hydrocracking of Long Chain N-Alkanes on Pt/Amorphous Sio2–Al2o3 Catalyst". *Applied Catalysis A: General*, 2000. 190(1-2): p. 207-218.

[46] A. Kubatova, J. St'avova, W.S. Seames, Y. Luo, S.M. Sadrameli, M.J. Linnen, G.V. Baglayeva, I.P. Smoliakova, and E.I. Kozliak, "Triacylglyceride Thermal Cracking: Pathways to Cyclic Hydrocarbons". *Energy & Fuels*, 2012. 26(1): p. 672-685.

[47] Y. Luo, I. Ahmed, A. Kubatova, J. St'avova, T. Aulich, S.M. Sadrameli, and W.S. Seames, "The Thermal Cracking of Soybean/Canola Oils and Their Methyl Esters". *Fuel Processing Technology*, 2010. 91(6): p. 613-617.

[48] A. Kubatova, Y. Luo, J. St'avova, S.M. Sadrameli, T. Aulich, E. Kozliak, and W. Seames, "New Path in the Thermal Cracking of Triacylglycerols (Canola and Soybean Oil)".*Fuel*, 2011. 90(8): p. 2598-2608.

[49] D. Kubicka and L. Kaluza, "Deoxygenation of Vegetable Oils over Sulfided Ni, Mo and Nimo Catalysts".*Applied Catalysis A: General*, 2010. 372: p. 199-208.

[50] S. Kovács, T. Kasza, A. Therneszb, I.W. Horváth, and J. Hancsók, "Fuel Production by Hydrotreating of Triglycerides on Nimo/Al2o3/F Catalyst". *Chemical Engineering Journal*, 2011. 176-177: p. 237-243.

[51] O.V. Kikhtyanin, A.E. Rubanov, A.B. Ayupov, and G.V. Echevsky, "Hydroconversion of Sunflower Oil on Pd/SAPO-31 Catalyst". *Fuel*, 2010. 89(10): p. 3085-3092.

[52] T.M. Sankaranarayanana, M. Banub, A. Pandurangana, and S. Sivasanker, "Hydroprocessing of Sunflower Oil–Gas Oil Blends over Sulfided Ni–Mo–Al–Zeolite Beta Composites". *Bioresource Technology*, 2011. 102(22): p. 10717-10723.

[53] R. Sotelo-Boyas, Y. Liu, and T. Minowa, "Renewable Diesel Production from the Hydrotreating of Rapeseed Oil with Pt/Zeolite and Nimo/Al$_2$O$_3$ Catalysts". *Industrial & Engineering Chemistry Research*, 2010. 50: p. 2791-2799.

[54] M. Krár, S. Kovács, D. Kalló, and J. Hancsók, "Fuel Purpose Hydrotreating of Sunflower Oil on Como/Al$_2$O$_3$ Catalyst". *Bioresource Technology*, 2010. 10(23): p. 9287-9293.

[55] O.İ. Şenol, T.-R. Viljava, and A.O.I. Krause, "Effect of Sulphiding Agents on the Hydrodeoxygenation of Aliphatic Esters on Sulphided Catalysts". *Applied Catalysis A: General*, 2007. 326(2): p. 236-244.

[56] J. Myllyaja, P. Aalto, and E. Harlin, *Process for the Manufacture of Diesel Range Hydrocarbons*, WPTO, Editor. 2007, Neste Oil OYJ: Finland.

[57] J. Lane. *Renewable Diesel Roundup*. 2012 May 9, accessed August 1, 2012; Available from: http://www.altenergystocks.com/archives/2012/05/renewable_diesel_roundup_1.html.

[58] eco-business.com. *Singapore: Exxonmobil to Raise 'Green' Diesel Output*. 2010 November 10, accessed August 1, 2012; Available from: Nov. 24 2010, The Business Times,

http://www.eco-business.com/news/singapore-exxonmobil-raise-green-diesel-output/.

[59] J. McFarlane, J.A. Gluckstein, M. Hu, M. Kidder, C. Narula, and M. Sturgeon, "Investigation of Catalytic Pathways and Separations for Lignin Breakdown into Monomers and Fuels". *Separation Science and Technology*, 2012. in press.

[60] B. Ozgen, "New Biodegradable Fibres, Yarn Properties and Their Applications in Textiles: A Review". *Industrial Textila*, 2012. 63(1): p. 3-7.

[61] S. Huda, N. Reddy, D. Karst, W.J. Xu, W. Yang, and Y.Q. Yang, "Nontraditional Biofibers for a New Textile Industry". *J. Biobased Materials and Bioenergy*, 2007. 1(2): p. 177-190.

[62] Y. Zhang, S. Ghasemzadeh, A.M. Kotliar, S. Kumar, S. Presnell, and L.D. Williams, "Fibers from Soybean Protein and Poly(Vinyl Alcohol)". *J. Appl. Polymer Sci.*, 1999. 71(1).

[63] Y. Li, N. Reddy, and Y. Yang, "A New Crosslinked Protein Fiber from Gliadin and the Effect of Crosslinking Parameters on Its Mechanical Properties and Wataer Stability". *Polymer International*, 2008. 57: p. 1174-1181.

[64] N. Reddy and Y. Yang, "Natural Cellulose Fibers from Soybean Straw". *Bioresource Technology*, 2009. 100(14): p. 3593-3598.

[65] B. Karki, D. Maurere, S. Box, T.H. Kim, and S. Jung, "Ethanol Production from Soybean Fiber, a Co-Product of Aqueous Oil Extraction, Using a Soaking in Aqueous Ammonia Pretreatment". *Journal of the American Oil Chemists Society*, 2012. 89(7): p. 1345-1353.

[66] A.A. Boateng, C.A. Mullen, N.M. Goldberg, K.B. Hicks, T.E. Devine, I.M. Lima, and J.E. McMurtrey, "Sustainable Production of Bioenergy and Biochar from the Straw of High-Biomass Soybean Lines Via Fast Pyrolysis". *Environmental Progress & Sustainable Energy*, 2010. 29(2): p. 175-183.

[67] D.A. Babb, *Polyurethanes from Renewable Resources*, in *Synthetic Biodegradable Polymers*, B. Rieger, et al., Editors. 2012. p. 315-360.

[68] B.K. Ahn, S. Kraft, D. Wang, and X.S. Sun, "Thermally Stable, Transparent, Pressure-Sensitive Adhesives from Epoxidized and Dihydroxyl Soybean Oil". *Biomacromolecules*, 2011. 12(5): p. 1839-1843.

[69] G. Lligadas, J.C. Ronda, M. Galia, and V. Cadiz, "Plant Oils as Platform Chemicals for Polyurethane Synthesis: Current State-of-the-Art". *Biomacromolecules*, 2010. 11(11): p. 2825-2835.

[70] L. Hojabri, X.H. Kong, and S.S. Narine, "Fatty Acid-Derived Diisocyanate and Biobased Polyurethane Produced from Vegetable Oil: Synthesis, Polymerization, and Characterization". *Biomacromolecules*, 2009. 10(4): p. 884-891.

[71] P. Tran, D. Graiver, and R. Narayan, "Biocomposites Synthesized from Chemically Modified Soy Oil and Biofibers".*J. Appl. Polymer Sci.*, 2006. 102(1): p. 69-75.

[72] R. Vendamme, K. Olaerts, M. Gomes, M. Degens, T. Shigematsu, and W. Eevers, "Interplay between Viscoelastic and Chemical Tunings in Fatty-Acid-Based Polyester Adhesives: Engineering Biomass toward Functionalized Step-Growth Polymers and Soft Networks". *Biomacromolecules*, 2012. 13(6): p. 1933-1944.

[73] G.H. Teng, J.R. Wegner, G.J. Hurtt, and M.D. Soucek, "Novel Inorganic/Organic Hybrid Materials Based on Blown Soybean Oil with Sol-Gel Precursors". *Progress in Organic Coatings*, 2001. 42(1-2): p. 29-37.

[74] A. Biswas, B.K. Sharma, J.L. Willett, S.Z. Erhan, and H.N. Cheng, "Soybean Oil as a Renewable Feedstock for Nitrogen-Containing Derivatives". *Energy & Environmental Science*, 2008. 1(6): p. 639-644.

[75] K.V. Camarda and P. Sunderesan, "An Optimization Approach to the Design of Value-Added Soybean Oil Products". *Industrial & Engineering Chemistry Research*, 2005. 44(4361-4367).

[76] K.D. Maher and D.C. Bressler, "Pyrolysis of Triglyceride Materials for the Production of Renewable Fuels and Chemicals". *Bioresource Technology*, 2007. 98: p. 2351-2368.

[77] T. Danuthai, S. Jongpatiwut, T. Rirksomboon, S. Osuwan, and D.E. Resasco, "Conversion of Methylesters to Hydrocarbons Over an H-ZSM5 Zeolite Catalyst H-Zsm5 Zeolite Catalyst". *Applied Catalysis A: General*, 2009. 361: p. 99-105.

[78] B. Donnis, R.G. Egeberg, P. Blom, and K.G. Knudsen, "Hydroprocessing of Bio-Oils and Oxygenates to Hydrocarbons. Understanding the Reaction Routes". *Topics in Catalysis*, 2009. 52(3): p. 229-240.

[79] G.W. Huber, S. Iborra, and A. Corma, "Synthesis of Transportation Fuels from Biomass: Chemistry, Catalysts, and Engineering". *Chemical Reviews*, 2006. 106(9): p. 4044-4098.

[80] T. Sooknoi, T. Danuthai, L.L. Lobban, R.G. Mallinson, and D.E. Resasco, "Deoxygenation of Methylesters over CsNaX". *Journal of Catalysis*, 2008. 258: p. 199-209.

[81] J.A. Melero, M.M. Clavero, G. Calleja, A. Garcia, R. Miravalles, and T. Galindo, "Production of Biofuels Via the Catalytic Cracking of Mixtures of Crude Vegetable Oils and Nonedible Animal Fats with Vacuum Gas Oil". *Energy & Fuels*, 2010. 24: p. 707-717.

[82] R.W. Hurn and H.M. Smith, "Hydrocarbons in the Diesel Boiling Range". *Industrial & Engineering Chemistry*, 1951. 43(12): p. 2788-2793.

Chapter 8

Soybean and Isoflavones – From Farm to Fork

Michel Mozeika Araújo,
Gustavo Bernardes Fanaro and
Anna Lucia Casañas Haasis Villavicencio

Additional information is available at the end of the chapter

1. Introduction

1.1. General

Soybean (*Glycine max* L. Merrill) were first grown as a crop in China about 5000 years ago [1] and have been widely consumed as folk medicines in China, India, Japan and Korea for hundreds of years [2]. Today is a major source of plant protein (70%) and oil (30%) and become a globally important crop [3,4]. Its nutrients become basic for humans consumption, beyond its by-products, that offer great diversities of products to the food industry. Soybean oil is highly consumed world-wide and soy milk is often used as a milk substitute to people who have lactose intolerance [5]. In addition soybean has phytoestrogens which can be used in replacement to women hormone [6]. Usage oil can be reused in several forms, including as a fuel source [7,8].

There are many kinds of soybean cultivars with different biological composition and economic values. According to the consensus recommendations of the Organization for Economic Cooperation and Development, soybean nutrients (such as amino acids, fatty acids, isoflavones) and antinutrients (such as phytic acid, raffinose and stachyose) are important markers in assessment the nutritional quality of soybean varieties [9].

Soybean production has expanded to most of continents and 90% of the world's soybean production is concentrated in tropical and semi-arid tropical regions which are characterized by high temperatures and low or erratic rainfall. In tropics, most of the crops are near their maximum temperature tolerance [10].

USA is the largest soybean producer in the world, following by Brazil and Argentina. World soybean harvest production reached 264.25 million metric tons in 2010/2011. USA, Brazil and Argentina were responsible to about 92.75% of all production [11].

Soybeans have been appreciated by consumers as a health-promoting food [12]. Soybeans and soy products are wide consumed by Asian populations and are encouraged for western diets because of its nutritional benefits. Soy products are abundant in traditional Asian diets which daily intake is from 7–8 g/day in Hong-Kong and China, up to 20–30 g/day in Korea and Japan. Most of Europeans and North Americans, however, consume less than 1 g/day [13].

The soybean consumption will depend on grains characteristics, for example large seed sizes with high sucrose content are desirable for the production of vegetable soybean, which is harvested at immature stage, also called edamame. On the other hand, cultivars with small seed size and low calcium contents are desirable for natto, a traditional fermented soy food in Japan with a firmer texture. For soymilk and tofu production, soybeans with light hilum color, large seed size, high water absorption, high protein and sucrose contents and low oligosaccharides contents are desirable [14].

1.2. Composition

Soybean´s unique chemical composition place this food products as one of the most economical and valuable agricultural commodity. Among cereal and other legume species, it has the highest protein content (around 40%) and the second highest (20%) oil content among all food legumes, after peanuts. Other valuable components found in soybeans include phospholipids, vitamins and minerals. Furthermore, soybeans contain many minor substances known to be biologically active including oligosaccharides, phytosterols, saponins and isoflavones. The actual composition of the whole soybean and its structural parts depends on many factors, including varieties, growing season, geographic location and environmental stress [12,15].

1.2.1. Macronutrients

On the average, oil and protein together constitute about 60% of dry soybeans. The remaining dry matter is composed of mainly carbohydrates (about 35%) and ash (5%) [15].

Soybeans store their lipids in an organelle known as oil bodies or lipid-containing vesicles. Their lipids are mainly in the form of triglycerides or triacylglycerols, with varying fatty acids as part of their structure. Triglycerides are neutral lipids, each consisting of three fatty acids and one glycerol that bound to three acids. Both saturated and unsatured can occur in the glycerides of soybean oil, however the fatty acids of soybean oil are primarily unsaturated [15,16]. The highest percentage of fatty acid in soybean oil is linoleic acid, following in decreasing order by oleic, palmitic, linolenic and stearic acid. Soybean oil contains some minor fatty acids, including arachidic, behenic, palmitoleic and myristic acid [16].

Protein content varies between 36% and 46% depending on the variety [17-19]. This component is present in the greatest amount in soybeans. Seed proteins are usually classified based on biological function in plants (metabolic proteins and storage proteins) and on solubility patterns. According to their functionality, metabolic proteins (such as enzymatic and structural) are concerned in normal cellular activities, including even the synthesis of storage proteins. Storage proteins, together with reserve of oils, are synthesized during soybean seed development and provide a source of nitrogen and carbon skeletons for the development seedling. In soybeans, most of proteins are storage type. A solubility pattern divides proteins into those soluble in water (albumins) and in salt solution (globulins). Globulins are further divided in legumins and vicilins. Under this classification system, most of soy protein is globulin. Certain soy proteins have their trivial names, as glycinin (legumins) and conglycinin (vicilins). Others, particularly those with enzymatic function, are based on the biological function of proteins themselves. Examples include hemagglutinin, trypsin inhibitors and lipoxygenases. A solubility classification can pose problems because an association with other proteins can change their solubility profile. Thus, a more precise means of identifying proteins has been based on approximate sedimentation coefficient using ultracentrifugation to separate seed proteins. Under appropriate buffer conditions, soy proteins exhibit four fractions after centrifugation. These fractions are designed as 2, 7, 11 and 15S. The major portion of the protein component is formed by storage proteins such as 7S globulin (β-conglycinin) and 11S globulin (glycinin), which represent about 80% of the total protein content [17]. Other proteins or peptides present in lower amounts include enzymes such as lipoxygenase, chalcone synthase and catalase.

On average, moisture-free soybeans contain about 35% of carbohydrates and defatted dehulled soy grits and flour contain about 17% soluble and 21% insoluble carbohydrates. Therefore, they are the second largest component in soybeans. However, the economical value of soy carbohydrates is considered much less important than soy protein and oil. A limited use of soybeans in human diet is due the flatucence produced by soluble carbohydrates such as raffinose and stachyose. Humans lack the enzymes to hydrolyze the galactosidic linkages of raffinose and stachyose to simpler sugars, so the compounds enter the lower intestinal tract intact, where they are metabolized by bacteria to produce flatus [16]. As a result, relatively fewer efforts have been made to study soy carbohydrates and their potential utilization. The principal use of soybean carbohydrate has been in animal feeds (primarily ruminant because they can digest the compound better than monogastric animals) where it contributes calories to the diet. Food processing can alter the carbohydrates composition making them more digestible to human organism. Although the presence of these oligosaccharides is generally considered undesirable because of their flatus activity, some studies shown some beneficial effects of dietary oligosaccharides in humans, mainly due to a increasing population of indigenous bifidobacteria in colon, such as: suppressing the activity of putrefactive bacteria by antagonist effect; preventing pathogenic and autogenous diarrhea; anti-constipation due to the production of high levels of short-chain fatty acids; producing nutrients such as vitamins. Other positive effects are: toxic metabolites and detrimental enzymes reduction; protecting liver function due to reduction of toxic metabolites; blood pressure reduction; anticancer effects [15,16].

1.2.2. Micronutrients

Soybean also contains a wide range of micronutrients and phytochemicals including minerals, vitamins, phytic acid (1.0–2.2%), sterols (0.23–0.46%) and saponins (0.17–6.16%) [20].

The primary inorganic compounds of the soybeans are minerals. Potassium is found in the highest concentration, followed by phosphorus, magnesium, sulfur, calcium, chloride and sodium, which vary in concentration to the variety, growing location and season. Minor minerals include silicon, iron, zing, manganese, copper and other [15,16].

Both water-soluble and oil-soluble vitamins are present in soybeans. Water-soluble vitamins in soybeans include thiamin, riboflavin, niacin, panthotenic acid and folic acid. They are not substantially lost during oil extraction and subsequently toasting of flakes. The oil-soluble vitamins present in soybeans are vitamins A and E. Vitamin E is especially unstable during soybean processing. During solvent extraction of soybeans, vitamin E goes with oil [15,16].

Phytate is the calcium-magnesium-potassium salt of inositol hexaphosphoric acid commonly known as phytic acid. As in the most seeds, phytate is the principal source of phosphorus in soybeans. His content depends on not only variety, but also growing conditions and assay methodology [15,16].

One important group of minor compounds present in soybean that has received considerable attention is a class of phytoestrogen called the isoflavones. Phytoestrogens are non-steroidal compounds that bind to and activate estrogen receptors (ERs) α and β, due to the fact that they mimic the conformational structure of estradiol. Phytoestrogens are naturally occurring plant compounds found in numerous fruits and vegetables, and are categorized into three classes: the isoflavones, lignans and coumestans [21,22]. Isoflavone compounds have been considered as nonnutrients, because they neither yield any energy nor function as vitamins. However, they play significant roles in the prevention of several diseases, so they may considered health-promoting substances.

Isoflavones belong to a group of compounds that share a basic structure consisting of two benzyl rings joined by a three-carbon bridge, which may or may not be closed in a pyran ring. This group of compounds is known as flavonoids, which include by far the largest and most diverse range of plant phenolics [15,16].

Isoflavones are present in just a few botanical families, because of the limited distribution of the enzyme chalcone isomerase, which converts 2(R)-naringinen, a flavone precursor, into 2-hydroxydaidzein. The soybean is unique in that it contains the highest amount of isoflavones, being up to 3 mg.g-1 dry weight [15,23].

The isoflavones in soybeans and soy products are of three types, with each type being present in four chemical forms. Therefore there are twelve isomers of isoflavones.

Isoflavones in soybean are mainly found as aglycones (daidzein, genistein, glycitein) (Figure 1), β-glucosides (daidzin, genistin, glycitin), malonyl-β-glucosides (6''-O-malonyldaidzin, 6''-O-malonylgenistin, 6''-O-malonylglycitin) and acetyl-β-glucosides (6''-O-acetyldaidzin, 6''-O-acetylgenistin, 6''-O-acetylglycitin) (Figure 2) [24].

Compounds	R₁	R₂
Daidzein	H	H
Glycitein	H	OCH₃
Genistein	OH	H

Figure 1. Chemical structure of soy aglycones.

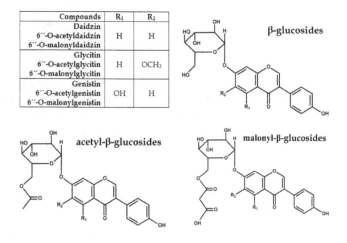

Compounds	R₁	R₂
Daidzin		
6''-O-acetyldaidzin	H	H
6''-O-malonyldaidzin		
Glycitin		
6''-O-acetylglycitin	H	OCH₃
6''-O-malonylglycitin		
Genistin		
6''-O-acetylgenistin	OH	H
6''-O-malonylgenistin		

Figure 2. Chemical structure of β-glucosides soybeans isoflavones.

Aglycones are flavonoid molecules without any attached sugars or other modifiers. Among the different forms of isoflavones, aglycones are especially important because they are readily bioavailable to humans [24]. Flavonoid β-glucosides also may carry additional small molecular modifiers, such as malonyl and acetyl groups. Sugar-linked flavonoids are called glycosides. The term glucoside only applies to flavonoids linked to glucose [25]. The malonyl-β-glucosides are the predominant form in conventional raw soybean [24].

Isoflavones concentration and composition vary greatly with structural parts within soybean seed. The concentration of total isoflavones in soybean hypocotyl is 5.5-6 times higher than that in cotyledons. Glycitein and its three derivatives occur extensively in the hypocotyl. Isoflavones are almost absent in seed coats [15]. The isoflavone content varies among soybean varieties, but most varieties contain approximately 100–400 mg per 100 g dry basis [26].

1.3. Isoflavones importance in human health

Isoflavones have received much attention because of their weak estrogenic property and other beneficial functions [24]. A large number of researchers have reported the positive aspects of isoflavones on human health, such as the ability to reduce the risk of cardiovascular, atherosclerotic and haemolytic diseases; alleviation of osteoporosis, menopausal and blood-cholesterol related symptoms; inhibition of the growth of hormone-related human breast cancer and prostate cancer cell lines in culture; and increased antioxidant effect in human subjects [27-30]. Isoflavones are also known for having anticancer activity and an effect on cell cycle and growth control [31-33].

Isoflavones in glycosides forms are poorly absorbed in the small intestine, due to their higher molecular weight and hydrophilicity. However bacteria in the intestine wall can biologically activate by action of β-glucosidase to their corresponding bioactive aglycone forms. Once hydrolyzed, aglycone forms are absorbed in the upper small intestine by passive diffusion. Nevertheless, pharmacokinetic studies confirm that healthy adults absorb isoflavones rapidly and efficiently. The average time to ingested aglycones reach peak plasma concentrations is about 4–7 h, which is delayed to 8–11 h for the corresponding β-glycosides. Despite the fast absorption, isoflavones or their metabolites are also rapidly excreted [13].

1.4. Food processing effect in isoflavones

Several investigations have been performed during the last years on soybeans consumption and their benefits, however the effects of seed processing and soybean processing into foods on the distribution of soy isoflavones are sparse. Processing significantly affects the retention and distribution of isoflavone isomers in soyfoods. The conversion and loss of isoflavones during processing significantly affect the nutraceutical values of soybean. Post-harvest changes in isoflavones in soybeans are influenced by processing methods however genotype has an effect on isoflavones profiles during seed development [34] and the environment has a greater e • ect than the genotype [26,35-37]. Distribution of isoflavones in soybean can be altered during various processing steps including fermentation, cooking, frying, roasting, drying and storage [24]. These effects in isoflavones can be even accelerated by heat, acid, alkaline and enzymes [38-40] (Figure 3). The effects of several food processing techniques on soybeans isoflavones will be reviewed and discussed in the following section.

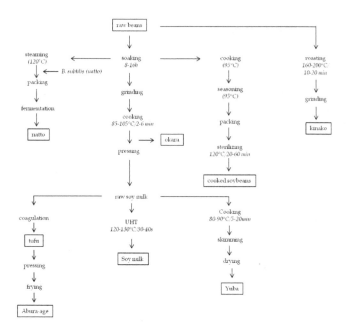

Figure 3. Flow diagram illustrating the processing of soybeans to commercial soybeans products. Adapted from Toda *et al.* [41].

1.4.1. Harvesting

Harvesting soybean seeds after their development and maturation is a critical step in profitable soybean production. Although most soybeans are harvested at the dry mature stage, a very small portion is harvested at the immature stage in certain regions. Immature soybeans refer to soybeans harvested at 80% maturity. The immature seed is used as a vegetable or as ingredient recipes. Soybeans are considered dry mature when seed moisture reduces to less than 14% in the field [15,42]. At this stage, seeds are ready for harvesting. The exact harvesting date depends on the variety, growing regions, plating date and local weather conditions [15].

1.4.2. Drying

Drying is an important process to extend the shelf life or to prepare food, including soybean, for subsequent production [43]. After harvest, if moisture content is more than 14%, soybeans need to be dried immediately in order to the following reasons: meet the quality standard of soybean trading; retain maximum quality of the grain; reach a level of moisture that does not allow the growth of bacteria and fungi and; prevent germination of seeds. Drying could be done naturally or artificially. Sun drying, or natural drying, soybeans are spread on the threshing for 2-3 days with frequently turning between top to bottom layers.

Once dried, seeds are transferred to storage facilities. Sun drying is not suitable for large quantities of soybeans or under humid and cloudy weather conditions. Artificially drying is carried out with various mechanical driers, including low temperature driers, on-the-floor driers, in-bin driers, medium temperature driers, tray driers, multiduct ventilated driers, countercurrent open-flame grain driers and solar driers. Regardless of which driers are used, caution is required so as to avoid too rapid drying; rapid drying hardens outlayers of seeds and seals moisture within the inner layer. Although the temperature of soybeans must be raised sufficiently to achieve the desired moisture content during drying process, excess heating (not exceed 76%) should be avoided to protect beans from discoloration and beans proteins from denaturation [15].

It is well known that drying significantly affects the quality and nutrients of dried food products. Drying temperatures affect the activity and stability due to chemical and enzymatic degradation which can for example alter significantly the isomeric distribution of the 3 aglycones [44]. All forms of isoflavones are generally lost due to thermal degradation and oxidation reactions during processing [45]. Conventional thermal treatment decreases malonyl derivatives into β-glucosides via intra-conversion while aglycones have higher heat resistance. Dry heat treatment such as frying, toasting, or baking process increases the formation of acetyl derivatives of isoflavones through decarboxylation from malonyl derivatives [24,41].

A comparison between freeze-drying and drum-drying of germ flour demonstrated that the former contained higher isoflavone aglycones than the latter. However, isoflavone glucoside contents in freeze-dried germ flour were lower than those of drum-dried germ flour. The content of isoflavone glucosides was significantly lower in processed (drum- and freeze-dried) germ flours compared with that of unprocessed germ flour because of conversion of isoflavone glucosides to isoflavone aglycones. Total isoflavone contents of drum-dried and freeze-dried germ flours were comparable but more than that of unprocessed germ flour [46].

Different drying methods (hot-air fluidized bed drying, HAFBD; superheated-steam fluidized bed drying, SSFBD; gas-fired infrared combined with hot air vibrating drying, GFIR-HAVD) were compared at various drying temperatures (50, 70, 130 and 150 °C). Higher drying temperatures led to higher drying rates and higher levels of β-glucosides and antioxidant activity, but to lower levels of malonyl-β-glucosides, acetyl-β-glucosides and total isoflavones. Comparing different drying methods to each other, at the same drying temperature GFIR-HAVD resulted in the highest drying rates and the highest levels of β-glucosides, aglycones and total isoflavones, antioxidant activity as well as α-glucosidase inhibitory activity of dried soybean. A drying temperature of 130 °C gave the highest levels of aglycones and α-glucosidase inhibitory activity in all cases [47].

Dehydration also helps to achieve longer shelf life and easier transportation and storage, enabling wider distribution of a product. Fermentation of soymilk with microorganisms with β-glucosidase activity promotes the biotransformation of isoflavone glucosides into bioactive aglycones, brings down the contents of aldehydes and alcohols responsible for the beany flavor in soy milk [48] and reduces the content of indigestible oligosaccharides. A spray-drying technology has been also applied to fermented soymilk. Daidzein and genistein retention were of about 87.6% and 85.3%, respectively. Retention was better at the lower inlet air temperatures. Coarser droplets formed at higher feed rates helped in higher retention of isoflavones because it reduced the incidence of direct exposure of isoflavones to higher temperature [49].

1.4.3. Storage

Soybeans are stored at farms, elevators and processing plants in various types of storage structures (steel tanks or concrete silos) before being channeled to next destination, and finally to processing. Loss in quality of soybeans during storage results from the biological activities of seeds themselves, microbiological activities and attacks of insects, mites and rodents. Quality loss is characterized by reduced seed viability and germination rate, coloration, reduced water absorption, compositional changes and ultimately reduced quality of protein and oil. Heat damage is a major cause of quality loss. Characterized by darkening of seed color coat, it results mainly from the improper control of temperature and moisture during storage and transportation. The excess presence of foreign matter can also cause heat buildup. Thus, cleaning soybeans before drying minimizes heat damage. Although minor losses are inevitable, major losses can be prevented by carefully control of storage temperature and humidity. Any biological activity requires a certain level of moisture present. Higher moisture content (or high moisture humidity) not only promotes bacteria and mold infection but also speeds up biological activity of seed themselves. Excessive moisture may also leed to seed germination. Generally, moisture content of 13.5% or below is considered to ensure storage stability of soybeans over reasonably long periods. However, this is true only when temperatures are kept below certain levels [15].

Storage conditions have also an important effect in the composition of soybeans, including isoflavone, anthocyanin, protein, oil and fatty acid.

Variation in isoflavone contents of different soybean cultivars were evaluated under different location and storage duration. Total isoflavone contents of soybeans stored for 1 year were only slightly higher than those of soybeans stored for 2 or 3 years. However, the concentrations of individual isoflavones, especially 6''-O-malonyldaidzin and 6''-O-malonylgenistin, decreased markedly in soybeans stored for 2 or 3 years, probably due to high temperatures during storage and oxidative reactions which transformed malonylated type to glycoside and aglycone groups, increasing their amounts with longer storage. They also pointed that the effect of crop year seemed to have a much greater influence on the variation in isoflavone content than did the location because of weather differences from one year to another. Differences among cultivars have been already expected [35]. Similar results were found comparing cropping year and storage for 3 years of soybeans seeds. Malonyldaidzin and malonylgenistin concentrations also decreased and the concentration of glucosides increased slightly over the 3 years [50]. Storage effect in soybeans isoflavones have been exhaustedly studied, and a markedly decreased in isoflavones content is proportional to storage periods, whereas protein, oil, and fatty acid of black soybeans showed a slight decrease over storage at room temperature [51].

A combined effect of storage temperature and water activity was evaluated on the content and profile of isoflavones of soy protein isolates and defatted soy flours. Storage for up to 1 year of soy products, at temperatures from -18 to 42 °C, had no effect on the total content of isoflavones, but the profile changed drastically at 42 °C, with a significant decrease of the percentage of malonyl glucosides with a proportional increase of β-glucosides. A similar effect was observed for soy protein isolates stored at aw = 0.87 for 1 month. For defatted soy flours, however, there was observed a great increase in aglycons (from 10 to 79%), probably due to the action of endogenous β-glucosidases [52].

1.4.4. Fermentation

Processing and fermentation of soybean has been reported to influence the forms isoflavones take. Studies have shown that the fermentation process of soybean promotes changes in the phytochemical compounds, causing changes in the isoflavone forms, hydrolysing the proteins and reducing the antinutritional factors, by reducing trypsin inhibitor content [53-55]. Fermentation process of soy leads to manufacturing different soy fermented foods, such as tempeh, soy extract, miso and natto.

Differences on isoflavones content between non-fermented and fermented soybean products have been extensively describe in literature in the last years. Isoflavone glucosides were the major components in soybean and non-fermented products, while isoflavone aglycones were abundant in sufu and partially in miso of soybean fermented products [56].

Tempeh is a traditional fermented soybean food product from Indonesia. It is normally consumed fried, boiled, steamed or roasted. A way to processing soybeans into tempeh occurs by fungus mediated fermentation. Several authors have already studied the effect of fermentation on isoflavones content. It was reported an increase of aglycones amount with fermentation time of tempeh, approximately two-fold higher after 24 h fermentation [57]. Later on, similar results were found by Haron et al. [58] who reported higher values of aglycone forms in raw tempeh. In addition, these researchers showed that fried tempeh had its total isoflavone content reduced in almost 50% during frying processing (reduction from 205 to 113 mg in 100 g of fried tempeh). A combined process of fermentation and refrigeration was evaluated by Ferreira et al. [25]. They quantified isoflavones content of two different soybean cultivars (low isoflavone content versus high isoflavone content cultivar) during processing of tempeh combined to refrigeration at 4 ºC for 6, 12, 18, and 24 hours. After 24 hours fermentation, isoflavone glucosides were 50% reduced, and the aglycone forms in the tempeh from both cultivars was increased. The malonyl forms reduced 83% after cooking. Refrigeration process up to 24 hours did not affect the isoflavone profile of tempeh from either cultivar. The fermentation process improves the nutritional value of tempeh by increasing the availability of isoflavone aglycones. Fermented soy foods, which are usually prepared by mixing soy with other components such as barley, rice and wheat, contained isoflavones at lower concentrations. In addition these fermented soy foods contain predominantly isoflavone aglycones [15]. Fermentation with microorganisms or natural products containing high β-glucosidase activity converts β-glucosides into corresponding aglycones by breaking the carbohydrate bond [59,60]. These aglycones exist in smaller amounts in other nonfermented soy products such as tofu and soymilk [37,61]. Other soybean products or by-products showed a similar behavior. Soy pulp is generated as a by-product during tofu or soymilk production and is sometimes used as animal food but is mainly burnt as waste. Fermentation increased isoflavone aglycone contents in black soybean pulp. Genistein concentrations in black soybean pulp were 6.8 and 7.2 fold higher than controls respectively after 12 h and 24 h of fermentation with *L. acidophilus*. Fermentation with *B. subtilis* showed a similar genistein concentration increase [62]. The effect of fermentation of whole soybean flour was investigated and also showed a conversion of isoflavone glycosides to the aglycone form [63].

Sufu, a fermented tofu product, showed ambiguous changes in isoflavone contents during manufacturing. Sufu manufacturing procedure promoted a significant loss of isoflavone content mainly attributed to the preparation of tofu and salting of pehtze. The isoflavone composition was altered during sufu processing. The initial fermentation corresponded to the fastest period of isoflavone conversion. Aglycones levels increased while the corresponding levels of glucosides decreased. The changes in the isoflavone composition were significantly related to the activity of β-glucosidase during sufu fermentation, which was inhibited by the NaCl content [64]. These influence of processing and NaCl supplementation on isoflavone contents was also investigated during douchi manufacturing. Douchi is a popular Chinese fermented soybean food. These results indicated that 61% of the isoflavones in raw soybeans were lost when NaCl content was 10%. Indeed, changes in isoflavone isomer distribution were found to be related to β-glucosidase activity during fermentation, which was affected by NaCl supplementation [65].

Other fermented soybean products are miso and soy sauce. Their production involves the application of pressurized steaming following to fermentation by bacteria for a lengthy period. During the fermentation of miso and soy sauce, β-glucosides are also reported to be hydrolyzed to aglycons by β-glucosidase of bacteria [41].

Fermentation shows the same pattern on isoflavones transformation, no matter the fermented soybean product. The fungi-fermented black soybeans (koji) also contained a higher content of aglycone, the bioactive isoflavone, than did the unfermented black soybeans [66]. However, it was later realized that the contents of various isoflavone isomers in black soybean koji may reduce during after 120 days of storage. Although the retention of isoflavone varied with storage temperature, packaging condition enabled black soybean koji to retain the highest residual of isoflavone [67].

1.4.5. Non-fermentation processing

Tofu is a popular nonfermented soy food. Processing of tofu involves soaking and heating procedures as well as the addition of protein coagulants such as calcium sulfate to soymilk to coagulate to make tofu. This soybean product has been also target of several studies during its manufacturing. Results of the stability of isoflavone during processing of tofu showed that the concentrations of the three aglycones increased with increasing soaking temperature and time, while a reversed trend was found for the other nine isoflavones. Tofu produced with 0.3% calcium sulfate was found to contain the highest total isoflavones yield (2272.3 µg/g) whereas a higher level (0.7%) of calcium sulfate resulted in a lower yield

(1956.6 μg/g) of total isoflavones in tofu. In the same study, authors showed that during processing of soymilk, an increase of concentration for β-glucosides and acetyl genistin, whereas malonyl glucosides exhibited a decreased tendency and the aglycones did not show significant change [68]. Previous reports have already demonstrated that during soaking of soybean malonyl glucosides can be converted to acetyl glucosides, which can be further converted to glycosides or aglycones depending on soaking temperature and time [34,69].

Regarding soymilk isoflavones, a research was done on the transformation of isoflavones and the β-glucosidase activity in soymilk during fermentation. Regardless of employing a lactic acid bacteria or a bifidobacteria as starter organism, fermentation causes a major reduction in the contents of glucoside, malonyl glucoside and acetyl glucoside isoflavones along with a significant increase of aglycone isoflavones content. Indeed, the increase of aglycones and decrease of glucoside isoflavones during fermentation coincided with the increase of β-glucosidase activity observed in fermented soymilk [70].

1.4.6. Heating processing

Distribution of isoflavones in soybeans and soybeans products are significantly affected by the method, temperatures and duration of heating.

Toasted soy flour and isolated soy protein had moderate amounts of each of the isoflavone conjugates. Pananum et al. [44] evaluated the effect of longer toasting of defatted soy flakes at 150 °C. They stated that toasting led to higher aglycone concentration, which increased the total phenolic recovery. Apparently, malonyl glucoside conjugates are thermally unstable and are converted to their corresponding isoflavone glycosides at high temperatures [15]. The chemical modification of isoflavones in soy foods during cooking and processing studies showed interesting results on isoflavones stability. Baking or frying of textured vegetable protein at 190 °C and baking of soy flour in cookies did not alter total isoflavone content. However, there was a steady increase in β-glucoside conjugates at the expense of 6''-O-malonylglucosides conjugates [71]. The de-esterifying reaction was presumably a result of transesterification of the ester linkage between the malonate or acetate carboxyl group and the 6''-O-hydroxyl group of the glucose moiety, yielding methyl malonate or methyl acetate and the isoflavone glucoside [15].

Roasting has been used to deactivate anti-nutritional components in soybeans and to give characteristic flavour and brown color to final products [43]. Kinako is a soybean product produced by roasting raw soybeans around 200 °C for 10-20 min, following by grinding. Roasting also promotes changes in isoflavones contents. Soybeans roasted upon at 200 °C without prior soaking in water showed a change in isoflavones profile. It was found that at the first 10 min of roasting caused an increase in 6''-O-acetyl-β-glucosides while 6''-O-malonyl-β-glucosides decreased drastically. Continued roasting showed a slightly decreased proportion of the 6''-O-acetyl-β-glucosides. These authors proposed that most 6''-O-malonyl-β-glucosides were decarboxylated and changed to 6''-O-acetyl-β-glucosides when roasted at a high temperature. Besides, β-glucosides and aglycons also increased gradually over time [41]. A similar trend was found roasting soybeans at 200 °C for 7, 14 and 21 min. Malonyl derivatives decreased drastically and acetyl-β-glucosides and β-glucosides increased significantly [24].

Similar to the roasting process, explosive puffing caused a significant decrease in malonyl derivatives and significant increase in acetyl derivatives and β-glucosides through 686 kPa explosive puffing treatment. Otherwise, aglycones did not increase during the explosive puffing process. This fact was suggested due that temperature of explosive puffing may not be high enough to cleave glycosidic linkage between β-glucopyranose and aglycones [24].

Some soy products are prepared by frying process. Abura-age is one of them being produced by frying tofu in oil. As pointed by Toda *et al.* [41], In comparison with tofu, heating in oil at a high temperature to produce abura-age resulted in a smaller proportion of 6''-O-malonyl-β-glucosides and a higher proportion of 6''-O-acetyl-β-glucosides.

Simonne *et al.* [34] evaluated the retention and changes of soy isoflavones in immature soybeans seeds during processing and found total isoflavone retention percentages means of 46% after boiling, 53% after freezing and 40% after freeze-drying. They assumed that probably the loss of isoflavones could be due to their leaching into the cooking water, however these authors did not analyze isoflavones in the cooking water. In the same study, they noted that boiling process also caused a substantial increase in daidzin, genistin, and genistein. In a previous work on soy milk production and cooking of dry soy products, it was proposed that hydrolysis of the malonyl and acetyl glucosides during boiling probably contributes to the conversion of isoflavone forms [71]. Changes in isoflavone compositions of different soybean foods during cooking process were performed later and supported this proposition [41]. Concerning to a possible leaching effect of isoflavones during blanching or boiling, Wang *et al.* [72] reported about 26% retention of isoflavones during the production of soy protein isolate. This soybean product is made by extracting soy flour under slightly alkaline pH, followed by precipitation, washing, and drying. Soy protein isolate showed a different isoflavone profile in comparison to soy flour. The former contained much more aglycones (genistein and daidzein), while the latter had almost none [12,72]. The high content of aglycones in soy protein isolate was probably due to the hydrolysis of glycosides. The percentages of total isoflavones lost during extraction, precipitation, and washing were 19, 14, and 22%, respectively. Washing was the step where most isoflavones were lost [72]. This statement was supported by another study on thermal processing of tofu. It was demonstrated a significant total isoflavone content decrease, most likely due to leaching of isoflavones into the water [69].

An approach on the stability of isoflavones in soy milk stored at temperatures ranging from 15 to 90 °C showed that genistin in soy milk is labile to degradation during storage. Although the loss rate was low at ambient temperatures, authors highlighted the potential loss of genistin when estimating shelf life of soy milk products [73].

Influence of thermal processing such as boiling, regular steaming and pressure steaming were also investigated in yellow and black soybeans. Again, all thermal processing caused significant increases in aglycones and β-glucosides of isoflavones, but caused significant decreases in malonyl glucosides of isoflavones for both kinds of soybeans. The malonyl glucosides decreased dramatically with an increase in β-glucosides and aglycones after thermal processing [74].

1.4.7. Irradiation

Food irradiation is a process in which food is exposed to ionizing radiations such as high energy electrons and X-rays produced by machine sources or gamma rays emitted from the radioisotopes ^{60}Co and ^{137}Cs. Depending on the absorbed radiation dose, various effects can be achieved resulting in reduced storage losses, extended shelf life and/or improved microbiological and parasitological safety of foods [75,76]. Food irradiation is one of the most effective means to disinfect dry food ingredients. Disisfestation is aimed at preventing losses caused by insects in stored grains, pulses, flour, cereals, coffee beans, dried fruits, dried nuts, and other food products. The dosage required for insect control is fairly low, of the order of 1 kGy or less [77,78].

Soybeans have been processed by ionizing radiation in order to improve their properties. An important improvement in microbiological properties, such as insect disinfestation and microbial contamination, can be achieved by radiation treatment. Physical properties are also enhanced, such as reduction of soaking and cooking time. Indeed, higher radiation doses may break glycosidic linkages in soybean oligosaccharides to produce more sucrose and decrease the content of flatulence causing oligosaccharides [79].

Gamma irradiation at 2.5-10.0 kGy caused the reduction of soaking time in soybeans by 2-5 hours and the reduction of cooking time by 30-60% compared to non-irradiated control samples. The irradiation efficacy on physical quality improvement was also recognized in stored soybeans for one year at room temperature [80,81].

Influence of radiation processing on isoflavones content has been also studied in the last years. Gamma-radiated (0.5-5.0 kGy) soybeans showed a radiation-induced enhancement of antioxidant contents. Interestingly, a decrease in content of glycosidic conjugates and an increase in aglycons were noted with increasing radiation doses. These results suggested a radiation-induced breakdown of glycosides resulting in release of free isoflavones. Whereas the content of genistein increased with radiation dose, that of daidzein showed an initial increase at a dose of 0.5 kGy and then decreased at higher doses. Degradation of daidzein beyond 0.5 kGy could thus be assumed. Glycitein appears to be the least stable among the three aglycons as its content decreased at all of the doses studied [82]. Gamma irradiation also induced an enhancement in isoflavones content of varying seed coat colored soybean up to a radiation dose of 0.5 kGy. However, the genotypes showed decrease in total isoflavone content at a higher radiation dose of 2.0 and 5.0 kGy [83].

Such as gamma-irradiation, an enhanced effect on soy germ isoflavones was found after electron beam processing. Interestingly, this study showed that applied radiation doses ranging from 1.0 to 20.0 kGy showed an increase in the amount of both glucosides and aglycones simultaneously [84].

Acknowledgements

We thank CAPES, IPEN and CNPq.

Author details

Michel Mozeika Araújo[1,2], Gustavo Bernardes Fanaro[1] and
Anna Lucia Casañas Haasis Villavicencio[1*]

*Address all correspondence to: mmozeika@yahoo.com

1 Nuclear and Energy Research Institute (IPEN), Brazil

2 Strasbourg University (UdS), France

References

[1] Liu, X., Jin, J., Wang, G., & Herbert, S. J. (2008). Soybean yield physiology and development of high-yielding practices in Northeast China. *Field Crops Research*, 105,
157-171.

[2] Jeng, T. L., Shih, Y. J., Wua, M. T., & Sung, J. M. (2010). Comparisons of flavonoids
and anti-oxidative activities in seed coat, embryonic axis and cotyledon of black soybeans. *Food Chemistry*, 123, 1112-1116.

[3] Chan, C., Qi, C., Li, M. W., Wong, F. L., & Lam, H. M. (2012). Recent Developments
of Genomic Research in Soybean. *Journal of Genetics and Genomics*, 39, 317-324.

[4] Tacarindua, C. R. P., Shiraiwa, T., Homma, K., Kumagai, E., & Sameshima, R. (2012).
The response of soybean seed growth characteristics to increased temperature under
near-field conditions in a temperature gradient chamber. *Field Crops Research*, 131,
26-31.

[5] Rosenthal, A., Deliza, R., Cabral, L. M. C., Cabral, L. C., Farias, C. A. A., & Domingues, A. M. (2003). Effect of enzymatic treatment and filtration on sensory characteristics and physical stability of soymilk. *Food Control*, 14, 187-192.

[6] Baber, R. (2010). Phytoestrogens and post reproductive health. *Maturitas*, 66, 344-349.

[7] Diya'uddeen, B. H., Abdul, Aziz. A. R., Daud, W. M. A. W., & Chakrabarti, M. H.
(2012). Performance evaluation of biodiesel from used domestic waste oils: A review.
Process Safety and Environmental Protection, 90, 164-179.

[8]   Viola, E., Blasi, A., Valerio, V., Guidi, I., Zimbardi, F., Braccio, G., & Giordano, G. (2012). Biodiesel from fried vegetable oils via transesterification by heterogeneous catalysis. *Catalysis Today*, 179, 185-190.

[9]   Jiao, Z., Si, X. X., Zhang, Z. M., Li, G. K., & Cai, Z. W. (2012). Compositional study of different soybean (Glycine max L.) varieties by 1H NMR spectroscopy, chromatographic and spectrometric techniques. *Food Chemistry*, 135, 285-291.

[10] Thuzar, M., Puteh, A. B., Abdullah, N. A. P., Lassim, M. B. M., & Jusoff, K. (2010). The Effects of Temperature Stress on the Quality and Yield of Soya Bean [(Glycine max L.) Merrill.]. *Journal of Agricultural Science*, 2, 172-179.

[11] United States Department of Agriculture (USDA). (2012). World Agricultural Supply and Demand Estimates:. *WASDE- 504*.

[12] Shao, S., Duncan, A. M., Yanga, R., Marconec, M. F., Rajcand, I., & Tsaoa, R. (2009). Tracking isoflavones: From soybean to soy flour, soy protein isolates to functional soy bread. *Journal of Functional Foods I*, 119-127.

[13] Cederroth, C. R., Zimmermann, C., & Nef, S. (2012). Soy, phytoestrogens and their impact on reproductive health. *Molecular and Cellular Endocrinology*, 355, 192-200.

[14] Saldivar, X., Wanga, Y. J., Chen, P., & Hou, A. (2011). Changes in chemical composition during soybean seed development. *Food Chemistry*, 124, 1369-1375.

[15] Liu, K. (1997). Soybeans: chemistry, technology, and utilization. *New York: Chapman & Hall*.

[16] Erickson, D. R. (1995). Practical Handbook of soybean processing and utilization. *USA: AOCS*.

[17] Garcia, M. C., Torre, M., Marina, M. L., & Labord, F. (1997). Composition and characterization of soyabean and related products. *Critical Reviews in Food Science and Nutrition*, 361-391.

[18] Grieshop, C. M., & Fahey Jr, G. C. (2001). Comparison of quality characteristics of soybeans from Brazil, China and the United States. *Journal of Agricultural and Food Chemistry*, 49(5), 2669-2673.

[19] Grieshop, C. M., Kadzere, C. T., Clapper, G. M., Flickinger, E. A., Bauer, L. L., Frazier, R. L., & Fahey Jr, G. C. (2003). Chemical and nutritional characteristics of United States soybeans and soybean meals. *Journal of Agricultural and Food Chemistry*, 51(26), 7684-7691.

[20] Kang, J., Badger, T. M., Ronis, M. J., & Wu, X. (2010). Non-isoflavone phytochemicals in soy and their health effects. *Journal of Agricultural and Food Chemistry*, 58(14), 8119-8133.

[21] Kuiper, G. G. J. M., Carlsson, B., Grandien, K., Enmark, E., Häggbla, J., Nilsson, S., & Gustafsson, J. A. (1997). Comparison of the Ligand Binding Specificity and Transcript Tissue Distribution of Estrogen Receptors α and β. *Endocrinology*, 138(3), 863-870.

[22] Kuiper, G. G., Lemmen, J. G., Carlsson, B., Corton, J. C., Safe, S. H., van der Saag, P. T., van der Burg, B., & Gustafsson, J. A. (1998). Interaction of estrogenic chemicals and phytoestrogens with estrogen receptor beta. *Endocrinology*, 139(10), 4252-4263.

[23] Rostagno, M. A., Palma, M., & Barroso, C. G. (2004). Pressurized liquid extraction of isoflavones from soybeans. *Analytica Chimica Acta*, 522, 169-177.

[24] Lee, S. W., & Lee, J. H. (2009). Effects of oven-drying, roasting, and explosive puffing process on isoflavone distribution in soybeans. *Food Chemistry*, 112, 316-320.

[25] Ferreira, M. P., Oliveira, M. C. N., Mandarino, J. M. G., Silva, J. B., Ida, E. I., & Carrão-Panizzi, M. C. (2011). Changes in the isoflavone profile and in the chemical composition of tempeh during processing and refrigeration. *Pesquisa Agropecuária Brasileira*, 46(11), 1555-1561.

[26] Hoeck, J. A., Fever, W. R., Murphy, P. A., & Grace, A. W. (2000). Influence of genotype and environment on isoflavone concentrations of soybean. *Journal of Agricultural and Food Chemistry*, 40, 48-51.

[27] Kwak, C. S., Lee, M. S., & Park, S. C. (2007). Higher antioxidant properties of chungkookjang, a fermented soybean paste, may be due to increased aglycone and malonylglycoside isoflavone during fermentation. *Nutrition Research*, 27, 719-727.

[28] Lee-J, S., Kim-J, J., Moon-I, H., Ahn-K, J., Chun-C, S., Jung-S, W., et al. (2008). Analysis of isoflavones and phenolic compounds in Korean soybean [Glycine max (L.) Merrill] seeds of different seed weights. *Journal of Agricultural and Food Chemistry*, 56, 2751-2758.

[29] Scambia, G., Mango, D., Signorile, P. G., Anselmi, Angeli. R. A., Palena, C., Gallo, D., et al. (2000). Clinical effects of a standardized soy extract in postmenopausal women: A pilot study. *Menopause*, 7, 105-111.

[30] Zhang, X., Shu, X. O., Gao, Y. T., Yang, G., Li, Q., Li, H., et al. (2003). Soy food consumption is associated with lower risk of coronary heart disease in Chinese women. *Journal of Nutrition*, 133, 2874-2878.

[31] Gourineni, V. P., Verghese, M., & Boateng, J. (2010). Anticancer effects of prebiotics Synergy1® and soybean extracts: Possible synergistic mechanisms in caco-2 cells. *International Journal of Cancer Research*, 6, 220-233.

[32] Katdare, M., Osborne, M., & Telang, N. T. (2002). Soy isoflavone genistein modulates cell cycle progression and induces apoptosis in HER-2/neu oncogene expressing human breast epithelial cells. *International Journal of Oncology*, 21, 809-815.

[33] Kim, M. H., Gutierrez, A. M., & Goldfarb, R. H. (2002). Different mechanisms of soy isoflavones in cell cycle regulation and inhibition of invasion. *Anticancer Research*, 22(6C), 3811-3817.

[34] Simonne, A. H., Smith, M., Weaver, D. B., Vail, T., Barnes, S., & Wei, C. I. (2000). Retention and changes of soy isoflavones and carotenoids in immature soybean seeds

(Edamame) during processing. *Journal of Agricultural and Food Chemistry*, 48, 6061-6069.

[35] Lee, S. J., Chung, I. M., Ahn, J. K., Kim, J. T., Kim, S. H., & Hahn, S. J. (2003). Variation in isoflavones of soybean cultivars with location and storage duration. *Journal of Agricultural and Food Chemistry*, 51, 3382-3389.

[36] Lee, S. J., Yen, W., Ahn, J. K., & Chung, I. M. (2003). Effects of year, site, genotype, and their interactions on various soybean isoflavones. *Field Crop Research*, 81, 181-192.

[37] Wang, H. J., & Murphy, P. A. (1994). Isoflavone composition of American and Japanese soybeans in Iowa- Effects of variety, crop year, and location. *Journal of Agricultural and Food Chemistry*, 42(8), 1674-1677.

[38] Beddows, C. G., & Wong, J. (1987). Optimization of yield and properties of silken tofu from soybeans. I. *The water:bean ratio. International Journal of Food Science and Technology*, 22, 15-21.

[39] Riedl, K. M., Zhang, Y. C., Schwartz, S. J., & Vodovotz, Y. (2005). Optimizing dough proofing conditions to enhance isoflavone aglycones in soy bread. *Journal of Agricultural and Food Chemistry*, 53, 8253-8258.

[40] Vaidya, N. A., Mathias, K., Ismail, B., Hayes, K. D., & Corvalan, C. M. (2007). Kinetic modeling of malonylgenistin and malonyldaidzin conversions under alkaline conditions and elevated temperatures. *Journal of Agricultural and Food Chemistry*, 55, 3408-3413.

[41] Toda, T., Sakamoto, A., Takayanagi, T., & Yokotsuka, K. (2000). Changes in isoflavone compositions of soybean during soaking process. *Food Science and Technology Research*, 6, 314-319.

[42] Rubel, A., Rinne, R. W., & Canvin, D. T. (1972). Protein, oil, and fatty acid in developing soybean seeds. *Crop Science*, 12, 739-741.

[43] Im, M. H., Choi, J. D., & Choi, K. S. (1995). The oxidation stability and flavor acceptability of oil from roasted soybean. *Journal of Agricultural and Food Chemistry and Biotechnology*, 38(5), 425-430.

[44] Pananun, T., Montalbo-Lomboy, M., Noomhorm, A., Grewell, D., & Lamsal, B. (2012). High-power ultrasonication-assisted extraction of soybean isoflavones and effect of toasting. *Food Science and Technology*, 47, 1999-2007.

[45] Chien, J. T., Hsieh, H. C., Kao, T. H., & Chen-H, B. (2005). Kinetic model for studying the conversion and degradation of isoflavones during heating. *Food Chemistry*, 91, 425-434.

[46] Tipkanon, S., Chompreeda, P., Haruthaithanasan, V., Prinyawiwatkul, W., No, H. K., & Xu, Z. (2011). Isoflavone content in soy germ flours prepared from two drying methods. *International Journal of Food Science and Technology*, 46, 2240-2247.

[47] Niamnuy, C., Nachaisin, M., Laohavanich, J., & Devahastin, S. (2011). Evaluation of bioactive compounds and bioactivities of soybean dried by different methods and conditions. *Food Chemistry*, 129, 899-906.

[48] Blagden, T. D., & Gilliland, S. E. (2005). Associated with the "Beany" flavor in soymilk by Lactobacilli and Streptococci. *Journal of Food Science*, 70(3), M186-M189.

[49] Telang, A. M., & Thorat, B. N. (2010). Optimization of process parameters for spray drying of fermented soy milk. *Drying Technology*, 28, 1445-1456.

[50] Kim, S. H., Jung, W. S., Ahn, J. K., & Chung, I. M. (2005). Analysis of isoflavone concentration and composition in soybean [Glycine max (L.)] seeds between the cropping year and storage for 3 years. *European Food Research and Technology*, 220, 207-214.

[51] Lee, J. H., & Cho, K. M. (2012). Changes occurring in compositional components of black soybeans maintained at room temperature for different storage periods. *Food Chemistry*, 131, 161-169.

[52] Pinto, M. S., Lajolo, F. M., & Genovese, M. I. (2005). Effect of storage temperature and water activity on the content and profile of isoflavones, antioxidant activity, and in vitro protein digestibility of soy protein isolates and defatted soy flours. *Journal of Agricultural and Food Chemistry*, 53, 6340-6346.

[53] Kim, H., Peterson, T. G., & Barnes, S. (1998). Mechanism of action of the soy isoflavone genistein: emerging role for its effects via transforming growth factor signaling pathways. *American Journal of Clinical Nutrition*, 68, 1418S-1425S.

[54] Molteni, A., Brizio-Molteni, L., & Persky, V. (1995). In-vitro hormonal effects of soybean isoflavones. *Journal of Nutrition*, 125, 751S-756S.

[55] Zhu, D., Hettiarachchy, N. S., Horax, R., & Chen, P. (2005). Isoflavone contents in germinated soybean seeds. *Plant Foods for Human Nutrition*, 60, 147-151.

[56] Chen, T. R., & Wei, Q. K. (2008). Analysis of bioactive aglycone isoflavones in soybean and soybean products. *Nutrition and Food Science*, 38, 540-547.

[57] Nakajima, N., Nozaki, N., Ishihara, K., Ishikawa, A., & Tsuji, H. (2005). Analysis of isoflavone content in tempeh, a fermented soybean, and preparation of a new isoflavone-enriched tempeh. *Journal of Bioscience and Bioengineering*, 100, 685-687.

[58] Haron, H., Ismail, A., Azlan, A., Shahar, S., & Peng, L. S. (2009). Daidzein and genestein contents in tempeh and selected soy products. *Food Chemistry*, 115, 1350-1356.

[59] Murphy, P. A., Song, T., Buseman, G., Barua, K., Beecher, G. R., Trainer, D., & Holden, J. (1999). Isoflavones in Retail and Institutional Soy Foods. *Journal of Agricultural and Food Chemistry*, 47(7), 2697-2704.

[60] Yang, S. O., Chang, P. S., & Lee, J. H. (2006). Isoflavone distribution and beta-glucosidase activity in Cheonggukjang, a traditional Korean whole soybean-fermented food. *Food Science and Biotechnology*, 15(1), 96-101.

[61] Astuti, M., & Dalais, F. S. (2000). Tempeh, a nutritious and health food from Indonesia. *Asia Pacific Journal of Clinical Nutrition*, 9, 322-325.

[62] Hong, G. E., Mandal, P. K., Lim, K. W., & Lee-H, C. (2012). Fermentation increases isoflavone aglycone contents in black soybean pulp. *Asian Journal of Animal and Veterinary Advances*, 6, 502-511.

[63] Silva, L. H., Celeghini, R. M. S., & Chang, Y. K. (2011). Effect of the fermentation of whole soybean flour on the conversion of isoflavones from glycosides to aglycones. *Food Chemistry*, 128, 640-644.

[64] Li-Jun, Y., Li-Te, L., Zai-Gui, L., Tatsumi, E., & Saito, M. (2004). Changes in isoflavone contents and composition of sufu (fermented tofu) during manufacturing. *Food Chemistry*, 87, 587-592.

[65] Wang, L., Yin, L., Li, D., Zou, L., Saito, M., Tatsumi, E., & Li, L. (2007). Influences of processing and NaCl supplementation on isoflavone contents and composition during douchi manufacturing. *Food Chemistry*, 101, 1247-1253.

[66] Lee, I. H., & Chou, C. C. (2006). Distribution profiles of isoflavone isomers in black bean kojis prepared with various filamentous fungi. *Journal of Agricultural and Food Chemistry*, 54, 1309-1314.

[67] Huang-Y, R., & Chou-C, C. (2009). Stability of isoflavone isomers in steamed black soybeans and black soybean koji stored under different conditions. *Journal of Agricultural and Food Chemistry*, 57, 1927-1932.

[68] Kao, T. H., Lu, Y. F., Hsieh, H. C., & Chen, B. H. (2004). Stability of isoflavone glucosides during processing of soymilk and tofu. *Food Research International*, 37, 891-900.

[69] Grün, I. U., Adhikari, K., Li, C., Li, Y., Lin, B., Zhang, J., & Fernando, L. N. (2001). Changes in the profile of genistein, daidzein, and their conjugates during thermal processing of tofu. *Journal of Agricultural and Food Chemistry*, 49, 2839-2843.

[70] Chien-L, H., Huang-Y, H., & Choua-C, C. (2006). Transformation of isoflavone phytoestrogens during the fermentation of soymilk with lactic acid bacteria and bifidobacteria. *Food Microbiology*, 23, 772-778.

[71] Coward, L., Smith, M., Kirk, M., & Barnes, S. (1998). Chemical modification of isoflavones in foods during cooking and processing. *American Journal of Clinical Nutrition*, 68, 1486S-1491S.

[72] Wang, C., Ma, Q., Pagadala, S., Sherrard, M. S., & Krishnan, P. G. (1998). Changes of isoflavones during processing of soy protein isolates. *Journal of the American Oil Chemists' Society*, 75, 337-341.

[73] Eisen, B., Ungar, Y., & Shimoni, E. (2003). Stability of Isoflavones in Soy Milk Stored at Elevated and Ambient Temperatures. *Journal of Agricultural and Food Chemistry*, 51, 2212-2215.

[74] Xu, B., & Chang, S. K. C. (2008). Total phenolics, phenolic acids, isoflavones, and anthocyanins and antioxidant properties of yellow and black soybeans as affected by thermal processing. *Journal of Agricultural and Food Chemistry*, 56, 7165-7175.

[75] Farkas, J. (2006). Irradiation for better foods. *Trends in Food Science and Technology*, 17, 148-152.

[76] Farkas, J., & Mohácsi-Farkas, C. (2011). History and future of food irradiation. *Trends in Food Science and Technology*, 22, 121-126.

[77] World Health Organization (WHO). (1994). Safety and Nutritional Adequacy of Irradiated Food. *Geneva, Austria*, 23-39.

[78] Hayashi, T., Takahashi, Y., & Todoriki, S. (1998). Sterilization of foods with low-energy electrons ("Soft-Electrons". *Radiation Physics and Chemistry*, 52, 73-76.

[79] Dixit, A. K., Kumar, V., Rani, A., Manjaya, J. G., & Bhatnagar, D. (2011). Effect of gamma irradiation on lipoxygenases, trypsin inhibitor, raffinose family oligosaccharides and nutritional factors of different seed coat colored soybean (Glycine max L.). *Radiation Physics and Chemistry*, 80, 597-603.

[80] Byun-W, M., Kwon-H, J., & Mori, T. (1993). Improvement of physical properties of soybeans by gamma irradiation. *Radiation Physics and Chemistry*, 42, 313-317.

[81] Pednekar, M., Das, A. K., Rajalakshmi, V., & Sharma, A. (2010). Radiation processing and functional properties of soybean (Glycine max). *Radiation Physics and Chemistry*, 490-494.

[82] Variyar, P. S., Limaye, A., & Sharma, A. (2004). Radiation-induced enhancement of antioxidant contents of soybean (Glycine max Merrill). *Journal of Agricultural and Food Chemistry*, 52, 3385-3388.

[83] Dixit, A. K., Bhatnagar, D., Kumar, V., Rani, A., Manjaya, J. G., & Bhatnaga, D. (2010). Gamma irradiation induced enhancement in isoflavones, total phenol, anthocyanin and antioxidant properties of varying seed coat colored soybean. *Journal of Agricultural and Food Chemistry*, 58, 4298-4302.

[84] Park, J. H., Choi, T. B., Kim, S. W., Hur, M. G., Yang, S. D., & Yu, K. H. (2009). A study on effective extraction of isoflavones from soy germ using the electron beam. *Radiation Physics and Chemistry*, 78, 623-625.

Chapter 9

Acrylated-Epoxidized Soybean Oil-Based Polymers and Their Use in the Generation of Electrically Conductive Polymer Composites

Susana Hernández López and
Enrique Vigueras Santiago

Additional information is available at the end of the chapter

1. Introduction

1.1. Electrical properties on polymer compounds with carbon black making emphasis in the polymer matrix for controlling them

Polymer composites with electrical properties modified by the amount of conductive filler particles emerge at a half of the XX Century. This type of materials are conformed by two components, the first is usually majority and continuous component known as matrix and the second is the minority and discontinuous frequently named conductive filler. Both, matrix and filler could be constituted by one or many different materials. For example, the matrix could be only one polymer or it may well be constituted by a polymer blend, the same way the conductive component. The polymer composites are characterized for being multiphase materials in which is possible to distinguish the two phases: polymer and conductive filler. The own properties of these composites are determined by the synergic coupling between properties of the matrix and the conductive particles, this means the electrical conductivity on these composites and in turn on their practical applications are determined by the chosen polymer matrix-conductive filler couple.

Modification of the electrical properties in polymer composites based on carbon black (CB) particles could reach until 10^{11} orders of magnitude. This characteristic has allowed designing materials with applications for electrostatic protection and electromagnetic shields. It is well known that electrical properties are modified by external factors making them useful for applications as vapor, gases, toxic or inert substances sensors as well as detectors of temperature and external mechanical strengths. For this reason from the 70′s the massive develop of polymer compounds plus conductive particles becomes important [1] and from the beginning it has been used all type of oleo-polymers and conductive particles being of major interest the carbon particles and nanoparticles. These carbonaceous particles allow obtaining polymer composites highly conductive, at great scale, at low cost, with the inherent polymer properties as lightness, processability into different shapes, sometimes reinforced in mechanical and thermal properties. The CB particles have very good conductivity thanks to a series of structural characteristics as the graphitic composition, the structure of the fundamental aggregate that could be low, middle or high, being a high structure [1] the required in order to render lower critical volume or weight percent (wt%) fractions. The particle surface is another important characteristic to take in account for having a good dispersion into the polymer. Some CB surfaces are oxidized and contain many functional groups as hydroxyls, carbonyls, esters, etc [1] which facilitate the interaction with the polymer matrix having a very good dispersion in it. Those and other characteristics as their easy synthesis at great scale make them the most studied conductive particles for preparing conductive polymer composites.

If the difference in the electrical conductivity between conductive particles and the matrix is at least six orders of magnitude then the electrical conductivity of polymer compounds in terms of the conductive particles load could be explained very well by the percolation concept [2]. According to this concept the electrical conductivity is achieved when the conductive particles reach a critical fraction which is the smallest amount of particles required to build a conductive network. This is established by the relationship 1.

$$\sigma(X) \approx (X - X_c)^\beta \tag{1}$$

X represents the volumetric fraction of the conductive filler, X_c the amount of conductive particles in the percolation threshold and is a critical exponent. The critical parameters Xc and β determine the transition from the dielectric to the conductive state in those polymer composites. This make an important difference in the electrical properties respect to the conventional materials in which is possible to determine the type of material according to the order of magnitude in its electrical conductivity. In composite materials the electrical conductivity is achieved when a volume or weight filler fraction is reached and it is characterized because the particles are in electric contact building conductive paths. In this conductive network the electrical interconnection is favored when the transit of electrons is allowed from one particle to another; transitions which are governed by the mechanical quantum laws [3]. The percolation model given by equation 1 fits very well to the experimental results allowing to explain the mechanism for which these composites could be considered as intelligent materials, useful for the detection of temperature gradients, pressure, displacement, solvents, gases, etc. For all those reasons the percolation model has been accepted as the well way to explain the electric conduction in terms of conductive particles fraction. However it has the important disadvantage to no predict the percolation threshold for a specific polymer-conductive particles couple.

Such disadvantage emerges from the fact that it only considers the particles interconnection probability without taking in account the polymer matrix nature and the possible interactions between particles and polymer.

The critical fraction is also associated with the potential applications of the polymer composite. In polymer composites with high critical fraction the mechanical properties could be influenced due the secondary forces between polymer chains are modified by the presence o a high fraction of particles and as a consequence the mechanical properties of the composite could detriment. On another hand in solvent sensing applications it has been demonstrate that as lower conductive particles fraction the sensibility to the detection is considerably increased due to the best interaction area between the polymer and the solvent, vapor or gas [4]. This is the reason for which one of the goals in the preparation of conductive composites is to diminish the critical fraction of conductive particles in conductive polymer composites. However there is not a generalized rule which allow predicting it considering both the polymer and the conductive particles properties. It is known that properties of the matrix polymer as density, viscosity at molten state, dielectric constant and those of the conductive particles as oxidative surface, size, shape and structure, have an important influence in the electrical composite properties in junction with the preparation method [1, 5-17]. The conditions in preparation method are especially important to achieve a homogeneous disaggregation and dispersion of the conductive particles and in turn to reach the electrical properties at a low fraction of conductive particles. A preferential and controlled distribution of those particles contributes to build conductive networks at a low critical fraction and it also guarantees a better reproducibility. Some strategies to obtain a preferential distribution have used polymer matrix derived from two or more immiscible polymers (blends) [10] in which the particles are in the interphase; the *in situ* synthesis using the emulsion polymerization [11] in which particles prefer to surround the polymer matrix nanoparticles. The polymer matrix viscosity [8] is very important to take in account for the conductive particles dispersion. In a high viscose polymer are required high shear forces in order to produce a disaggregation of the particles but also it could produce a breaking of the CB particles structure. This last effect is not convenient due are necessary more particles to interconnect them and form the conductivity network, increasing the critical fraction. Usually the strength distribution is not uniform, not a good dispersion is achieved and the conductivity is not reached a low fraction of conductive particles. In summary the polymers with low viscosity tend to reach the percolation threshold at a lower critical fraction in comparison with polymers highly viscose. The weight molecular mass [12] also influences the process of conductive particles distribution and is directly associated with the viscosity. In according with the latest discussion, as higher the molecular weight the polymer has a higher viscosity and viceversa, in such a way that conductive particles have less problems to disperse in a low molecular weight matrix. The polymer morphology [13] determines in some way the preparation process and the critical fraction of the conductive particles in the percolation threshold. It has been well established that particles tend to disperse uniformly in the amorphous region. As a consequence in semicrystalline matrixes is possible to have a preferential distribution of the conductive particles in the amorphous and interphases zones diminishing the critical fraction. In other cases the polarity of the polymer matrix has taken in account for dispersing CB particles and there are many opinions respect to its influence on the critical fraction [9, 14,15]. It has been mentioned that a major polarity the critical fraction increases [9]. However other authors claim [14, 15] that a major polarity the particles dispersion is better due to the interactions between the groups on the surface of the CB particles and those on or pending on the polymer chain favoring their dispersion. Recently in a systematic study [16] has been demonstrated that critical fraction of CB particles included in some polystyrene-based polymers with different dielectric constant, tend to diminish as the polarity (dielectric constant) of the polymer matrix increases.

In other recently studies, has been evaluated that polymer derivate from vegetable oils as the soybean and linseed oil and some derivates have a very good compatibility not only with some natural fillers as fibers, henequen, cellulose, bamboo, etc, but also with CB particles and carbon nanotubes producing composites with a huge gamma of mechanical and thermal properties, similar to those offered for oleo-polymers, with a great potential to use them as engineering applications. Much less research has been focus to the electrical properties however those studies shown that critical fraction in composites based on acrylated-epoxidized soybean oil and epoxidized linseed oil with CB are much lower (lower than 4 wt% CB) than those exhibited by common oleo-polymers (usually higher than 8 wt % CB) [17]. This fact has been explained in terms of the functional groups of the polymer matrixes and on the ability to crosslink drawing paths in which the CB particles are favorable dispersed. In order to understand this phenomena, the next section will be devoted to describe the structure, characteristics and uses of vegetable oils, specifically focus to one derivate of soybean oil, the acrylated-epoxidized soybean oil (AESO) a derivate well studied as monomer and comonomer in the preparation of conductive composites with different CB particles.

1.2. Acrylated-epoxidized soybean oil: Some relevant advances and achievements

Polymeric materials prepared from renewable natural resources have become increasingly important because of their low cost, ready availability, and potential biodegradability. Vegetable oils represent one of the cheapest and most abundant biological feedstocks available in all around the world in large quantities, and their use as starting materials offers numerous advantages, such as low toxicity, biocompatibility, inherent biodegradability as well as certain excellent frictional properties e.g. good lubricity, low volatility, high viscosity index, solvency for lubricant additives, and easy miscibility with other fluids, and more recently in useful polymers and polymer composites [18-21].

In addition to their application in medicine, cosmetics, surfactants, nutrition, food industry, lubricants and as an alternative fuel for diesel, natural oils also have been used quite extensively to produce coatings, inks [22] plasticizers, lubricants, and agrochemical [23]. Structurally vegetable oils are predominantly constituted of triglyceride molecules. Triglycerides in turn are constituted of three fatty acids joined at a glycerol central structure. Most common oils contain fatty acids that vary from 14 to 22 carbons in length, with 0 to 3 double bonds per fatty acid. Because of the many different fatty acids present, these oils are composed of many different types of triglycerides with numerous levels of unsaturation. A high degree of multiple unsaturations (-C=C-) in the fatty acid (FA) chain of many vegetable oils causes poor thermal and oxidative stability and confines their use as lubricants to a modest range of temperature [24]. Although they possess double bonds, which could be used as reactive sites, they are of low reactivity and cross-linked polymers could be produced under a specific and limited reaction conditions [25-29]. Usually as triglycerides are made up of aliphatic chains, the produced materials lack of the necessary rigidity and strength required for some applications. To reach a higher level of molecular weight and crosslink density is necessary to copolymerize [30-34] or to incorporate chemical functionalities [35] easily to polymerize and that are known to improve the mechanical properties of the polymer networks. The double bonds are usually used to functionalize the triglyceride with polymerizable chemical groups, for example to convert the unsaturation to epoxy group is the most common reaction. There are a lot of published studies devoted to establish [36-40], control [41, 42] and scale [43] many different reaction conditions for partial [44-46] or complete epoxidization of the trygliceride's double bonds. By themselves the epoxidized oils are polymerizable under temperature [47-49] ultraviolet radiation [50-53], or by opening epoxy ring reaction. However is well know that the diamines [54, 55], anhydrides [56, 57], dicarboxylic acids [56], dioles [58, 59] are also used for curing epoxidized oils rendering cured resins with a gamma of properties controlled by the stoichiometry, the structure of the crosslinker and the cure degree.

Epoxidation in general is a commercially important reaction due to the high reactivity of that functional group that makes it to be readily transformed into other important functional and polymerizable groups. The most studied and important polymerizable groups are hydroxyl by complete or partial ring opening reaction of epoxy group [60] making the triglyceride capable of reaction via addition polymerization for producing polyurethanes [61-63] or hydroxylated polyesters for example with maleic anhydride which could further being cured by free radical reaction. Another important functional group is the acrylate inserted by reaction of epoxidized vegetable oils with acrylic acid [18, 64]. This last modification provides monomers which can then be blended with reactive diluents and cured by free radical polymerization. There are natural oils comprise fatty acids with these types of functionalities as the vernonia oil which contain epoxy groups and castor oil that comprise hydroxyl groups.

Soybean oil is an exemplar model of how the natural oils can be used for producing polymer and polymer composites useful in some applications. About 80% of the soybean oil produced each year is used for human food. Another 6% is exploiting for animal feed, while the remainder (14%) has nonfood uses (soap, fatty acids, lubricants, coatings, etc.). Structurally (Figure 1), soybean oil usually contains in average 11% palmitic, 4% stearic, 23% oleic, 54% linoleic and 8% linolenic acids. Palmitic acid contains 16 length carbons and no unsaturations, the remanent acids are all of them are 18 length carbons and are unsaturated acids: oleic (1 double bond), linoleic (2 unsaturations) and linolenic (3 double bonds). Soybean oil contain in average 4.5 double bonds which are thermally polymerizable under the special conditions above mentioned.

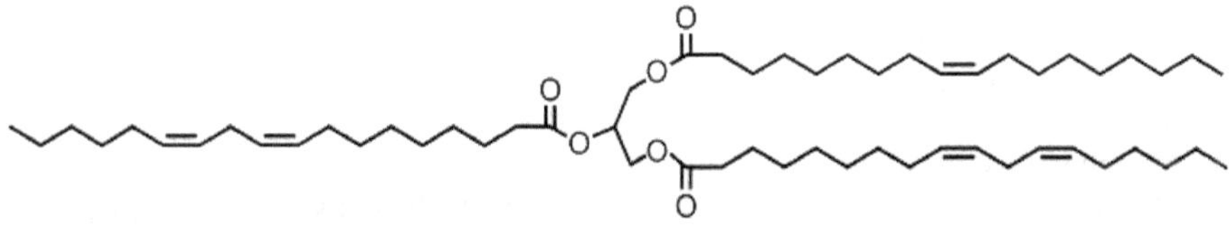

Figure 1. Representative structure of SBO showing three 18 carbon-length chains with 1 and two double bonds

For example, cationic polymerization of the soybean oil with divinylbenzene comonomer initiated by boron trifluoride diethyl etherate produces crosslinked polymers with properties from the rubbers (soft) to the thermosets (rigid) [34, 65]. These properties are dependent on the stoiquiometry of both comonomers. Some others copolymers prepared also by cationic polymerization [66] showed shape-memory effect which refers to the ability to remember a specific shape after deformation. Polymerized soybean oils have already been employed in printing inks and paints and much effort has been devoted to converting soybean oils into solid polymeric materials which now usually possess viable mechanical properties and thus may be useful as structural materials in a variety of specific applications [67]. In this context, one of the most important derivate is the epoxidized soybean oil (ESO). The epoxidized soybean oil is widely used as plasticizer instead of phtalates in the plastic industry to increase flexibility, stability and processability in PVC products. For this application the higher the epoxy degree more efficient is the stabilizer ability [41] to heat and UV-radiation. It was also studied as potential source of high-temperature lubricant [24]. It shows a low thermal and oxidative deteriorate compared with raw soybean oil and other oils due to the absence of allyl hydrogens which are the responsible of those unwanted process. Crosslinked polyesters were prepared by curing ESO with different dicarboxylic acid anhydrides [56]. The mechanical and thermal properties were evaluated and they were dependent on the type of anhydride, the type and wt% of catalyst and on the ratio ESO/anhydride. A broad range of glass transition temperatures, T_g's, from -5 to 75°C were shown and flexural modulus from 520 to 980 MPa. Another example of ESO cured with different acid anhydrides [57] in presence of tertiary amines as initiators. The dynamical mechanical properties were studied in terms of the type of anhydride, initiator and epoxidation level. Composites based on ESO resin and chicken feathers fibers (CFFs) were studied as potential applications in electronic devices such as printed circuits and boards (PCBs) [68]. Equipment based on PCBs requires high electrical resistance, relatively low electric constant and loss. Properties were compared with those of the prepared composites using the standard epoxy resin for PCBs and CFFs as reinforcement. The resistivity of CFFs composites was two to four orders of magnitude higher than E-glass fiber composites indicating that CFFs have better insulating effect. The CFF composite can be potentially used as PCBs industry. In other study ceramer coatings based on ESO were prepared using three sol-gel titanium (IV) and zirconium precursors and several coating properties were evaluated, exhibiting excellent flexibility and hardness. Ceramer coatings (inorganic/organic hybrid materials) have potential applications in protecting optical and electronic devices. In this study the properties of ELO as low volatile organic content, low price and viscosity were taken in advantage in order to prepare the mentioned ceramers [69].

Another important derivate of soybean oil (Figure 2), in which this chapter will be devoted, is the acrylated-epoxidized soybean oil (AESO), produced in two steps from soybean oil. The first one is the epoxidation of soybean oil by any described method. After, this epoxy intermediate is reacted with acrylic acid [21, 64] in presence of an acid as catalyst, being the level of acrylatation very important for the mechanical properties as studied in [70]. T_g increases linearly with the number of acrylates per triglyceride from -50 to 92°C for 0.6 to 5.8 acrylates per triglyceride and it is possible to obtain soft and rigid polymers as the level of acrylatation increases.

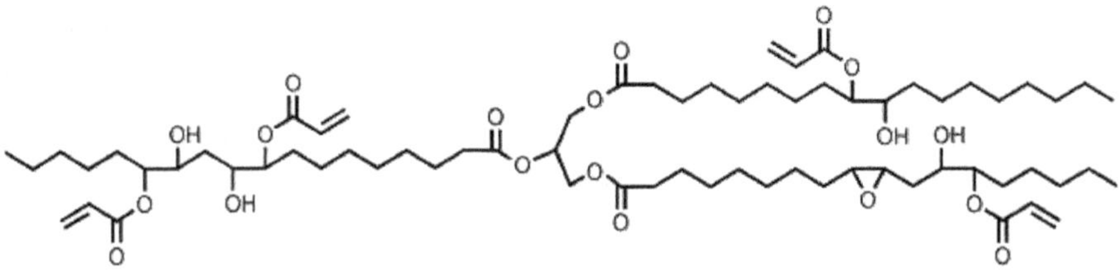

Figure 2. Representative structure for AESO

This molecule is interesting due the possibility of use the double bonds from acrylate functional group in order to polymerize/copolymerize easily via free radicals reaction under several initiator systems as thermal initiator decomposition, photoinitiators and UV or visible radiation, and high energy radiations as gamma rays. However, pure AESO polymer (poly-AESO) does not display important mechanical properties by itself. It looks like an amorphous crosslinked rubber without possibility of processing in useful shapes. However these properties were taken in advantage for modify the mechanical and tribological properties of goat leather [71]. Elastic modulus, tensile deformation, friction and wear were evaluated for goat leather before and after to graft AESO monomer by each and both faces of the leather. The grafting was made by free radicals using a photonitiator and ultraviolet light as initiator system [72]. Those properties were dependent on the irradiation dosages (360, 720 and 2700 J/m^2) being the changes more evident at higher dosages and on the grafted face (inner or outer). In general friction and tensile deformation decreased (17% and 39%) respect to ungrafted leather whereas elastic modulus and wear decreasing as the dosage increased (19% and 39%, respectively).

However for engineering applications polyAESO do not have enough mechanical properties. An attempt to improve the mechanical properties of polyASEA was made using gamma radiation as initiator system [73]. In this work AESO was successfully polymerized and the polyASEA was obtained in one transparent and homogeneous piece depending on the used mold for polymerization. That piece could be cut in different shapes. Its properties as friction and scratch were evaluated respect to the dosage irradiation which was directly related to the crosslink degree. Some probes were made for preparing composites with carbon black under the same conditions, not rendering the same successfully results. In this case those composites were obtained as an unshaped mass without the possibility to processing them into specific shapes.

It is because AESO use to be copolymerized with other vinyl comonomers. The most used is the styrene comonomer which imparts stiffness and as the same time is useful as diluent in order to reduce the viscosity which is of great help when some filler is used for preparing polymer composites. In this context, a lot of interesting works have been published. These resins have proven to be comparable to commercial, oleo-based thermosetting resins. Also, AESO has been combined with several natural fibers [74] and other fillers as glass or carbon fibers and clays in order to produce new economical composites and very useful in many fields like agricultural, automotive, infrastructures, housing, and construction. Some examples are cited. Network blends of polyurethane network films were prepared using polyester urethane acrylate (PUA) having terminal double-bond functional groups and acrylated epoxidized soybean oil by a simultaneous thermal polymerization process. The weight ratios of PUA/ AESO affected the thermal and mechanical properties; with an increase in AESO content the glass-transition temperature of the networks decreased from 40 to -4.8 °C, tensile strength increased from 1.2 to 9.8 MPa, and elongation at break decreased from 470 to 70% [75]. Copolymers of AESO and styrene mixed with butyrated kraft lignin as compatibilizing agent for natural fiber reinforced thermoset composites were prepared and evaluated [76]. An improvement of adhesion of the resin to the fibers was achieved and the flexural strength increased 40% for a 5 wt% butyrate lignin. The effect of different lignin concentrations on the mechanical properties of composite made of flax and wheat straw fibers, was investigated. The ultimate goal was a lignin-based, soluble additive that gives rise to a very strong resin-natural fiber interface. For construction structural panels and unit beams were manufactured from AESO and natural fibers (flax, cellulose, pulp, recycled paper and chicken feathers) as natural fiber reinforcements (20-55 wt%), E-glass fiber and closed cell structural foam achieve good mechanical strength. The goal of this work [77] was to develop monolithic structural panels that would be suitable for use as the (load-bearing) roof, floors or walls of a home. It was compared the mechanical properties. In summary, the results show that the recycled paper beam with chicken feathers and the recycled paper beam with corrugated cardboard have flexural rigidities for a commercial wood product. The beam with some E-glass is a stiff as stiff as other three of the wood commercial references. All composites beams are stronger than the woods references. In conclusion the composite beams made for recycled paper have strengths and stiffness that make them suitable for use in structural applications where wood members would normally used. Other important works are they where AESO was copolymerized with styrene in presence of cellulose fibers or corrugated cardboard boxes [78] in order to manufacture composites structures useful for residential roof construction, rendering successfully results according to the evaluated mechanical properties.

AESO/Styrene in a 65/35 wt ratio and 3 wt% of SWNT composites were prepared [79] by sonication procedure in order to study the mechanical properties. It was found that the addition of NTC results in an increasing of the flexural modulus about in a 44% and the yield strength in 9% as well as the T_g in 8%. In other work AESO/styrene (65/35 wt ratio) also has been reinforced with impure multi walled carbon nanotubes (containing 60-70% of carbon soot) with an aspect ratio of 33.3 [80]. Impure MWNT were added in 1, 3 and 5 wt% to the monomers mix and the dispersion conditions were evaluated until obtain a stable dispersion. Mixes were polymerized via free radicals using tert-butyl peroxy benzoate as initiator and after their mechanical properties were analyzed. Modulus was increased respect to the same resin composition without MWNT in a 30% for samples having 1wt% MWNT. In both cases electrical properties were not evaluated.

However, electrical properties of polymer composites based on AESO and AESO copolymers with carbon black were studied [17]. As mentioned before, crosslinked polyAESO is not adequate to processes the respective composites into convenient shapes for evaluate electrical and other properties. Nevertheless, something that was very interesting is that the percolation concentration for this type of composite was very low (4.1 wt% CB) in comparison with those obtained with the traditional oleo-polymers based composites with carbon black as polystyrene (15 - 9.4) [16, 81], low density polyethylene (17%), polybutyl methacrylate (14%) [17], etc. This behavior is one evidence that polymer matrix in conjunction with the carbon black characteristics, is another parameter that has a very important influence in reach the percolation threshold at low critical fraction of conductive particles. A possible explanation is that AESO helps to low the concentration percolation due to CB particles disperse very well thanks to the presence of the carbonyl groups on the acrylic functional groups that participate in the crosslinking of the AESO monomer, drawing for the carbon black particles a tridimensional conductive network in which they are well distributed and dispersed. To overcome the inconvenient of processing the polyAESO composites one option was to copolymerize the AESO with other acrylic monomer as butylmethacrylate (BMA). Polymer from BMA is not a rigid polymer as polystyrene or poly(metylmethacrylate) but it contains a carbonyl group as ASEA and a short alkyl chain enough to be miscible with ASEA. After studying the electrical behavior and evaluating the critical fraction of the polyAESO and polyBMA both with CB particles, rendered values of 4 and 14%CB respectively. It was established to copolymerize both monomers, ASEA and BMA, in order to improve the processing properties and to evaluate the changes in the electrical properties respect the composites based on the respective homopolymers. Considering the great difference in the critical fraction values of polyASEA and polyBMA composites it was expected that this value was dependent on the comonomers proportion. In order to establish one only proportion it was necessary to do a scanning on the composition vs resistivity property maintaining constant the CB load which was chosen as 10 wt% CB. Once the percolation curve (Figure 3 in reference 17) was built it could be appreciate that as increased the ASEA load in the range of 0 to 10%, the resistivity decreased sharply (seven orders of magnitude) until it reached a minimum and constant value. From these results it was chosen a copolymer composition of 30:70 wt% of ASEA/MAB, taking in account that a higher quantity of ASEA produces processing problems. The results will be discussed later in section 2.1.2.

Another interesting option was to copolymerize the ASEA with another commercial modified natural-derivate comonomer named (Acrylamidomethyl) cellulose acetate butyrate (ACAB). This semicrystalline polymer is a cellulosic product soluble in organic solvents and also it is crosslinkable by free radical reaction. It is well know that cellulose is the most abundant naturally occurring organic substance which is very insoluble and its products are extensively used in non-food and industrial applications mainly as textile, paper, thickeners, flocculants, etc. Many derivates have been synthesized in order to synthetic polymer replacements, being hydroxyl groups in cellulose the most available source of chemical modification that has been exploited. For example, cellulose esters and ethers, which usually are water-insoluble, have been used as plastics for molding, in modern coatings, extrusion, laminates, controlled release of actives, biodegradable plastics, composites, optical films, and membranes and related separation media [82, 83]. ACAB was synthesized in order to have not only esters and ether functionalities, but also polymerizable vinyl functionalities by react the hydroxyl cellulosic groups with N-metilolacrilamide [84].

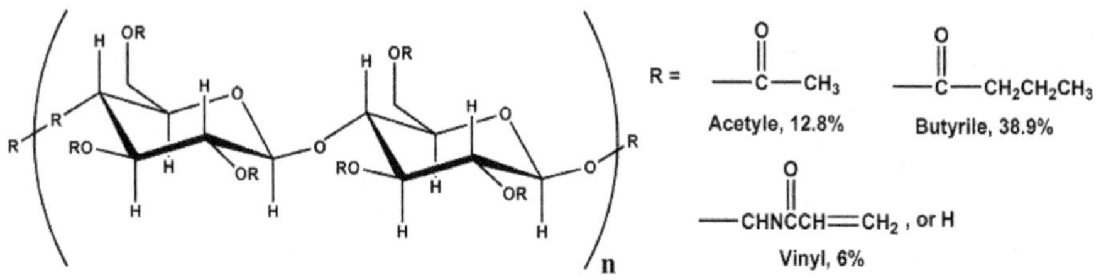

Figure 3. Structural representation and composition of commercial ACAB.

Polymerization and copolymerization could carry out by thermal, ultraviolet, gamma or electron beam radiation. It contains in average the next composition (Figure 3): 38.9% of butyril groups (~1.8 mol butyrate per mol cellulose), 12.8% of acetyl groups (~1.0 mol acetate per mol cellulose), and 0.6% of vinyl ones (~0.1 mol vinyl per mol cellulose). It is a white powder useful for produce crosslinked and insoluble resistant copolymers, hard, durables as adhesives and composites in the textile industry, to provide articles which should retain a permanent shape. Coatings are effective on wood, paper, metal, plastics and they are useful to produce containers, furniture, floors, appliances, trucks, pipe, boats, paper products, among others. Recently ACAB was used to produce columns with higher separation efficiencies of aminoacids and peptides by capillary electrochromatography [85].

Considering the good processing properties of ACAB, the miscibility with AESO, the great number of polar groups and the presence of a vinyl crosslinkable functional group, were the reasons to establish this study in which the main goal is to obtain processable conductive composites and with low critical fraction. In this study three types of CB particles were used. The properties and methodology are described in section 2.2.

The possibility to generate polymer composites from natural sources monomers with modified electrical properties is of high interest due the mentioned reasons and for the possibility of reduce the critical fraction. As mentioned before the preparation methodology is very important in order to obtain a very good CB dispersion and reproducibility of the electrical properties. The next section will be devoted to describe the preparation of the AESO-co-BMA and AESO-co-ACAB copolymers and the respective composites with CB's as well as the characterization of products by FT-IR spectroscopy, Differential Scanning Calorimetry (DSC) and Thermogravimetric Analysis (TGA).

2. Detailed description for obtained copolymers and their structural and thermal characterization

2.1. System AESO-co-BMA

2.1.1. Reactants and equipment

The used reactants in this section are described: AESO containing in average 3.4 acrylates per molecule, and contains 4,000 ppm of monomethyl ether hydroquinone (MEHQ) as inhibitor; density 1.04 g/mL at 25°C. BMA (99%) comonomer contains 10 ppm of MEHQ; bp. 162-165°C, density 0.894 g/mL at 25°C. Potassium Bromide, KBr was spectroscopic degree; Tetrahydrofurane (THF) (99.7%), bp. 70°C; and Acetone (99%) both as solvents. All of them were purchased from Sigma-Aldrich, Co. Dibenzoyl peroxide (BPO) (> 97%), mp 102-105°C, was purchased from Merck Co; and Carbon black Vulcan X-C72 was kindly donated by Cabot Co. Before use, ASEA and BMA were surpassed by an inhibitor-remover packing (from Sigma-Aldrich, Co) in order to remove the MEHQ. Other reactants were used as received.

For Infrared characterization a FT-IR Nicolet Avatar 360 was used. All samples were mixed with previously dried KBr and pressured at 8 Tons in order to obtain transparent plates. 32 scans were recorder from 4500 to 400 cm^{-1} at a resolution of 8 cm^{-1}.

For calorimetric characterization, a SDT Q600 modulus (TGA coupled to a DSC) from TA Instruments was used. Experiments were recorder at a heating rate of 10°C/min, under nitrogen atmosphere (100 mL/min), from 23 to 500°C.

2.1.2. Synthesis and characterization of pure AESO-co-BMA copolymer, 30/70 wt%.

The pure copolymer is obtained following the next methodology, using as example 1g of total prepared compound: 0.3 g of ASEA is weighed into a 150 mL round bottom flask and 10 mL of THF are added. The mix is sonicated in an ultrasonic processor (Ultrasonik™ 28X 50/60Hz) at 5-8°C for 10 min until ASEA is completely solved. Then 0.7g (0.78mL) of BMA is added in junction with 0.05 mL of 0.5M PBO in THF and again the mix is sonicated for 2 min only, at same temperature. After, a condenser with inlet and outlet nitrogen ultrapure gas (from Infra, Co) is adapted to the flask. It was left the low nitrogen flux for 5 min and finally the temperature is increased from room temperature (RT) to 70°C in order to copolymerize. Copolymerization takes around 6 hr and it can be detected because the copolymer starts to precipitate due to the crosslinking reaction. Once the product is precipitated as several pieces, the temperature is lowered to RT, the nitrogen is turn off and the transparent lightly yellow product is filtered and washed twice with acetone to remove the possible residual monomers and initiator. The copolymer is well dried under vacuum for 24 hr. This copolymer is characterized by FTIR spectroscopy, detecting specific signals for both comonomers. Finally, the greatest piece was cut into a cylinder shapes (1.0 cm diameter x 0.2 cm thickness) and the faces were covered with silver paint in order to measure the resistivity.

For a structural characterization of the copolymer by FT-IR spectroscopy, it is necessary to know the main absorption bands of the monomers and homopolymers, in this case for MAB, polyMAB, AESO and polyAESO. For BMA, the next bands are identified: vinyl bond (=C-H) from acrylic group is found at 3100 cm^{-1}, methyl and methylene groups (-CH3-, -CH2-) between 2900 cm^{-1} and 2800 cm^{-1}, carbonyl (C=O) from ester group at 1743 cm^{-1}, polymerizable double bond (-C=C-) from acrylic group is sited at 1640 cm^{-1} and another vibration related with end double bonds (H2C=) also from acrylic group is sited at 950 cm^{-1}. When BMA is polymerized (Figure 3), the main changes on the respective spectrum are: the diminishing of the bands corresponding to the double bonds at 3100 cm^{-1}, 1650 cm^{-1}, and 950 cm^{-1}. These absorption bands diminish in intensity as the polymerization take place and finally they disappear when polymerization has been completed.

For AESO monomer (Figure 4) the most important identified bands are: Hydroxyl (-OH) groups at 3500 cm^{-1}, vinyl group (=C-H) from acrylic one at 3100 cm^{-1}, methyl and methylene groups (-CH3-, -CH2-) between 2900 cm^{-1}, 2800 cm^{-1}, carbonyl vibration (C=O) from esters groups is sited at 1730 cm^{-1}, double bonds (C=C) from the acrylic pendant groups are found at 1650 cm^{-1}. Finally, at 750 cm-1 is sited the typical band due to four or more continuous methylenes (-CH2CH2CH2CH2-...) from the fat acids chains. The main changes observed when ASEA is polymerized to polyASEA (Figure 4) are those related with the acrylic double bonds at 3100 cm^{-1} and 1650 cm^{-1}. As the same as BMA polymerization, those bands decrease in intensity as polymerization takes place; at the end of polymerization they are no detected. All the other signals practically do not change.

For copolymer ASEA-co-MAB (Figure 4) the main information obtained from the FT-IR spectra are that signals of both monomers are seen overlaid and the main ones are those corresponding to the acrylic double bonds. It is logic that they are not detected by copolymerization; however there are still small signals corresponding to acrylic double bonds of the BMA at 1632 cm^{-1}, even after washing the product with THF. BMA monomer was not detected in the filtered THF after remove the solvent. That indicates that it does not correspond to an incomplete reaction of BMA (residual monomer). The small double bonds signals seen on the spectrum are attributed to a higher load than ASEA, generating segments of BMA units with disproportion ends chains.

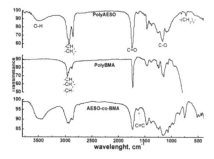

Figure 4. Comparison of the IR absorption bands for polyAESO, polyBMA and AESO-co-BMA

The FT-IR study is not determinant to corroborate if there was a copolymerization, however the thermal properties indicates that there is. DCS/TGA analysis also were made in order to found a possible transition and to determine the decomposition temperature considered it at which de polymer lost the 10 wt% (T_{10}) of its original weight percent. By DSC, ASEA monomer shows an exothermal crosslink curve with a maximal temperature at 349°C and immediately the decomposition is observed. TGA shows that ASEA´s T_{10} is detected at 360°C in one step. Once the ASEA is polymerized, the DSC curve does not show any transition: no residual exothermal curve due to crosslinking, neither T_g indicating a complete polymerization under the described conditions. There is difference in the T_{10} in 7°C lower for polyASEA in comparison with ASEA monomer. This decreasing is explained in terms on that crosslink decreases the free movement of the fat acids arms on the polyASEA in such a way that heating tends to break that inter-chain tension at lower temperatures. The T_{10} of polyBMA was detected at 230°C. The T_{10} of the copolymer is 302°C which is lightly next to the polyASEA value (Figure 5) due to the crosslinking reaction between both monomers, even BMA is in major proportion. Decomposition of the copolymer occurs in only one step, indicating the product is only one compound (copolymer) without byproducts such as homopolymers or oligomers.

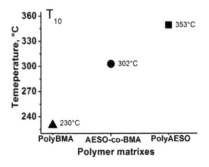

Figure 5. Decomposition temperature (T_{10}) for AESO-co-BMA copolymer and its respective homopolymers: polyAESO and polyBMA

2.1.3. Preparation and characterization of AESO-co-BMA copolymer composites with carbon black

For preparing composites based on ASEA-co-BMA and CB particles, it is followed the next methodology which is similar as the described for pure copolymer synthesis. Composites preparation starts with the dissolution of 0.3 g of ASEA monomer in 20 mL of THF and then the addition of 0.78 mL of MAB monomer. At this homogeneous mix is added the corresponding amount of CB depending on the composition to prepare. In this case were prepared the next CB loads: 0.5, 1.0, 1.2, 1.3, 1.4, 1.5, 2.0, 2.5, 3.0, 3.5, 4.0, 6.0, 8.0, 9.0, 10.0, 15.0 and 20.0 wt% CB. The mix lefts to disperse by sonication for 2 hr at 8°C. Finally the same milliliters of initiator solution is added, nitrogen flux is adapted and the composite is left to polymerize at 70°C for 4 hr and then at 95°C for 2 hr. The composites precipitate as a black unshaped mass; they are washed three times with acetone and filtered in order to remove the probable residual monomers and/or initiator. Finally they are dried under vacuum for 48 hr.

Respect to the composite characterization by FT-IF spectroscopy, the spectra are very similar to the pure copolymer with only one difference, the signals corresponding to the graphitic structure of CB, composed by double bonds C=C sited at 1633 cm^{-1} increases subtle as the CB load do. Referent to the thermal properties, T_{10} of the composites based on ASEA-co-BMA depends on the CB load. For example, for 0.5 wt% CB a T_{10} of 342°C is detected, whereas for 6 wt% CB the T_{10} is 335°C. Is a fact that CB acts as thermal reinforcing filler in this matrix however, that reinforcement decreases as the CB load increases. This is one reason for the interest in produce composites with low critical fraction, usually as the CB load increases some other properties as mechanical or processing ability tend to decrease as well the potential applications as sensing.

It was not a successfully improvement of the processability by copolymerize AESO with BMA, copolymer and the respective composites were insoluble, unmelted, and the thermomolding process was not possible, nevertheless it was obtained a very low critical fraction (1.2 wt% CB)[17].

2.2. System AESO-co-ACAB

In this section the electrical properties of polymer composites based on (Acrylamidomethyl) cellulose acetate butyrate, ACAB, and its copolymer with acrylate-epoxidized soybean oil (AESO) (50:50 wt%) are discussed. Three different commercial carbon black (CB) particles were used in this study: Raven 5000, Vulcan XC-72 and N660 which have different properties as size, structure (ramification percent), oxidize surface and conductivity (see Table 1).

2.2.1. Reactants and equipment

Commercial (Acrylamidomethyl) cellulose acetate butyrate, ACAB has a Mn of 10,000, mp 155-165 °, Tg 118°C, density 1.31 g/mL a 25°C, it is soluble on a variety of acrylic monomers. Benzophenone photoinitiador, BPN (> 99%), mp 47-51°C, both were purchased from Sigma-Aldrich, Co. Reaven 5000 and N660 CB's particles were kindly donated by Professor Wiltold Brostow from UNT, USA. All of them were used as received. For electric contacts silver paint SPI 18DB70X was purchased from Electron Microscopy Sciences. Some properties of the carbon blacks particles are shown in Table 1.

Carbon particles have many different properties according to the raw reactants and on the synthesis method. They are into the 50 chemical products more produced in all around the world and actually, CB particles are the result of an incomplete combustion from hydrocarbons. 90% of CB is useful as reinforcement particles in elastomers for fabrication of wells, 9% as pigment and the rest 1% in other specific applications as paints and electric devices [86]. Additional to their stain power, their electric or charge action provide UV radiation protection to the polymers at low cost. The CB particles are constituted from almost spherical particles with a graphitic structure (electrically conductive) and colloidal dimensions. Their structure consists in many primary nanoparticles (from 10 to 100 nm) bonded into a bunch named "aggregate" with dimensions from 50 to 500 nm. Those aggregates use to form macro-agglomerates with more than 1μm of size [1]. CB aggregates have different shapes, they could be 1) spherical, 2) elispsoids, 3) lineal or 4) branched. Chemically are constituted from 83% to 99% of elemental carbon and according to the fabrication method it could be founded many oxygenated functional groups as phenols, quinolics, carboxylics, etc. on the surface. This is related to the ash content which is usually less than 5%. For electrical properties, are required that CB particles have branched structures (high structure), an oxidized surface which allows to disperse them into several polymer matrix, a high diameter and good intrinsic electrical conductivity. Branched structures as in CB Vulcan X-C72 allow a large contact number or electronic sits for build the conductive paths with a lower load of conductive particles. Opposite to this, CB N660 has the lowest structure and superficial area but a particle diameter of 50 nm and a structure manly lineal. CB Raven 5000 particles show a low structure and superficial area but have a highly oxidized surface in comparison with the others. Their resistivities [see section 3.1] are not so different making interesting to study how the other mentioned properties have influence in reaching the percolation threshold at low critical fractions.

CB Particles	Average diameter size, nm	Volatile % (oxidized surface)	Branched structure, %	Resistivity (cm)
Raven 5000 Ultra II [87]	8	10.5	45.5	1.1
Vulcan XC 72 [86, 88]	32	< 2	77.3	0.08
N660 [89]	50	2.5	31.7	0.19

Table 1. Characteristics of the three used carbon black particles

The FTIR spectra for the three different types of CB, shows the same superficial functional groups, being the only difference the intensity of the signals (Figure 6). Those characteristic absorption bands are: hydroxyls groups (O-H) from 3416 cm^{-1} a 3440 cm^{-1}, methylene (C-H) from the amorphous CB composition at 2923 cm^{-1} y 2850 cm^{-1}; carbonyl bond (C=O) due to different groups as esters, carboxylic acids and quinones are seen from 1709 cm^{-1} to 1740 cm^{-1}; typical vibrations of the double bonds (C=C) that constitute the graphitic composition of the different CB's are sited from 1578 cm^{-1} to 1632 cm^{-1}; this due there are some double conjugated bonds with carbonyl group from quinones. A combination of vibrations due to the C-O and O-H from esters and phenols, respectively are founded at 1380 cm^{-1} and 1216 cm^{-1}; finally are seen a combination of vibrations for C-O bond from hydroxyl groups in phenols at 1064 cm^{-1} y 1066 cm^{-1}. At 680 cm^{-1} is sited an out of the plane deformation vibration for O-H bond which is only seen in CB N660.

2.2.2. Synthesis and characterization of crosslinked ACAB ($_{cross}$ACAB)

This reaction was neccesary to know and stablish the copolymerization reaction conditions and the processing properties. For pure crosslinked ACAB, 1 g of ACAB is weighed into a 150 mL round bottom flask, 10 mL of THF are added and the mix is sonicated for 15 min until the

Acrylated-Epoxidized Soybean Oil-Based Polymers and Their Use in the Generation of Electrically
Conductive Polymer Composites

191

complete dissolution of ACAB. At this point 0.05 mL of a 2.0M THF solution of PBO is added and shaking another 2 min. The condenser and the nitrogen atmosphere are adapted to the flask and the temperature starts to increase from RT to 70°C as a first step. The crosslinked ACAB ($_{cross}$ACAB) starts to precipitate. After 5 hr of reaction, the temperature increases from 70 to 88-90°C for 4hr in order to finish the crosslinking reaction. Heating and nitrogen gas are turn off and at RT and the transparent one piece polymer is washed twice with acetone en order to remove residual ACAB and/or initiator. Finally it was well dried under vacuum for 24 hr. The polymer was cut in small pieces and they were processed by thermo mechanic technique. This allowed preparing cylinders of 1.0 cm diameter x 0.2-0.3 mm thickness in order to measure the resistance.

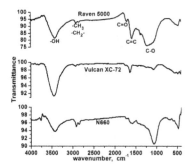

Figure 6. FI-IR spectra if the three types of used CB particles: Raven 5000, Vulcan XC-72, N660

The characterization of ACAB by FTIR spectroscopy allows distinguishing the next important signals: Hydroxyls and amide (–O-H, –N-H) vibrations appear at 3700-3250 cm^{-1}; methyl and methylene vibrations (C-H) from of the polysaccharide ring are sited at 2970 and 2879 cm^{-1}; carbonyl vibration (C=O) from ester groups (acetate and butyril) are found at 1750 cm^{-1} and at 1679 cm^{-1} it is possible to distinguish the carbonyl vibration from amidomethyl group. Double bond band (C=C) is sited at 1630 cm^{-1}, and the corresponding terminal double bond (C=CH$_2$) is at 919 cm^{-1}. Signals monitored in order to found the time in which ACAB completely crosslinks are those corresponding to the vinyl group, 1630 and 919 cm^{-1}. The $_{cross}$ACAB is characterized by FTIR spectroscopy detecting the disappearance of those mentioned signals (Figure 7).

Concerning with the thermal properties, it was interesting to note that the decomposition temperature, in this case taken as the onset temperature (T$_{ons}$) due some little lost before decomposition temperature due to humidity or some remnant solvents in case of $_{cross}$ACAB. Those temperatures were very similar with a difference of 2°C higher for $_{cross}$ACAB respect to the ACAB. This slightly difference could be due to the small amount of crosslink vinyl groups (only 0.6%) which not do much difference in that property. The DSC thermogram for both compounds does not show differences, it is possible to distinguish an exothermal curve with a maximal temperature at 145°C that encloses both the T$_g$ and the T$_m$ reported transitions. This is one reason for which polyACAB shows processability under temperature and pressure and is possible to soften or melt in order to mold it to any required shape, which was something expected.

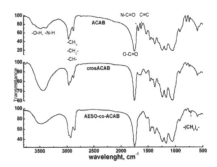

Figure 7. FT-IR spectra for ACAB, _cross_ACAB and AESO-co-ACAB polymer matrixes

2.2.3. Synthesis and characterization of the pure copolymer, AESO-co-ACAB (50/50 wt%)

For preparing 1g of AESO-co-ACAB (50:50 wt%), 0.5g of ACAB are weighed into a round bottom flask, 5 mL of THF are added and it was solved into an ultrasonic processor for 15 min. Then 0.5 g of ASEA and 5mL more of THF are added to the last mix and again homogenized into the ultrasonic processor for 10 min. The same quantity of PBO initiator as the described polymerizations is added and all mix is copolymerized following the same procedure described for _cross_ACAB. The reaction for this system needs 7 hr for polymerizes at 70°C, and then the temperature is increased to 88-90°C for 4 hr. The product as a one translucent and lightly yellow homogeneous piece which is filtered and washed three times with acetone and finally dried under vacuum for 24 hr. This product, softer than _cross_ACAB also shows processability by thermo molding being one of the main goal of this proposed system.

The FT-IR spectrum shows absorption bands of the two comonomers and the most characteristics are: 1535 cm^{-1} (C-H and N-H), 755 cm^{-1} (N-H) from the amide group. 1601 cm^{-1} (C=O) it is a vibration from all esters in both monomers. The most representative band for ASEA is at 723 cm^{-1} which is typical for more than four continuous methylene groups. There is not band sited at 1634 cm^{-1} corresponding to double bonds indicating a complete copolymerization between both monomers.

The TGA analysis of the copolymer indicates two lost, one of 2% before the 150°C due to remnant solvent and humidity and a second at 351°C which corresponds to its decomposition. Figure 8 shows the decomposition temperature of the three pure polymer matrix. As we can see, the decomposition temperature for the copolymer is in the middle of the homopolymers. Taking in account that the decomposition occurs in only one step, it becomes and evidence that copolymerization between ACAB and ASEA took place.

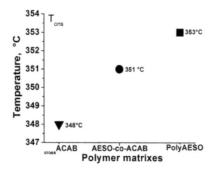

Figure 8. Decomposition temperature of the three polymer matrix: $_{cross}$ACAB, AESO and AESO-co-ACAB (50:50 wt%) copolymer

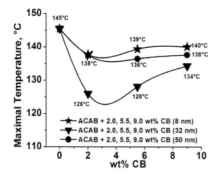

Figure 9. Changes on maximal T_{gm} according to the type and amount of CB particles

The DSC curve showed a diminishing curve due to the melting point of ACAB matrix (145°C), but it does not show any residual exothermal curve due to the ASEA cure. These events suggest that the copolymer could be processable by thermomolding thanks to the thermoplastics properties imparted by the ACAB polymer.

2.2.4. Preparation and characterization of AESO + CB particles

The CB compositions for composites based on ACAB with CB, independently of the type of CB, are: 1.0, 2.0, 2.5, 3.0, 4.0, 4.5, 5.5, 7.5 and 9.0 wt% CB. An example is described for preparing 1g of composite with 1 wt% CB. 0.99 g (90 wt%) of ACAB is solved in 30 mL of THF by sonication for 10 min into a 150 mL round bottom flask and 0.01g (1.0 wt%) of CB is added. The mix is sonicated for 2 hr until a homogeneous solution is observed. At this moment the solvent is distilled with a rotovapor and finally the solid mix is well dried under vacuum for 24 hr until a black powder is obtained. For the other compositions the methodology was the same, the only difference is the dispersion time: 2 hr for 1.0 to 4.0 wt% CB and 4hr to 4.5 to 9.0 wt% CB.

When CB is dispersed in ACAB or into de copolymer, the detection of CB particles by FTIR is the same in any case. Here is exhibited the analysis only with the copolymer and CB Vulcan XC72: the absorption band a 1632 cm⁻¹ which correspond to graphitic CB zone (C=C) or 1582 cm⁻¹ for 8nm-CB, is subtle increased as the CB load increases. There are other bands which overlayer the composite spectra. At 3416 cm⁻¹ - 3440 cm⁻¹ we have a vibration for (O-H) hydroxyl functional group, at 2923 cm⁻¹ and 2850 cm⁻¹ are the vibrations corresponding to the amorphous zone of the CB's (C-H). From 1709 cm⁻¹ to 1740 cm⁻¹ appear the vibrations of C=O bond corresponding to esters groups; at 1578 cm⁻¹ a 1632 cm⁻¹ are the vibrations of the graphitic composition of the CB's particles which correspond to C=C conjugated with carbonyl group from quinones. At 680 cm⁻¹ we have a bending vibration of the O-H bond out of plane which is only observed in CB N660.

In Figure 9 was plotted a temperature (T_{gm}) which encloses the Tg and maximal T_m for ACAB matrix in composites with 2, 5.5, 9 wt% CB, identifying the next information: All ACAB-based composites show a diminishing on the T_{gm} respect to pure ACAB. An explanation is that CB particles act as impure or lubricant, abating both temperatures, mainly the T_m as observed for other authors [90]. The higher effect is shown when CB N660 is used.

On the other hand, respect to the decomposition temperatures, shown in Figure 10, it can be seen that the addition of any CB used particles tends to increases the decomposition temperature respect to the ACAB matrix (346°C). In Figure 10 only T_{ons} for three compositions are plotted: 2, 5.5 and 9 wt% CB. The smallest CB particles (8nm) give the composites with the higher decomposition temperatures around 351°C, followed by those based on 50nm-CB (N660) which have T_{ons} around 348°C and finally those particles that almost do not have a great effect on the decomposition temperature were the 32nm-CB particles, maintaining it around 346.5°C.

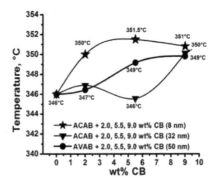

Figure 10. Decomposition temperatures for ASEA+CB composites as a function of the type and amount of the CB particles

The incorporation of the CB particles increases the decomposition temperature of the compo-site, it could be due to that intermolecular interaction seems to be favourable between the functional groups of the polymer matrix and those of the CB particles [1]. It is important to take in account that 8nm-CB particles have a higher oxidized superficial area implying a great number of hydroxyl and carboxyl groups that could render hydrogen bridges with those on the ACAB polymer. It is because composites prepared with 8nm-CB have a higher resistance to decompound thermally. However 50nm-CB particles have scarce superficial functional groups and less superficial area and in turn negligible interaction with the polymer matrix and a less decomposition temperatura for the respective composites. An intermediate case is for ASEA+ 32nmCB composites.

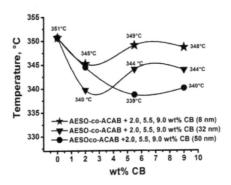

Figure 11. Decomposition temperature for the copolymer composites base don AESO-co-ACAB as a function on the type and amount of CB particles

2.2.5. Preparation and characterization of AESO-co-ACAB + CB's composites

The preparation of these composites also is also via free radicals. For 32 and 50 nm CB's, thermal initiator decomposition was used, but for CB Raven 5000 (8nm) was necessary to use the photochemical method via benzophenone/UV radiation as initiator system. The reason is that was not possible to obtain composites via thermal decomposition. A possibility is that some superficial groups on CB Raven 5000 act as inhibitors or quenching free radicals consuming them. In case of UV radiation, it is possible to reactivate the apparently inhibited free radicals. This could be supported by the fact that those copolymer composites were cured under air atmosphere, not under nitrogen flux as in thermal copolymerization.

The CB compositions prepared for these systems were the same as those described for ACAB: 1.0, 2.0, 2.5, 3.0, 4.0, 4.5, 5.5, 7.5 and 9.0 wt% CB. In the copolymer composites,the CB load was calculated considering 1 g of comonomers in which 0.5 g of ASEA and 0.5 g of ACAB are used for synthesize the copolymer matrix. The first stage of the preparation is the same independ-ently of the type of CB used: First 0.5 g of ACAB are weighted and solved in 30 mL of THF by ultrasonic shake for 10 min. 0.5g of ASEA are added and sonicated again for 10 min in order to get a homogeneous clear solution. Then the respective amount of CB is added (depending on the composition) and the mix is sonicated for 2 or 4 hr (as same as ACAB composites). Finally, depending on the initiator system, when 8nm-CB is added, 0.01 mL of 0.2M BPN in acetone is dropped and sonicated 2 min more. The mix is poured into a glass mold covering the bottom. The solvent is slowly evaporated inside an extraction bell in such a way that composite mix forms a homogeneous layer. The glass mold is put into a CL-1000 ultraviolet-crosslinker UVP (with a maxima wavelength of 254 nm) and copolymerized at 720 J/cm² for 4 hr. The composites are obtained as a solid layer with a shining black aspect, which is insoluble in water and organic solvents. The composite layer is collected and reserved.

When CB of 32 and 50nm are used, 0.05 mL of 2.0M PBO in THF is added, and following the same methodology as in pure copolymer, the mix is cured at 70°C for 5h and then a 90°C for 2 hr under nitrogen flux. The cured composite precipitate as black solid pieces which are filtered and washed twice with acetone and dried under vacuum for 24 hr.

By FT-IR spectroscopy, as the same to the aforementioned for the ACAB composites, the only effect is the increasing of the band at 1632 nm which corresponds to the double bonds of the graphitic composition in CB particles. The same effect is observed in all copolymers and that signal seems to increase as the wt% CB does.

Concerning with the thermal properties of copolymer composites only was detectable the decomposition temperature, it was very difficult to detect some T_m surely due to the addition of CB particles which influence the possibility to crystallize and T_g was not evident. But TGA analysis shows interesting results, in the case of the copolymer matrix the effect on the T_{ons} is opposite to that analyzed for the ASEA polymer (Figure 10). In copolymer composites at 2, 5.5 and 9wt% CB's, the T_{ons} is diminished with the CB particles addition respect to the copolymer matrix. 8nm-CB particles seem not to have an important effect on the decomposition temperature, except for 2 wt% CB in which the temperature is reduced from 351°C (copolymer) to 345°C. At this same CB composition, 32 nm particles show the stronger effect on T_{ons} diminishing it at 340°C, but with the increasing of the CB load, the reached value for T_{ons} was in average 344°C. The 50nm-CB particles had its more critical effect at 5.5 wt% CB decreasing the T_{ons} from 351 to 338°C.

3. Electrical properties evaluation

3.1. Electrical properties of carbon black particles

In order to compare the electrical properties of the different prepared and characterized polymer composites, the electrical properties of the three types of carbon black were determined under the same conditions [91, 92]. Each type of CB was hydrostatically compressed into a system which consists of a cylinder-piston couple, as shown in Figure 12 and the electrical resistivity was calculated using the relationship 2.

$$\rho = R\frac{A}{l} \tag{2}$$

Where the electrical resistance R was measured with a LRC720 Stanford Research System. The length of the compacted particles l, and the transversal section area A were measured with a micrometer and a Vernier respectively. The same amount (100 mg) and pressure (20 Kg/cm^2) were applied to the three CB particles. Cupper electrodes of 6 mm diameter, 1cm length for the inferior and 2.5 cm length for the superior one were used. The electrical resistivities are shown on Table 1. As it can be see, the Vulcan XC-72 have the lowest resistivity and Raven 5000 the highest indicating that the most conductive CB is the Vulcan XC-72.

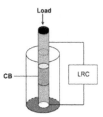

Figure 12. Design for measure the resistance of the CB particles

3.2. Percolation curves and percolation threshold calculus for composites

For composites based on AESO-co-BMA, the methodology for electrical measurements is well detailed in reference [17] and the most interesting results were the low critical fraction of both composites based on polyAESO and on the AESO-co-BMA matrixes, 4.0 and 1.2 wt% CB, respectively. These results will be discussed on section 3.3.

For the other systems (ACAB, AESO-co-ACAB) the electrical treatment is described: Polymers composites were processed by thermo-molding technique at 1.5 MPa in order to get samples of 1.2 cm diameter and 2 mm thickness [81, 93]. The processing conditions of temperature and time are summarized in Table 2 and those conditions depend on the polymer matrix. The bulk resistivity was determined using the two-points technique and silver paint (SPI de Electron Microscopy) as electric contacts were put on the parallel faces of the composite samples.

The voltage-current relationship is measured with a Keithley 6517A electrometer [17, 81, 93]. For each composition the resistivity was calculated via the relation 2 building the respective percolation curves in a width range of compositions. As an example, the percolation curves for AESO-co-ACAB are exhibited in Figure 13. From these curves and using Origin 6.0 software the percolation threshold and the critical fraction are numerically calculated fixing the experimental data to the equation 1. As we can see, the critical fraction for polymer composites based on ACAB and AESO-co-ACAB is modified due to the CB particles as well as the polymer matrix.

Polymer composite	Temperature/ °C	Compression time /min
ACAB	122	30
ASEA/ACAB	118	40

Table 2. Processing conditions of ACAB and ACAB-co-AESO composites

For the ACAB composites the critical fraction varies from 2.8 to 7.5 wt% CB whereas the AESO-co-ACAB copolymer composites the respective values are from 2.4 to 3.2 wt% CB (Figure 14). This demonstrates that polymer matrix influences the electrical properties of the final composite. However in terms of CB properties, the decreasing of the critical fraction with a increasing of the CB structure converges very well with the percolation theory. As we discussed before, when are used aggregates with small particles and low structure it tend to need more particles in order to interconnect them and to build the electrically conductive paths, in comparison with high structure and bigger particles (as Vulcan XC-72 respect to Raven 5000). Also, other CB properties as intrinsic resistivity are evident. For both systems the CB particles with the highest resistivity (Raven 5000, 8 nm) are those with a higher critical fraction.

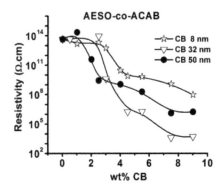

Figure 13. Percolation threshold of the AESO-co-ACAB composites for different carbon black particles

3.3. Analysis of electrical properties as a function of carbon black particles and polymer matrix nature

From the results we could discuss that the chemical nature of the polymer matrix as well the CB properties contribute to the disaggregation, dispersion and preferential distribution of the conductive particles into the polymer matrix. However it is possible to control the preferential build of conductive paths and diminishes as a consequence the critical fraction? As was mentioned before, there are some successfully attempts for controlling it but only oleo-polymer has been used for that and critical fraction lower that 3% are not obtained yet. However, AESO has become a question in to decrease the critical fraction under copolymerization. For the system AESO-co-BMA, a very low critical fraction of 1.2 wt% CB was calculated. Something

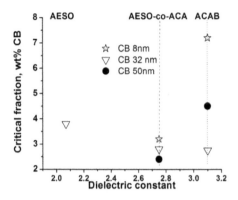

Figure 14. Critical fractions respect to the type of CB and to the polymer matrix

similar is observed for the AESO-co-ACAB composites. If we analyze the critical fractions on Figure 15, we could appreciate that copolymers composites have the lowest critical fraction independently on the type of CB particle. For oleo polymers, in a systematic work [16] was demonstrate that the dielectric constant of the polymer matrix is directly associated with the critical fraction. Due to the unprocessability of the AESO-co-BMA copolymer it was not possible to measure the dielectric constant but it was made for ACAB, AESO and the AESO-co-ACAB matrixes (Figure 15). For dielectric constant polymers were processed into disks of 2cm diameter x 0.9-1.1 mm thickness by compression molding. An Agilent 4991 A RF impedance/Material Analyzer at 450 MHz and room temperature [16].

We can realize that AESO has the lowest dielectric constant (2.1) whereas ACAB has the highest one (3.1). Even both molecules have polar groups; the AESO has large alkyl chains also, whereas ACAB has oxygen as constituent of different functional groups being carbonyl and hydroxyl the main. Unsurprising constant dielectric value for the copolymer was obtained in 2.7. The tendency with CB Vulcan XC-72 is the same as the reported results on [16]: at a highest dielectric constant, a diminishing in the critical fraction is observed. The amazing results are in the low critical fractions for the copolymer composites (from 2.4 to 3.2) with the three types of CB particles in comparison with the ACAB composite which possess the highest dielectric constant. These results claim that the dielectric constant is important to take in account however the role of the AESO is completely understood but definitively it tends to low the critical percolation. From the structural point of view, polyAESO is an amorphous and 3-D crosslinked polymer while AESO is a semicrystalline and lineal one. However when AESO crosslinks builds a bonds network with polar groups thanks to the reaction of the acrylic groups. This network draw preferential paths (in blue, Figure 16) for a good dispersion and distribution of the CB particles without have to fill the entire matrix. In conjunction with the polar functional groups of the other comonomer, very low critical fractions are reached under copolymerization with BMA and ACAB. The differences on the critical fraction will surely dependent on the CB properties as the size, structure, surface and the intrinsic conductivity.

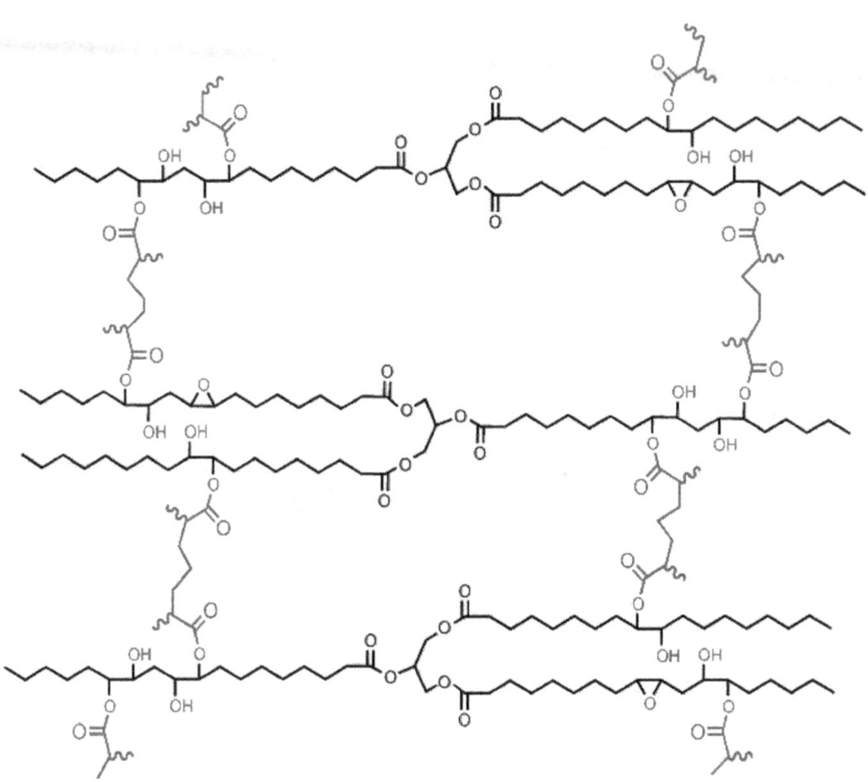

Figure 15. Representation of the 3D- polar paths (in blue) in which the CB particles surely disperse and preferentially distribute in all matrixes containing AESO comonomer.

4. Conclusions and remarks

It has been proved that polarity of the polymer matrix is very important in order to have a better distribution and dispersion of the conductive particles. However, AESO is a very interesting monomer that renders very low critical fractions under copolymerization with other polar comonomers as BMA and ACAB. The explanation we have is that AESO provides preferential distribution to the CB conductive particles due the 3D-crosslinking paths which are bonded to polar functional groups. This is the reason for thinking that conductive polymer composites based on AESO have a promising future in some applications as solvent, gases, vapors or pressure sensors. There are some previous results (unpublished) about the capacity for sensing solvents as well as pressure changes indicating they are very reliable and reproducible materials. The sensibility to pressure changes could be dependent on the crosslinking degree, the proportion and on the comonomer chemical structure. Finally is considered that polymers derived from renewable sources could be a better alternative for reducing the oleopolymers use in some specific applications as discussed in this chapter.

Acknowledgements

Authors thank to the Universidad Autónoma del Estado de México under the projects 3135/2012FS and 3198/2012U for the financial support.

Author details

Susana Hernández López[*] and Enrique Vigueras Santiago[*]

*Address all correspondence to: eviguerass@uaemex.mx

Research and Development of Advanced Materials Laboratory, Chemistry Faculty of the State of Mexico Autonomous University, Toluca, Mexico

References

[1] Donnet, J. B., Bansal, R. C., & Wang, MJ. (1993). Carbon Black. *Science and Technology.*, USA: Marcel Dekker,.

[2] Kirkpatrick, S. (1973). Percolation and Conduction. *Reviews of Modern Physics*, 45(4), 574-578.

[3] Balberg, I. (1987). Tunneling and Nonuniversal Conductivity in Composite Materials. *Physical Review Letters*, 59(12), 1305-1308.

[4] Sisk, B. C., & Lewis, N. S. (2006). Vapor Sensing Using Polymer/Carbon Black Composites in the Percolative Conduction Regime. *Langmuir*, 22(18), 7928-7935.

[5] Yu, J., Zhang, L. Q., Rogunova, M., Summers, J., Hiltner, A., & Baer, E. (2005). Conductivity of Polyolefins Filled with High-Structure Carbon Black. *Journal of Applied Polymer Science*, 98(4), 1799-1805.

[6] Gaurav, R., Kasaliwal, Goldel. A., Potschke, P., & Heinrinch, G. (2011). Influences of Polymer Matrix Melt Viscosity and Molecular Weight on MWCNT agglomerate dispersion. *Polymer*, 52(4), 1027-1036.

[7] Cheng, G. S., Hu, J. W., Zhang, M. Q., Li, M. W., & Xiao, Rong. M. Z. (2004). Electrical Percolation of Carbon Black Filled Poly (ethylene oxide) Composites in Relation to the Matrix Morphology. *Chinese Chemical Letters*, 15(12), 1501-1504.

[8] Sumita, M., Abe, H., Kayati, M., & Miyasaka, K. (2006). Effect of Melt Viscosity and Surface Tension of Polymers on the Percolation Threshold of Conductive-Particle-Filled Polymeric Composites. *Journal of Macromolecules Science*, 25(1-2), 171 EOF-184 EOF.

[9] Tchoudakov, R., Breuer, O., & Narkis, Siegmann. A. (1996). Conductive Polymer Blends with Low Carbon Black Loading: Polypropylene/Polyamide. *Polymer Engineering and Science*, 36(10), 1336-1346.

[10] Sumita, M., Sakata, K., Asai, S., Miyasaka, K., & Nakagawa, H. (1991). Dispersion of Fillers and the Electrical Conductivity of Polymer Blends Filled with Carbon Black. *Polymer Bulletin*, 25(2), 265-271.

[11] Moriarty, G. P., Whittemore, J. H., Sun, K. A., Rawlins, J. W., & Grunlan, J. C. (2011). Influence of Polymer Particle Size on the Percolation Threshold of Electrically Conductive Latex-Based Composites. *Journal of Polymer Science*, Part B: Polymer Physics, 49(21), 1547-1554.

[12] Kasaliwal, G. R., Goldel, A., Potschke, P., & Heinrinch, G. (2011). Influences of Polymer Matrix Melt Viscosity and Molecular Weight on MWCNT Agglomerate Dispersion. *Polymer*, 52(4), 1027-1036.

[13] Cheng GS, Hu JW,. Zhang MQ, Li MW, Xiao DS, Rong MZ. (2004). Electrical Percolation of Carbon Black Filled Poly (Ethylene Oxide) Composites in Relation to the Matrix Morphology. *Chinese Chemical Letters*, 15(12), 1501-1504.

[14] Carmona, F. (1989). Conducting Filled Polymers. *Phisica A.*, 157(1), 461-469.

[15] Li, Y., Wang, S., Zhang, Y., & Zhang, Y. (2005). Electrical Properties and Morphology of Polypropylene/Epoxy/Glass Fiber Composites Filled with Carbon Black. *Journal of Applied Polymer Science*, 98(3), 1142-1149.

[16] Castro-Martínez, M., Hernández, López. S., & Vigueras, Santiago. E. (2012). Relationship Between Polymer-Chemical Structure and Electrical Properties in Conductive Carbon Black Composites. Submitted to e-Polymers.

[17] Hernández-López, S., Vigueras-Santiago, E., Mercado-Posadas, J., & Sánchez-Mendieta, V. (2006). Electrical Properties of Acrylated-Epoxidized Soybean Oil Polymers-Based Composites. *Advances in Technology of Materials and Materials Processing Journal*, 8(2), 214-219.

[18] Guner, F. S., Yagci, Y., & Erciyes, A. T. (2006). Polymer from Triglyceride Oils. *Progress in Polymer Science*, 31(7), 633-670.

[19] Sharma, V., & Kundu, P. P. (2006). Addition Polymers from Natural Oils- A Review. *Progress in Polymer Science*, 31(11), 983-1008.

[20] Meier, M. A. R., Metzger, J. O., & Shubert, U. S. (2007). Plant Oil Renewable Resources as Green Alternatives in Polymer Science. *Chemical Society Reviews*, 36(11), 1788-1802.

[21] Khot, S. N., Lascala, J. J., Can, E., Morye, S., Williams, G. I., Palmese, G. R., Kusefoglu, S. H., & Wool, R. P. (2001). Development and Application of Triglyceride-based Polymers and Composites. *Journal of Applied Polymer Science*, 82(3), 703-723.

[22] Blayo, A., Gandini, A., & Le Nest-F, J. (2001). Chemical and Rheological Characterizations of Some Vegetable Oils Derivates Commonly Used in Printing Inks. *Industrial Crops and Products*, 14(2), 155-167.

[23] Hill, K. (2000). Fats and Oils as Oleochemical Raw Materials. *Pure and Applied Chemistry*, 72(7), 1255-1264.

[24] Adhvaryu, A., & Erhan, S. Z. (2002). Epoxidized Soybean Oil as a Potential Source of High-Temperature Lubricants. *Industrial Crops and Products*, 15(3), 247-254.

[25] Tuman, S. J., Chamberlain, D., Scholsky, K. M., & Soucek, . (1996). Differential Scanning Calorimetry Study of Linseed Oil Cured with Metal Catalysts. *Progress in Organic Coatings*, 28(4), 251-258.

[26] Refvik, M. D., & Larock, R. C. (1999). The Chemistry of Metathesized Soybean Oil. *Journal of the American Oil Chemists' Society*, 76(1), 99-102.

[27] Stemmelen, M., Pessel, F., Lapinte, V., Caillol, S., Habas-P, J., & Robin-J, J. (2011). A Fully Biobased Epoxy Resin from Vegetable Oils: From the Synthesis of the Precursors by Thiol-ene Reaction to the Study of the Final Material. *Journal of Polymer Science*, Part A: Polymer Chemistry, 49(11), 2434-2444.

[28] Erhan, S. Z., Bagby, M. O., & Nelsen, T. C. (1997). Drying Properties of Metathesized Soybean Oil. *Journal of the American Oil Chemists' Society*, 74(6), 703-706.

[29] Montero de, Espinosa. L., & Meier, M. A. R. (2011). Plants Oils: The Perfect Renewable Resource for Polymer Science. *European Polymer Journal*, 47(5), 837-852.

[30] Teng, G., Wegner, J. R., Hurtt, G. J., & Soucek, MD. (2001). Novel Inorganic/Organic Hybrid Materials Based on Blown Soybean Oil with Sol-Gel Precursors. *Progress in Organic Coatings*, 42(1), 29-37.

[31] Andjelkovic, D. D., Vakverde, M., Henna, P., Li, F., & Larock, R. C. (2005). Novel Thermosets Prepared by Cationic Copolymerization of Various Vegetable Oils-Synthesis and Their Structure-Property Relationships. *Polymer*, 46(23), 9674-9685.

[32] Cakmakli, B., Hazer, B., Tekin, I. O., Kizgut, S., Koksal, M., & Menceloglu, Y. (2004). Synthesis and Characterization of Polymeric Linseed Oil Grafted Methyl Methacrylate or Styrene. *Macromolecular Bioscience*, 4(7), 649-655.

[33] Rybak, A., & Meier, M. A. R. (2008). Cross-Metathesis of Oleyl Alcohol with Methyl Acrylate: Optimization of Reaction Conditions and Comparison of Their Environmental Impact. *Green Chemistry*, 10(10), 1099-1104.

[34] Li, F., & Larock, R. C. (2000). Novel Polymeric Materials from Biological Oils. *Journal of Polymers and the Environment*, 59 EOF-67 EOF.

[35] Biermann, U., Friedt, W., Lang, S., Luhs, W., Machmuller, G., Metzger, J. O., Rusch, gen., Klaas, M., Schafer, H. J., & Schneider, M. P. (2000). *Angewandte Chemie International Edition*, 39(13), 2206-2224.

[36] Rusch, gen., Klaas, M., & Warwel, S. (1997). Lipase-Catalyzed Preparation of Peroxy Acids and Their Use for Epoxidation. *Journal of Molecular Catalysis: A Chemical*, 117(1-3), 311-319.

[37] La Scala, J., & Wool, R. P. (2002). Effect of FA Composition Kinetics of TAG. *Journal of the American Oil Chemists' Society*, 9(4), 373-378.

[38] Okieimen FE, Bakare OI, Okieimen CO. (2002). Studies on the Epoxidation of Rubber Seed Oil. *Industrial Crops and Products*, 15(2), 139-144.

[39] Guidotti, M., Ravasio, N., Psaro, R., Gianotti, E., Coluccia, S., & Marchese, L. (2006). Epoxidation of Unsaturated FAMEs Obtained from Vegetable Source Over Ti(IV)-grafted Silica Catalysts: A Comparison between Ordered and Non-Ordered Mesoposous Materials. *Journal of Molecular Catalysis A: Chemical*, 250(1-2), 218-225.

[40] Campanella, A., Baltanas-Sánchez, Capel., Campos-Martín, M. C., Fierro, J. M., & , J. L. G. (2004). Soybean Oil Epoxidation with Hydroperoxide Using an Amorphous Ti/Si Catalysts. Green Chemistry , 6(7), 330-334.

[41] Parreira, T. F., Ferreira, M. M. C., Sales, H. J. S., & de Almeida, W. B. (2002). Quantitative Determination of Epoxidized Soybean Oil using Near-Infrared Spectroscopy and Multivariate Calibration. *Applied Spectroscopy*, 56(12), 1607-1614.

[42] Vlcek, T., & Petrovic, Z. S. (2006). Optimization of the Chemoenzymatic Epoxidation of Soybean Oil. *Journal of the American Oil Chemists' Society*, 83(3), 247-252.

[43] Hilker, I., Bothe, D., Pruss, J., & Warnecke-J, H. (2001). Chemo-Enzimatic Epoxidation of Unsatured Plant Oils. *Chemical Engineering Science*, 56(2), 427-432.

[44] Rusch, gen., Klaas, M., & Warwel, S. (1999). Complete and Partial Epoxidation of Plant Oils by Lipase-Catalyzed Perhydrolysis. *Industrial Crops and Products*, 9(2), 125-132.

[45] López-Téllez, G., Vigueras-Santiago, E., & Hernández-López, S. (2009). Characterization of Linseed Oil Epoxidized at Different Percentages. *Superficies y Vacío*, 22(1), 5-10.

[46] Muturi, P., Wang, D., & Dirlikov, S. (1994). Epoxidized Vegetable Oils as Reactive Diluents I. Comparison of Vernonia, Epoxidized Soybean and Epoxidized Linseed Oils. *Progress in Organic Coatings*, 25(1), 85-94.

[47] Tan, S. G., & Chow, W. S. (2010). Biobased Epoxidized Vegetable Oils and Its Greener Epoxy Blends: A Review. *Polymer-Plastics Technology and Engineering*, 49(15), 1581-1590.

[48] Jin F-L, Park S-J. (2007). Thermal and Rheological Properties of Vegetable Oil-Based Epoxy Resins Cured with Thermally Latent Initiator. *Journal of Industrial Engineering of Chemistry*, 13(5), 808-814.

[49] Uyama, H., Kuwabara, M., Tsujimoto, T., Nakano, M., Usuki, A., & Kobayashi, S. (2003). Green Nanocomposites from Renewable Resources: Plant Oil-Clay Hybrid Materials. *Chemistry of Materials*, 15(13), 2492-2494.

[50] Samuelsson, J., Sundell-E, P., & Johansson, M. (2004). Synthesis and Polymerization of a Radiation Curable Hyperbranched Resin Based on Epoxy Functional Fatty Acids. *Progress in Organic Coatings*, 50(3), 193-198.

[51] Chakrapani, S., & Crivello, J. V. (1998). Synthesis and Photoinitiated Cationic Polymerization of Epoxidized Castor Oil and Its Derivates. *Pure and Applied Chemistry A*, 35(1), 1-20.

[52] Crivello, J. V. (2008). Effect of Temperature on the Cationic Photopolymerization of Epoxides. *Journal of Macromolecular Science, Part A: Pure and Applied Chemistry*, 45(8), 591-598.

[53] Cheong, M. Y., Ooi, T. L., Amhad, S., Wan, M. Z. W. Y., & Kuang, D. (2009). Synthesis and Characterization of Palm-Based Resin for UV Coating. *Journal of Applied Polymer Science*, 111(5), 2353-2361.

[54] López-Téllez, G., Vigueras-Santiago, E., & Hernández, López. S. (2008). Synthesis and Thermal Cross-Linking Study of Partially-Aminated Epoxidized Linseed Oil. *Designed Monomers and Polymers*, 11(5), 435-445.

[55] Manthey, N. W., Cardona, F., Aravinthan, T., & Cooney, T. (2011). Cure Kinetics of an EpoxidIzed Hemp Oil based Bioresin System. *Journal of Applied Polymer Science*, 122(1), 444-451.

[56] Rosch, J., & Mulhaupt, R. (1993). Polymers from Renewable Resources: Polyester Resins and Blends upon Anhydride-Cures Epoxidized Soybean Oil. *Polymer Bulletin*, 31(6), 679-665.

[57] Gerbase, A. E., Petzhold, C. L., & Costa, A. P. O. (2002). Dynamical Mechanical and Thermal Behavior of Epoxy Resins Based on Soybean Oil. *Journal of the American Oil Chemists' Society*, 79(8), 797-802.

[58] Yue, S., Hu, J. F., Huang, H., Fu, H. Q., Zeng, H. W., & Chen, H. Q. (2007). Synthesis, Properties and Application of Novel Epoxidized Soybean Oil Toughened Phenolic Resins. *Chinese Journal of Chemical Engineering*, 15(3), 418-423.

[59] Munoz, J. C., Ku, H., Cardona, F., & Rogers, D. (2008). Effects of Catalysts and Post-Curing Conditions in the Polymer Network of Epoxy and Phenolic Resins: Preliminary results. *Journal of Materials Processing Technology*, 486 EOF-492 EOF.

[60] Kiatsimkul, P-p., Suppes, G. J., Hsieh, F-h., Lozada, Z., & Tu-C, Y. (2008). Preparation of High Hydroxyl Equivalent Weight Polyols from Vegetable Oils. *Industrial Crops and Products*, 27(3), 257-264.

[61] Hofer, R., Daute, P., Grutzmacher, R. K., & Ga-H, A. A. W. (1997). Oleochemical Polyols-A New Raw Material Source for Polyurethane Coatings and Floorings. *Journal of Coatings Technology*, 69(869), 65-72.

[62] Zlatanić, A., Lava, C., Zhang, W., & Petrović, Z. S. (2004). Effect of Structure on Properties of Polyols and Polyurethanes Based on Different Vegetable Oils. *Journal of Polymer Science Part B: Polymer Physics*, 42(5), 809-819.

[63] Lligadas, G., Ronda, J. C., Galià, M., & Cádiz, V. (2010). Plant Oils as Platform Chemicals for Polyurethane Synthesis: Current State-of-the-Art. *Biomacromolecules*, 11(11), 2825-2835.

[64] La Scala, J., & Wool, R. P. (2002). The Effect of Fatty Acid Composition on the Acrylation Kinetics of Epoxidized Triacylglycerols. *Journal of the American Oil Chemistis' Society*, 79(1), 59-6.

[65] Li, F., Hanson, M. V., & Larock, R. C. (2001). Soybean Oil-Styrene-Divinylbenzene Thermosetting Polymers: Synthesis, Structure, Properties and Their Relationships. *Polymer*, 42(4), 1567-1579.

[66] Li, F., & Larock, R. C. (2002). New Soybean Oil-Styrene-Divinylbenzene Thermosetting Copolymers V: Shape-Memory Effect. *Journal of Applied Polymer Science*, 84(8), 1533-1543.

[67] Li, F., & Larock, R. C. (2002). Novel Polymeric Materials from Biological Oils. *Journal of Polymers and the Environment*, 59 EOF-67 EOF.

[68] Zhang, M., Wool, R. P., & Xiao, J. Q. (2001). Electrical Properties of Chicken Feather Fiber Reinforced Epoxy Composites. *Composites Part A: Applied Science and Manufacturing*, 42(3), 229-233.

[69] Teng, G., & Soucek, . (2000). Epoxidized Soybean Oil-Based Ceramer Coatings. *Journal of the American Oil Chemists' Society*, 77(4), 381-387.

[70] La Scala, J., & Wool, R. P. (2005). Property Analysis of Triglyceride-Based Thermosets. *Polymer*, 46(1), 61-69.

[71] Vigueras-Santiago, E., Hernández-López, S., Linares-Hernández, K., & Linares-Hernández, I. (2009). Mechanical Properties of Goat Leather Photo Grafted with Acrylate-Epoxidized Linseed Oil. *Advances in Technology of Materials and Materials Processing Journal*, 11(2), 43-48.

[72] Ureña-Núñez, F., Vigueras-Santiago, E., Hernández-López, S., Linares-Hernández, K., & Linares-Hernández, I. (2008). Structural, Thermal and Morphological Characterization of UV-Graft Polymerization of Acrylated-Epoxidized Soybean Oil onto Goat Leather. *Chemistry and Chemical Technology*, 2(3), 191-197.

[73] Hernández-López, S., Martín-López del, Campo., Sánchez-Mendieta, E., Ureña-Núñez, V., Vigueras, F., & Santiago, E. (2006). Gamma Irradiation Effect on Acrylated-Epoxi-

dized Soybean Oil: Polymerization and Characterization. *Advances in Technology of Materials and Materials Processing Journal*, 8(2), 220-225.

[74] Williams, G. I., & Wool, R. P. (2000). Composites from Natural Fibers and Soy Oil Resins. *Applied Composites Materials*, 7(5-6), 421 EOF-432 EOF.

[75] Oprea, S. (2010). Properties of Polymer Networks Prepared by Blending Polyester Urethane Acrylate with Acrilated Epoxidized Soybean oil. *Journal of Materials Science*, 45(5), 1315-1320.

[76] Thielemans, W., & Wool, R. P. (2004). Butyrated Kraft Lignin as Compatibilizing Agent for Natural Fiber Reinforced Thermoset Composites. *Composites Part A Applied Sciencie and Manufacturing*, 35(3), 327-338.

[77] Dweib, Hu. B., O', Donnell. A., Shenton, H. H., & Wool, R. P. (2004). All Natural Composite Sandwich Beams for Structural Applications. *Composite Structures*, 63(2), 147-157.

[78] Dweib, Hu. B., Shentin, I. I. I. H. W., & Wool, R. P. (2006). Bio-Based Composite Roof Structures: Manufacturing and Processing Issues. *Composite Structures*, 74(4), 379-388.

[79] Panhuis, M., in, H., Thielemans, W., Minett, A. I., Leahy, R., Le Foulgoc, B., Blau, W. J., & Wool, R. P. (2003). A Composite from Soy Oil and Carbon Nanotubes. *International Journal of Nanoscience*, 2(3), 185-194.

[80] Thielemans, W., Mc Aninch, I. M., Barron, V., Blau, W. J., & Wool, R. P. (2005). Impure Carbon Nanotubes as Reinforcements for Acrylated Epoxidized Soy Oil Composites. *Journal of Applied Polymer Science*, 98(3), 1325-1338.

[81] San-Farfán, Juan., Hernández-López, R., Martínez-Barrera, S., Camacho-López, G., Vigueras-Santiago, M. A., & , E. (2005). Electrical Characterization of Polystyrene-Carbon Black Composites. *Physica Status Solidi*, 2-3762.

[82] Verbanac, F. (1985). Cellulosic Organic Solvent Soluble Products. Patent USA,(4)

[83] Edgar, K. J., Buchanan, CM, Debenham, JS, Rundquist, PA, Seiler, BD, Shelton, MC, & Tindall, D. (2001). Advances in cellulose ester performance and application. *Progress in Polymer Science*, 26(9), 1605-1688.

[84] Verbanac, F. (1984). Cellulosic Organic Solvent Soluble Products. Patent number 4,490,516, Dec 25,.

[85] Shediac, R., Ngola, S. M., Throckmorton, D. J., Anex, Shepodd. T. J., & Singh, A. K. (2001). Reversed-Phase Electrochromatography of Amino Acids and Peptides Using Porous Polymer Monoliths. *Journal of Chromatography*, 925(1-2), 251 EOF-63 EOF.

[86] Carbon Black User's Guide, (2006). International Carbon Black Association,. ., 28.

[87] Columbian Chemicals Company. (2010). http://www.columbianchemicals.com/ Portals/Markets%20and%20Applications/Industrial%20Carbon%20Black/Inks/ CCC_lores_Raven_0505.pdf, (accessed 1 August).

[88] Cabot Corporation:. (2010). http://www.cabot-corp.com/wcm/download/en-us/sb/VULCAN_XC72-English1.pdf, (accessed 1 August).

[89] Nhumo^{MR} . (2010). http://www.nhumo.com.mx/UploadedFiles/Adjunto-Standard_Carbon_Blacks-200671910496.pdf, (accessed 1 August).

[90] Brostow, W., Keselman, M., Mironi-Harpz, I., Narkis, M., & Peirce, R. (2005). Effects of Carbon Black on Tribology of Blends of Poly(Vinylidene Fluoride) with Irradiated an Non-Irradiated Ultrahigh Molecular Weight Polyethylene. *Polymer*, 46(14), 5058-5064.

[91] Marinho, B., Ghislandi, M., Tkalya, E., Koning, CE, & de With, G. (2012). Electrical Conductivity of Compacts of Graphene, Multiwalled Carbon Nanotubes, Carbon Black, and Graphite Powder. *Powder Technology*, 221(1), 351-358.

[92] Celzard, A., Marêché, J. F., Payot, F., & Furdin, G. (2002). Electrical Conductivity of Carbonaceous Powders. *Carbon*, 40(15), 2801-2815.

[93] Vigueras, Santiago. E., Hernández, López. S., & Mercado, Posadas. J. (2009). Modificación de propiedades eléctricas en compuestos poliméricos con negro de carbono. In: Camacho-López M.A, et al (ed), *Tópicos en Materiales,*, México: Universidad Autónoma del Estado de México;.

Chapter 10

Value - Added Products from Soybean: Removal of Anti-Nutritional Factors *via* Bioprocessing

Liyan Chen, Ronald L. Madl, Praveen V. Vadlani, Li Li and Weiqun Wang

Additional information is available at the end of the chapter

1. Introduction

Soybean is the second largest acreage crop in the United States (29%), right after corn (35%) according to the American Soybean Association [1]. Soybean is widely consumed in the world, particularly in Asian countries. The various soybean products could be separated into non-fermented and fermented soybean products. The non-fermented soybean products include soymilk, tofu, yuba, soybean sprouts, okara, roasted soybeans, soynuts and soy flour, immature soybeans, cooked whole soybeans, and the fermented oriental soybean products include soy paste (Jiang and Miso), soy sauce, Tempeh, Natto, soy nuggets (Douchi), sufu. In the United States, soy oil is often used for food and biodiesel production. The soybean processing process is shown in Figure 1. After the oil extraction, the residue – flaked soy meal, is usually produced into four products (textured soy flour, soy protein concentrate and soy protein isolate, 48% soy meal, soluble soy carbohydrate). The textured soy flour could be used in bakery products, meat products, infant food etc. Soy protein concentrate and isolate could be used in baby food, bakery products, cereals, lunch meat etc. SSPS (soluble soybean polysaccharides) functions as a dispersing agent, stabilizer, emulsifier, and has good adhesion properties [2]. The 48% soy meal is used for animal feed. The portions of different animal usages are poultry (48%), swine (26%), beef (12%), dairy (9%), pets (2%), others (3%) [1]. Poultry and swine usages account for74%.

The popular usage of soybean for food and feed is due to its nutritional profile. Soybean is a good protein source and the only dietary isoflavone source together with other legumes. Anti-nutritional factors in soybean, such as phytic acid, oligosaccharides, trypsin inhibitor etc, limit its usage. Fermentation with GRAS (generally recognized as safe) microorganisms has been used to help degrade these anti-nutritional factors. The nutritional value of fermented soybean and soy meal products with additional nutritional factors is then largely enhanced.

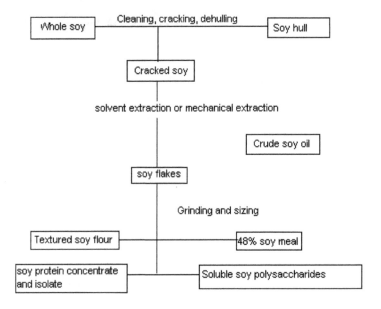

Figure 1. Soybean processing and products

2. Nutritional enhancement of soybean and soy meal via fermentation

2.1. Soy Protein

Protein content in soybean and soy meal are around 40%- 50% respectively. The high protein content makes soybean and soy meal a rich protein source for food and feed. As food source, the Protein Digestibility Corrected Amino Acid Score (PDCAAS), which is adopted by FDA and FAO/WHO, for isolated soy protein is 0.92, soy protein concentrate is 0.99, comparing with beef (0.92) and egg white (1.00). The human subject studies show that well-processed soy protein could serve as the sole source of protein intake for human beings [3]. FDA claims that diets containing 25 g of soy protein can reduce levels of low-density lipoproteins by as much as 10 percent and have considerable value to heart health. The specific reason for the heart protection function is unclear, for there are hundreds of protective compounds in soybean. As feed source, soy protein is high in lysine, but low in sulfur-containing amino acids, with methionine being the most limiting amino acid, followed by threonine [4]. The complementation of soy and corn for lysine and methionine makes them a valuable feed when combined.

2.1.1. Fermentation increases protein and amino acid content, and degrades protein into small functional peptides.

During fermentation, microorganisms digest the carbohydrates in soybean or soy meal and use for their own growth. The decreased dry matter and increased microorganisms weight ratio result in enhanced protein content[5-7]. In reference [5], fermented soy meal with *S.cerevisae* increased its protein level from 47% to 58%, while with *L. plantarum* and *B. lactis*, protein level increased to 52.08% and 52.14%. Microorganisms used for soybean fermentation have been reported to secret protease during fermentation [8-11]. In Cheonggukjang, the *Bacillus subtilis* fermented traditional soybean food in Korea, the acidic protease activity level could be as high as 590.24±2.92 µg/ml. Neutral protease activity level could achieve 528.13±3.11 µg/ml [9]. Because of protein degradation during fermentation, fermented soybean products are easier to digest.

Four parameters have been often used to evaluate the protein degradation of fermented soybean products. They are trichloroacetic acid (TCA) soluble nitrogen, degree of protein hydrolysis, SDS-PAGE profile, and amino acid content. Usually peptides having 10 or fewer amino acids would dissolve in TCA[12]. During fermentation, the degree of protein hydrolysis increases because of protease hydrolysis [13-14]. Meanwhile, TCA soluble nitrogen and peptide contents could also be enhanced [6, 9, 13-14]. SDS-PAGE analysis shows less large (>70 kDa) and medium (20-60 kDa) peptides and more small (<10 kDa) peptides in soy meal after fermentation of *Lactobacillus plantarum, Bifidobacterium lactis, Sccharomyces cerevisiae*, or *Aspergillus oryzae* [5, 7, 15]. Reference [13] showed that after 24 hr *Bacillus subtilis* fermentation, soy protein with molecular weight above 20 Kd disappeared from the electrophoretograms. The total amino acid content increased significantly (p<0.05) in fermented soy meal or soybean with *Lactobacillus plantarum, Bifidobacterium lactis, Sccharomyces cerevisiae, Bacillus subtilis, Aspergillus oryzae, Rhizopus oryzae, Actinomucor elegans, Rhizopus oligosporus* et al. [5, 7-8, 10-11, 13, 16]. *L.plantarum* fermentation of soy flour led to an increase in sulfur amino acids (Met plus Cys), Phe, Tyr, Lys, and Thr [16].

2.1.2. Functional biopeptide

Fermentation degrades large protein molecules into small peptides and amino acids. The biologically active peptides from soybean play an important role as angiotenisin converting enzyme (ACE) inhibitor [17] and as antioxidants [18]. In this section, we will discuss the ACE inhibitor. The antioxidant activity of biopeptides will be discussed in the antioxidant section.

Angiotensin I-converting enzyme (ACE, EC 3.4.15.1) is a dipeptidylcarboxypeptidase associated with the regulation of blood pressure as well as cardiovascular functions [19]. ACE-inhibitory substances are used to lower the blood pressure of hypertensive patients. Various ACE inhibitory peptides have been isolated from traditional fermented soybean foods, like natto, doujiang, soy sauce, and miso paste [20-23].

ACE inhibitory activity of peptides generated by protease is greatly dependent upon fermentation time. ACE inhibitory activity in Textured Vegetable Protein (TVP) fermented by

Bacillus subtilis for 24 and 72hr showed IC50 values of 2.20 and 3.80 mg/ml, respectively [24]. The initial fermentation of TVP resulted in production of effective peptides, but longer fermentation time produced less active peptides as ACE inhibitor. Peptide with ACE inhibitory activity consisted of low molecular weight. Molecular weight of 500-1,000 Da shows the highest ACE inhibitory activity [24]. In [25], oligopeptides generated from soy hydrolysate and fermented soy foods though endoprotease digestion, demonstrated a range of biological activities – angiotensin converting enzyme (ACE) inhibitory, anti-thrombotic, surface tension and antioxidant properties.

2.1.3. Fermentation decreases soy immunoreactivity

Soybean is defined as one of the "big 8" food allergens in the United States [16].The estimated prevalence of soybean allergies is about 0.5% of the total U.S. population [16]. Patients with soy allergy could react subjectively and objectively with 0.21 and 37.2 mg of soy protein, respectively [5]. The principle for food allergy is that epitopes in allergenic protein bind to the immunoglobulin E (IgE) molecules residing in the mast cells and basophils, causing them to release inflammatory mediators, including histamine. Alpha- (72 kDa) and beta- (53 kDa) conglycinin subunits, P34 fraction, and glycinin basic (33 kDa) and acidic (22 kDa) subunits, and trypsin inhibitor (20 kDa) are the main protein components causing plasma immunoreactivity [16]. Glycine was found to stimulate local and systemic immune responses in allergic piglets and had negative effects on piglet performance [26]. The severity of the immune reactions depends on the dose of glycinin; higher doses cause more severe symptoms. The effect of purified beta-conglycinin on the growth and immune responses of rats were investigated [27]. Results showed that purified beta-conglycinin possesses intrinsic immune-stimulating capacity and can induce an allergic reaction. Also, newly weaned pigs with limited stomach acid and enzymatic secretions in the small intestine can have difficulty digesting proteins with complex structures and large molecular weights [28].

Studies have confirmed degradation of soybean allergens during fermentation by microbial proteolytic enzymes in fermented soybean products, such as soy sauce, miso and tempeh [5, 29]. In the fermented soy products, soy protein has been hydrolyzed into smaller peptides and amino acids;therefore the structure of antigen epitopes might be altered, becoming less reactive. The IgE binding potential is reduced and therefore the immunoreactivity is lowered. In soy sauce, proteins are completely degraded into peptides and amino acids after fermentation and allergens are no longer present [29]. The reduction of immunoreactivity by nature and induced fermentation of soy meal with *Lactobacillus plantarum*, *Bifidobacterium lactis*, *Saccharomyces cerevisiae* were evaluated [5]. *S.cerevisiae*, *B.lactis* and *L. plantarum* reduced the immune response77.2%, 77.2%, 78.0%, when using 97.5 kUA/l human plasma and 88.7%, 86.3%, 86.9%, respectively, when the pooled human plasma was used. All three fermented soy meal products showed fewer large (>70 kDa) and medium (20-60 kDa) peptides, and more small (<10 kDa) peptides. Protein hydrolysis reduction of soy protein immunoreactivity was also confirmed through enzyme hydrolysis conducted in reference [30], which showed that after hydrolyzing with three food-grade proteases (Alcalase, Neutrase, Corolase PN-L), no residual antigenicity was observed in resulting soy whey.

Animal experiments have also confirmed the hypoallergenic properties of fermented soybean or soy meal products. With regard to the soybean allergy, fermentation of soy meal decreased the immune response to soy protein in piglets and the level of serum IgG decreased by 27.2% [31]. Antigenic soybean proteins in the diet of early weaned pigs provoke a transient hypersensitivity associated with morphological changes including villi atrophy and crypt hyperplasia in the small intestine [31]. All of these morphological changes can cause a malabsorption syndrome [26, 33], growth depression, and diarrhea [34, 35]. Differences of the villi condition in such pigs fed soy meal and fermented soy meal were investigated by using scanning electron microscopy [36]. Piglets fed soy meal had shorter, disordered, and broader villi, whereas piglets fed fermented soy meal had long, round, regular, and tapering villi that could better digest and absorb nutrients.

2.2. Isoflavone

One of the acknowledged bioactive compounds in soybean is isolavone. Isoflavones generally exist in soybeans and soy foods as aglycones (daidzein, genistein, and glycitein), beta-glycosides (daidzin, genistin, and glycitin), acetylglycosides (6''-O-acetyldaidzin, 6''-O-acetylgenistin, and 6''-O-acetylglycitin), and malonylglycosides (6''-O-malonyldaidzin, 6''-O —malonylgenistin, and 6''-O-mlonylglycitin). The structures of the 12 isomers are shown in figure 2.Isoflavones physiological effects include their estrogenic activity, antioxidant and antifungal activity, and more importantly, to act as anti-carcinogens. Isoflavones may also help to reduce blood serum cholesterol levels [4].

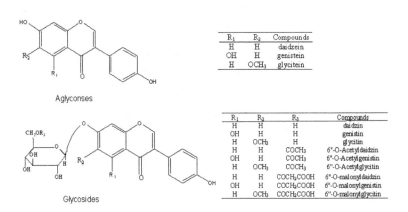

Figure 2. Chemical structures of 12 isoflavone isomers in soybean [4]

Among soy isoflavones, the relative abundance of genistein including respective derivatives, is the highest – about 50% of isolavone content, followed by daidzein (40%) and glycitein (10%). However, glycitein has been shown to be more bio-available than other isoflavones, followed by genistein [37]. Most of the isoflavones are associated with proteins in soy, with very little present in the lipid fraction. In their natural state, the majority of iso-flavones exists as inactive glycosides (genistin, daidzin, and glycitin) and the remaining as their active aglycone forms (genistein, daidzein, and glycitein, respectively). Glycoside forms are heat sensitive, being converted into malonyl-beta-glycosylated isoflavone when heated. The aglycone forms are quite stable at high temperature [38].

Isoflavones when ingested are metabolized extensively in the intestinal tract, absorbed, transported to the liver, and undergo enterohepatic recycling. Intestinal bacterial glucosidas-es cleave the sugar moieties and release the biologically active isoflavones as aglycones. Aglycones could be directly absorbed in the adult and these can be further biotransformed by microorganisms to the specific metabolites, equol, desmethylangolensin, and p-ethylphe-nol. All of these phyto-oestrogens are then eliminated, mainly by the kidney, and therefore share the physiological features and behavior of endogenous estrogens. Among the glyco-nes, beta-glucosides are easier to be hydrolyzed than 6''-O-malonylglucosides and 6''-O-ace-tylglucosides.

Some microorganisms have been reported to secrete beta-glucosidase, which could convert isoflavones to aglycones. The quantities of melonyl, acetyl, and glycosidyl isoflavonoids de-crease during fermentation but those of isoflavonoid aglycones, daidzein, and genistein in-crease by over 10 to 100 fold. In Meju (long term fermented soybeans), compared with unfermented samples, the total glycosides in 60 hr fermented samples decreased from 1827 ug/g to 487 ug/g, while total aglycones increased from 22 ug/g to 329 ug/g, with daidzein increasing from trace to 152 ug/g, genistein from 16 ug/g to 170 ug/g. However, the quantity of glycitein was not increased [39]. In another study about meju fermented by *Aspergillus or-yzae*, malonyl glycosides that initially accounted for 57.2-72.2% in the different soybeans markedly decreased to 7.4-19.9%, while aglycones originally accounted for only 1.1-2.8% in the soybeans, but markedly increased to 34.1-53.2% in miso [40]. In reference [38], total agly-cones increased from 12.27 ug/g in whole soy flour to 446.90 ug/g after 48 hr fermentation by *Aspergillus oryzae*(ATCC 22876). Its percentage of total isoflavones increased from 2.67% to 75.51%. Daidzein content increased from trace to 133.07 ug/g, glycitein from trace to 35.56 ug/g, genistein from 12.27 ug/g to 278.27 ug/g. In *Bacillus subtilis* fermented soy paste ChungGuklang (CGJ), about 85% of isoflavones were in the aglycone form in the CGJ, 14% in the glucoside form and acetylglucoside and malonylglucoside forms contributed less than 1% [41]. In *Bacillus pumilus* HY1 fermented Cheonggkjang, the beta glucosidase increased to 24.8 U/g until 36 h. The glycosides and malonylglycosides decreased throughout the fermen-tation to about 80% - 90% of their starting amount at 60 hr [42]. Part of isoflavonoid agly-cones are broken down into secondary metabolites, so the total quantity of isoflavonoids decreased [38, 42].

Aglycones could alleviate the symptoms of type2 diabetes, which the beta-glycosides, ace-tylglycosides and malonylglycosides forms are not able to do. Type 2 diabetes mellitus emerges from uncompensated peripheral insulin resistance that is associated with unregu-lated nutrient homeostasis, obesity, peripheral insulin resistance and progressive beta-cell failure. The effect of isoflavones in Meju on alleviating the symptoms of the type2 diabetes was investigated and four mechanismswere found [39]. Isoflavonoid aglycones could improve insulin-stimulated glucose intake. Also, they could induce PPAR-γ activation to increase insulin-stimulated glucose uptake. The PPAR-γ is the central regulator of insulin and glucose metabolism. It could help improve insulin sensitivity in type 2 diabetic patients and in diabetic rodent models. Besides, aglycones have strong effects for insulin/IGF – 1 signaling through IRS2, which plays a crucial role in beta-cell growth and survival. In this study, aglycones in meju increased GLP-1 secretion. GLP-1 is one of the incretins secreted from enteroendocrine L-cells that augments insulin secretion after the oral intake of glucose and free fatty acids. The induction of its secretion can prevent and/or relieve diabetic symptoms.

2.3. Antioxidant activity

Oxidative stress has been found to be the primary cause of many chronic diseases as well as the aging process itself. Antioxidants could help to delay or prevent oxidative stress. Epidemiological studies show that antioxidants could lower the risk of cardiovascular disease, cancer (overall risk reduction between 30 – 50%), diabetes, neurological diseases, immune diseases, eye diseases et al. [43]. Antioxidant compounds play an important role as a health protecting factor. It is beneficial to eat antioxidant enriched food. Products prepared through the fermentation of soybean including various traditional oriental fermented products of soybean such as natto, tempeh, miso and other fermented products, have been found to exhibit a significantly higher antioxidant activity than their respective non-fermented soybean substrate.

Di(phenyl)-(2,4,6-trinitrophenyl)iminoazanium(DPPH) radical scavenging activity, Fe2+ chelating activity and reducing property have been used to quantify antioxidant activity. DPPH is a stable nitrogen-centered, lipophilic free radical that is used in evaluating the antioxidant activities in a short time. Ferrous ionsare the most effective pro-oxidants in food system. Metal chelation agents prevent metal-assisted homolysis of hydroperoxides and block the formation of chain initiators. IC50 and the relative scavenging effect are the parameters to describe the DPPH radical scavenging and Fe2+ chelating abilities. IC50 is the inhibition concentration of extracts required to decrease initial DPPH radical or Fe2+ concentration by 50%. The relative scavenging effect is calculated by dividing the extract content with the IC50 of the respective extract, and then compare with the scavenging effect of control samples. The reducing property indicates that these compounds are electron donors and can reduce the oxidized intermediates, and therefore, can act as primary or secondary antioxidants. The reduction of Fe3+/ferricyanide complex to ferrous form in presence of antioxidants has been used to test the reducing activity of samples [44].

Antioxidant activity of methanol extracts of soybean koji fermented with *Asp. oryzae*, *Asp. sojae*, *Actinomucor taiwanensis*, *Asp. awamori*, and *Rhizopus* spp. have been investigated [44]. Methanol extracts of soybean koji, which mainly contained phenolic acid, have higher relative DPPH – scavenging effect and Fe2+ chelating effect than the unfermented steamed soybeans. The koji methanol extracts had a relative DPPH - scavenging effect of 2.3-8.9 compared with that of the non-fermented soybean, which was assigned as 1.0. Among them, the *Aspergillus awamori*-soybean koji exhibited the highest DPPH-scavenging effect, at a level approximately 9.0 fold than that exhibited by the non-fermented steamed soybean [44].The Fe^{2+} ion chelating ability of soybean increased by 2.1 – 6.7 fold after fermentation, depending on starter organism employed [44]. In Cheonggukjang fermented by *Bacillus pumilus HY1*, the level of DPPH radical scavenging activity increased from 54.5% to 96.2% by 60 hr [42]. In *Bacillus subtilis* fermented soybean kinema, when the methanol extract concentration was 50 mg/ml, 60% DPPH radical scavenging was observed. In the same product with a concentration of 10 mg/ml, the methanolic extract of kinema exhibited 64% metal chelation which was much higher than the activity shown by cooked non-fermented soybean (22%) [45]. Similar findings of enhanced reducing power of fermented bean and bean products were reported from *Bacillus subtilis* fermented soybean kinema[45] and from *Aspergillus oryzae* fermented soybean koji [44].

The enhanced antioxidant activities in fermented soybean products may be due to the increased phenolic compounds contents. Phenolic compounds have been demonstrated to exhibit a scavenging effect for free radicals and metal-chelating ability [46]. Most phenolic acids in cereals primarily occur in the bound form, as conjugates with sugars, fatty acids, or proteins [47].Isoflavones are the predominant phenolics in soybean, and the glucoside form of isoflavones represents 99% of the total isoflavones in soybean [45].This condition lowers the antioxidant activity since the availability of free hydroxyl groups in the phenolic structure is an important characteristic for the resonance stabilization of free radicals. The enhancement of bioactive phenolic compounds by enzymatic hydrolysis from different cereals was reported by Wojdylo [47] and Yuan [48]. Different enzymes during bacteria or fungi fermentation, such as alpha-amylase, alpha and beta-glucosidase, beta-glucuronidase, cellulase et al., have been reported to be involved in the lignin remobilization and phenolic compounds contents enhancement [49]. Fermented soybean products have higher amount of phenolic compounds [42, 44]. In Cheonggukjang fermented by *Bacillus pumilus HY1*, total phenolics increased markedly from the starting amount of 253 g/kg to 9586 g/kg at the end of fermentation (60hr) [42]. In *Bacillus subtilis* fermented kinema, total phenol content of kinema was 144% higher than that of cooked non-fermented soybean. The total phenol content was positively correlated ($p<0.001$) with radical- scavenging, reducing power, metal-chelating activity in *Bacillus subtilis* fermented kinema [45].

The other reason is the short chain peptides generated by fungi or bacterial protease. Antioxidant activities of peptides have been reported [50-52]. In [51], five different proteases were used to hydrolyze soybean β-conglycinin and the hydrolysates from three of them had significant enhanced antioxidative activities. Peptide antioxidant activity is related, but not limited, to the amino acid composition and its sequence. In reference [51], peptides isolated from the antioxidative beta-conglycinin hydrolysate contain histidine, proline or tyrosine residue in their sequences and hydrophobic amino acids, valine or leucine at the N terminus. The constituent amino acids had no antioxidant activity when mixed at the same concentration as the peptides. Anti-oxidative activities of peptides with different amino acid composition or sequences have different antioxidant mechanisms. Reference [52] studied the anti-oxidative properties of combinatorial tri-peptides. Tri-peptides Tyr-His-Tyr, Xaa-Xaa-

Trp/Tyr, and Xaa-Xaa-Cys (SH) had a strong synergistic effect with phenolic antioxidants, a high radical scavenging activity, and a high peroxynitrite scavenging activity, respectively.

2.4. Phytic acid

Phytate is the calcium-magnesium-potassium salt of inositol hexaphosphoric acid common-ly known as phytic acid [Figure 3]. Phytate and phytic acid are also referred to as phytin in some literature.

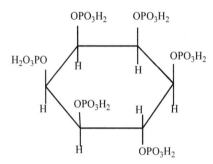

Figure 3. Basic structure of phytic acid [52]

Phytate is the main storage form of phosphorous in soybean. Its content in soybean ranged from 1.00 to 1.47% on a dry matter basis [4].Phytate is known to be located in the protein bodies, mainly within their globoid inclusions. Phosphorous in the phytate form could not be absorbed by monogastric animals, because they lack phytase, the digestive enzyme re-quired to release phosphorus from the phytate molecule. Phytic acid could form protein-phytate or protein-phytate-protein complexes; these have more resistance to digestion by proteolytic enzymes; thus, utilization of dietary protein is reduced. Also, phytic acid has a strong binding affinity to important minerals such as calcium, magnesium, iron, and zinc. When a mineral binds to phytic acid, it becomes insoluble, precipitates, and is not absorba-ble in the intestines. In food industry, the presence of phytic acid in high concentration is undesirable. In feed industry, the unabsorbed phytate passes through the gastrointestinal tract of monogastric animals, elevating the amount of phosphorus in the manure. Excess phosphorus excretion can lead to environmental problems such as eutrophication. With the pressure on the swine industry to reduce the environmental impact of pork production, it is important to use feed ingredients that can minimize this influence.

The ability of the molds for oriental fermented soybean food to produce phytase has been investigated. Phytase is the enzyme hydrolyzing phytic acid to inositol and phosphoric acid and thereby removing the metal chelating property of phytic acid. There are two strains of *Rhizopus oligosporus* used for tempeh fermentation, two strains of *Aspergillus oryzae* used for soy sauce and six strains of *Aspergillus oryzae* used for miso fermentation, all the ten strains were reported to be able to secret both extracellular and intracellular phytases [54]. The phytic acid content of soybeans was reduced by about one-third as a result of *Rhizopus oligo-sporus* NRRL 2710 fermentation [55-56]. That was from original 1.27% to 0.61% after 48 hr fermentation [56]. Animal test showed that fermentation of soy meal increased phosphorus availability [57-58] and zinc availability [59] and reduced phosphorus excretion without affecting growth of chicks. Using fermented soy meal as substitute for regular soy meal saved 0.2% of dietary inorganic phosphorous [60].

2.5. Oligosaccharides

Galacto-oligosacchrides (GOS) generally represent approximately 4 to 6% of soybean dry matter. In soy meal produced at 10 commercial processing plants in the United States, concentrations of stachyose, raffinose, and verbascose ranged from 41.0 to 57.2, 4.3 to 9.8, and 1.6 to 2.4 mg/g DM, respectively [61]. Oligosaccharides in the carbohydrate fraction, particularly raffinose and stachyose [figure 4], could lead to flatulence and abdominal discomfort for monogastric animals [62-63] because of the deficiency of alpha-galactosidase.

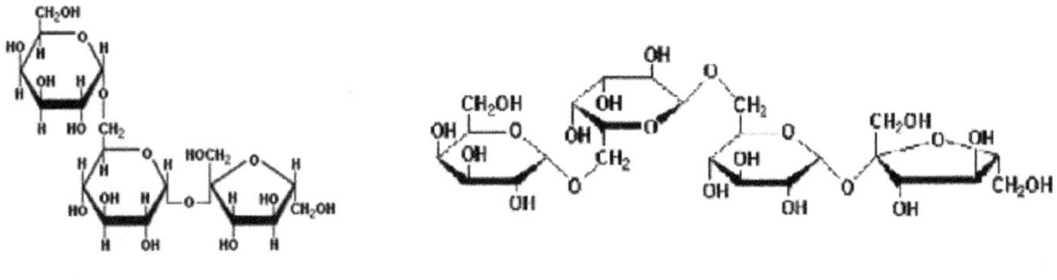

(1)   (2)

Figure 4. Structure of raffinose and stachyose (1) raffinose; (2) stachyose

Galacto-oligosacchrides are digested to some extent in the small intestine (76 to 88% for stachyose, 31 to 65% for raffinose, and 32 to 55% for verbascose), resulting in the production of carbon dioxide and hydrogen [64]. In some cases, the accumulation of flatulent rectal gas provokes gastrointestinal distress such as abdominal pain, nausea, and diarrhea. Weanling pigs fed a GOS-free diet supplemented with 2% stachyose or fed a diet containing soy meal had increased incidence of diarrhea compared with pigs fed a control diet [63]. Additionally, fermentation of GOS has been implicated to have negative effects on nutrient digestibilities and energy availability of soy meal. Roosters fed soy meal with low oligosaccharide concentrations had higher total net metabolizable energy values (2931 kcal/kg dm) than those fed conventional soy meal (2739 kcal/kg DM) [65]. The removal of polysaccharides from soy foods and feed is, therefore, a major factor in improving their nutritive value. To reduce non-digestible oligosaccharides, fermentation with fungi, yeast, and bacteria with alpha-galactosidase secreting ability has been attempted over the years.

The enzyme alpha-D-Gal (alpha-D-galactoside galactohydrolase, EC 3.2.1.22) is of interest for hydrolyzing the raffinose-type sugars found in soybeans. *Rhizopus oligosporus, Lactobacil-*

lus curvatus R08, Leuconostoc mesenteriodes, Lactobacillus fermentum, Bifidobacterium sp. et al. have been reported to be able to produce alpha-galactosidase [56, 66-70]. These microorganisms have been applied for soybean fermentation to reduce the oligosaccharides [56, 67-68, 71]. In [56],stachyose and raffinose decreased by 56.8% and 10%, respectively, in soybean during 48 hr fermentation by Rhizopus oligosporus. In Leu.mesenteriodes JK55 and L.curvatus R08 fermented soymilk, the non-digestive oligosaccharides were completely hydrolyzed after 18-24 h of fermentation [67].

Alpha-galactosidase has been isolated from plant and microbial sources, and its properties have been documented. In general, alpha-galactosidase acts upon gal-gal bonds in the tetra-saccharide stachyose, releasing galactose and raffinose, and also acts upon gal-glu bonds with the release of sucrose. Sucrose is, in turn, split by invertase, producing fructose and glucose [72]. So, α-galatosidase activity is noticeably dependent on the type of sugar. The type and concentration of the carbon source are known to be nutritional factors that regulate the synthesis of bacterial galactosidase. Reference [67] found that the existence of glucose and fructose inhibit the alpha-galatosidase expression both in Lactobacillus curvatus R08 and Leucomostoc mesenteriodes.

2.6. Trypsin inhibitor

Protease inhibitors constitute around 6% of soybean protein [73]. Two protein protease inhibitors have been isolated from soybeans. The Kunitz trypsin inhibitor has a specificity directed primarily toward trypsin and a molecular weight of about 21.5 kDa. The Bowman-Birk (BB) inhibitor is capable of inhibiting both trypsin and chymotrypsin at independent reactive sites and has a molecular weight of 7.8 kDa [74]. Trypsin and chymotrypsin, the two major proteolytic enzymes produced in the pancreas, belong to the serine protease class.

Trypsin inhibitors present in soybeans are responsible for growth depression by reducing proteolysis and by an excessive fecal loss of pancreatic enzymes rich in sulfur-containing amino acid which can not be compensated by dietary soy protein [75]. Trypsin inhibitors account for 30-50% of the growth inhibition effect [76]. Rats fed a raw soybean extract from which trypsin inhibitors had been inactivated showed improved growth performance when compared with control rats fed diets containing raw soybeans from which inhibitors had not been inactivated [77].

Diets with a trypsin inhibitor concentration of 0.77 mg/g and less did not reduce the growth of pigs according to reference [7]. And, research showed that after fermentation with Aspergillus oryzae GB-107, the trypsin inhibitor in soy meal was reduced from 2.70 mg/g to 0.42 mg/g [7]. In the in vivo experiment of reference [15], the activities of total protease and trypsin at the duodenum and jejunum of piglets fed with fermented soy meal increased because of the inactivation of trypsin inhibitor. Just as was mentioned above, protease produced during fermentation could degrade protein molecules into peptides and amino acids. The trypsin inhibitor may be degraded or modified during fermentation and lose its activity binding to trypsin.

2.7. Vitamin

The increased content of some vitamins or provitamins, both water-soluble and fat-soluble vitamins, such as riboflavin, niacin, vitamin B6, β-carotene et al., which are due to fungal metabolic activities,is one of the healthy and nutritional advantages of fermented soybean products and has been extensively examined.

Vitamin or provitamin formation during tempeh fermentation by *Rhizopus oligosporus*, *R.arrhizus*, *R. oryzae* and *R.slolonifer*, respectively, were studied [11, 78]. In ref [11], all of the fourteen Rhizopus strains used in the research could form riboflavin, vitamin B6, nicotinic acid, nicotinamide, ergosterol, with isolates of *R.oligosporus* the best vitamin formers. In ref [78], six of 14 Rhizopus strains were able to form β-carotene in significant amounts. Five of these six strains belonged to the species of *R.oligosporus* and one was identified as *R.stolonifer*. A newly fourfold increase in β-carotene from 0.6 ug/g dw to 2.2 ug/g dw could be detected between 34 and 48hr in fermentations with *R.oligosporus* strain. During this period the content of total carotenoids increased from 9.1 ug/g dw to 11.2 ug/g dw in the fermentations with *R.oligosporus* strain. Ergosterol is the vitamin D2 precursor. Vitamin D can be derived from two naturally occurring compounds: ergocalciferol (D2) and cholecalciferol (D3). Both forms have equal biological activities in humans. The fourteen strains could produce ergosterol. The ergosterol concentration could be up to 1610 ug/g dw after 96 hr fermentation.

Vitamin K is an essential cofactor for the posttranslational conversion of glutamic acid residues of specific proteins in the blood and bone into γ-carboxyglutamic acid (Gla). There are two naturally occurring forms of vitamin K, vitamins K1 and K2. Vitamin K1 (phylloquinone) is formed in plants. Vitamin K2 (menaquinone, MK) is primarily synthesized by bacteria [79].Menaquinone (MK) plays an important role in blood coagulation and bone metabolism [80]. Japanese fermented soybean product, Natto, has been regarded as a high content of MK source (about 6- 9 ug/g) and is found in everyday products. Natto produced by a mutated *B. subtilis* strain showed a much higher content of MK up to 12.98 ug/g [81].Aromatic amino acids (phenylalanine, tyrosine, and trypstophan) could slow down the MK synthesis rate in cheonggukjang by using *Bacillus amyloliquefaciens KCTC11712BP*, while the supplement of 4% glycerol could signicantly increase its yield [82].

3. Conclusion

After fermentation by GRAS microorganisms, the anti-nutritional factors in soybean or soy meal are totally degraded, including oligosaccharides, trypsin inhibitor and phytic acid. Fermentation could also degrade large soy protein into peptides and amino acids, therefore, removing the allergenic effect of soy protein. Nutritional factors are formed during fermentation along with removal of undesirable factors. Functional peptides, such as peptides with ACE inhibitory activity are created by protein degradation. Isoflavones are converted to their functional forms, the aglycones. Antioxidant activity is enhanced, contributed by the increase of short chain peptides and phenolic compounds. Certain vitamins or provitamins are formed such as riboflavin, β-carotene, vitamin K2 and ergosterol. Total nutritional profiles of soybean and soy meal are greatly enhanced by fermentation.

Acknowledgements

The authors are grateful to the Kansas Soybean Commission and the Department of Grain, Science and Industry, Kansas State University, for funding this project. This chapteris contribution no 13-041-B from the Kansas Agricultural Experiment Station, Manhattan, KS 66506

Author details

Liyan Chen[1], Ronald L. Madl[1], Praveen V. Vadlani[1*], Li Li[2] and Weiqun Wang[3]

*Address all correspondence to: vadlani@ksu.edu

1 Bioprocessing and Renewable Energy Laboratory, Department of Grain Science and Industry, Kansas State University, Kansas, USA

2 Department of Food Science, South China University of Technology, Guangzhou, China

3 Department of Human Nutrition, Kansas State University, Manhattan, Kansas, USA

References

[1] American Soybean Association. (2012). Soy Stats. http://www.soystats.com/2012, accessed 1st May 2012).

[2] Maeda, H., & Nakamura, A. (2009). Soluble soybean polysaccharide. In: Phillips GO, Williams PA., *Handbook of Hydrocolloids*, New York: CRC Press;, 693-709.

[3] Young, V. R. (1991). Soy protein in relation to human protein and amino acid nutrition. *Journal of American Dietetic Association*, 91(7), 828-835.

[4] Liu, K. (1997). Soybeans: Chemistry, Technology, and Utilization. New York: Chapman & Hall Press;.

[5] Song, Y. S., Frias, J., Martinez-Villaluenga, C., et al. (2008). Immunoreactivity reduction of soybean meal by fermentation, effect on amino acid composition and antigenicity of commercial soy products. *Food Chemistry*, 108, 571-581.

[6] Chen, C. C., Shih, Y. C., Chiou, P. W. S., et al. (2010). Evaluating nutritional quality of single stage-and two stage-fermented soybean meal. *Asian-Australasian Journal of Animal Science*, 23(5), 598-606.

[7] Hong, K. J., Lee, C. H., & Kim, S. W. (2004). Aspergillus oryzae GB-107 Fermentation Improves Nutritional Quality of Food Soybeans and Feed Soybean Meals. *Journal of Medical Food*, 7(4), 430-435.

[8] Kim, J., Hwang, K., & Lee, S. (2010). ACE Inhibitory and hydrolytic enzyme activities in textured vegetable protein in relation to the solid state fermentation period using Bacillus subtilis HA. *Food Science and Biotechnology*, 19(2), 487-495.

[9] Kim, M., Han, S., Ko, J., & Kim, Y. (2012). Degradation characteristics of proteins in Cheonggukjang (fermented unsalted soybean paste) prepared with various soybean cultivars. *Food Science and Biotechnology*, 21(1), 9-18.

[10] Omafuvbe, B. O., Abiose, S. H., & Shonukan, O. O. (2002). Fermentation of soybean (Glycine max) for soy-daddawa production by starter cultures of Bacillus. *Food Microbiology*, 19, 561-566.

[11] Bisping, B., Hering, L., Baumann, U., Denter, J., et al. (1993). Tempeh fermentation: some aspects of formation of γ-linolenic acid, proteases and vitamins. *Biotechnology Advances*, 11, 481-493.

[12] Low, A. G. (1980). Nutrient absorption in pigs. *Journal of the Science of Food and Agriculture*, 31, 1087-1130.

[13] Weng, T., & Chen, M. (2010). Changes of protein in Natto (a fermented soybean food) affected by fermenting time. *Food Science and Technology Research*, 16(6), 537-542.

[14] Weng, T., & Chen, M. (2011). Effect of two-step fermentation by Rhizopus oligosporusrus and Bacillus subtilis on protein of fermented soybean. *Food Science and Technology Research*, 17(5), 393-400.

[15] Feng, J., Liu, X., Xu, Z. R., et al. (2006). The effect of Aspergillus oryzae fermented soybean meal on growth performance, digestibility of dietary components and activities of intestinal enzymes in weaned piglets. *Animal Feed Science and Technology*, 134, 295-303.

[16] Frias, J., Song, Y. S., Martinez-Villaluenga, C., et al. (2008). Immunoreactivity and amino acid content of fermented soybean products. *Journal of Agricultural and Food Chemistry*, 56, 99-105.

[17] Zhang, J. H., Tatsumi, E., Ding, C. H., & Li, L. T. (2006). Angiotensin converting enzyme inhibitory peptides in douchi, a Chinese traditional fermented soybean product. *Food Chemistry*, 98, 551-557.

[18] Kim, S. K., Choi, Y. R., Park, P. J., et al. (2000). Purification and characterization of antioxidative peptides from enzymatic hydrolysate of cod teiset protein. *J. Korean Fish. Soc*, 33, 198-204.

[19] Saxena, P. R. (1992). Interaction between the renin-angiotensin-aldosterone and sympathetic nervous systems. *Journal of Cardiovascular Pharmacology*, 19, 6-8.

[20] Okamoto, A., Hanagata, H., Kawamura, Y., & Yanagida, F. (1995). Antihypertensive substances in fermented soybean, natto. *Plant Foods for Human Nutrition*, 47, 39-47.

[21] Shin, Z. I., Nam, H. S., Lee, H. J., Lee, H. J., & Moon, T. H. (1995). Fractionation of angiotensin converting enzyme (ACE) inhibitory peptides from soybean paste. *Korean Journal of Food Science and Technology*, 27, 230-234.

[22] Oka, S., & Nagata, K. (1974). Isolation and characterization of neutral peptides in soy sauce. *Agricultural and Biological Chemistry*, 38, 1185-1194.

[23] Teranaka, T., Ezawa, M., Matsuyama, J., Ebine, H., & Kiyosawa, I. (1995). Inhibitory effects of extracts from rice-koji miso, barley-koji miso, and soybean-koji miso on activity of angiotensin I converting enzyme. *Nippon Nogeik. Kaishi*, 69, 1163-1169.

[24] Kim, J. E., Hwang, K., & Lee, S. P. (2010). ACE inhibitory and hydrolytic enzyme activities in textured vegetable protein in relation to the solid state fermentation period using Bacillus subtilis HA. *Food Science and Biotechnology*, 19(2), 487-495.

[25] Gibbs, B. F., Zougman, A., Masse, R., & Mulligan, C. (2004). Production and characterization of bioactive peptides from soy hydrolysate and soy-fermented food. *Food Research International*, 37(2), 123-131.

[26] Sun, P., Li, D. F., Dong, B., et al. (2008a). Effects of soybean glycinin on performance and immune function in early weaned pigs. *Archives of Animal Nutrition*, 62, 313-321.

[27] Guo, P., Piao, X., Ou, D., et al. (2007). Characterization of the antigenic specificity of soybean protein b-conglycinin and its effects on growth and immune function in rats. *Archives of Animal Nutrition*, 61(3), 189-200.

[28] Kim, S. W. (2010). Bio-fermentation technology to improve efficiency of swine nutrition. *Asian-Australasian Journal of Animal Science*, 23(6), 825-832.

[29] Kobayashi, M. (2005). Immunological functions of soy sauce: hypoallergenicity and antiallergicantivity of soy sauce. *Journal of Bioscience and Bioengineering*, 100(2), 144-151.

[30] Penas, E., Restani, P., Ballabio, C., et al. (2006). Assessment of the residual immunoreactivity of soybean whey hydrolysates obtained by combined enzymatic proteolysis and high pressure. *European Food Research and Technology*, 222, 286-290.

[31] Liu, X., Feng, J., Xu, Z., et al. (2007). The effects of fermented soybean meal on growth performance and immune characteristics in weaned piglets. *Turkish Journal of Veterinary and Animal Sciences*, 31(5), 341-345.

[32] Dreau, D., & Lalles, J. P. (1999). Contribution to the study of gut hypersensitivity reactions to soybean proteins in preruminant calves and early-weaned piglets. *Livestock Production Science*, 60, 209-218.

[33] Gu, X., & Li, D. (2004). Effect of dietary crude protein level on villous morphology, immune status and histochemistry parameters of digestive tract in weaning piglets. *Animal Feed Science and Technology*, 114, 113-126.

[34] Dreau, D., Lalles, J. P., Philouzerome, V., et al. (1994). Local and systemic immune-responses to soybean protein ingestion in early-weaned pigs. *Journal of Aanimal Science*, 72, 2090-2098.

[35] Sun, P., Li, D. F., Li, Z. J., et al. (2008b). Effects of glycinin on IgE-mediated increase of mast cell numbers and histamine release in the small intestine. *The Journal of Nutritional Biochemistry*, 19, 627-633.

[36] Feng, J., Liu, X., Xu, Z. R., et al. (2007). Effect of fermented soybean meal on intestinal morphology and digestive enzyme activities in weaned piglets. *Digestive Diseases and Sciences*, 52, 1845-1850.

[37] Masilamani, M., Wei, J., & Sampson, H. A. (2012). Regulation of the immune response by soybean isoflavones. *Immunologic Research*, in press.

[38] Da, Silva. L. H., Celeghini, R. M. S., & Chang, Y. K. (2011). Effect of the fermentation of whole soybean flour on the conversion of isoflavones from glycosides to aglycones. *Food Chemistry*, 128, 640-644.

[39] Kwon, D. Y., Hong, S. M., Ahn, S., et al. (2011). Isoflavonoids and peptides from meju, long-term fermented soybeans, increase insulin sensitivity and exert insulinotropic effect in vitro. *Nutrition*, 27, 244-252.

[40] Yamabe, S., Kobayashi-Hattori, K., Kaneko, K., Endo, H., & Takita, T. (2007). Effect of soybean varieties on the content and composition of isoflavone in rice-koji miso. *Food Chemistry*, 100, 369-374.

[41] Chung, I., Seo, S., Ahn, J., et al. (2011). Effect of processing, fermentation, and aging treatment to content and profile of phenolic compounds in soybean seed, soy curd and soy paste. *Food Chemistry*, 127, 960-967.

[42] Cho, K. M., Hong, S. Y., Math, R. H., Lee, J. H., et al. (2009). Biotransformation of phenolics (isoflavones, flavanols and phenolic acids) during the fermentation of cheonggukjang by Bacillus pumilus HY1. *Food Chemistry*, 114, 413-419.

[43] Willcox, J. K., Ash, S. L., & Catignani, J. L. (2004). Antioxidants and prevention of chronic disease. *Critical reviews in Food Science and Nutrition*, 44, 275-295.

[44] Lin, C., Wei, Y., & Chou, C. (2006). Enhanced antioxidative activity of soybean koji prepared with various filamentous fungi. *Food Microbiology*, 23, 628-633.

[45] Moktan, B., Saha, J., & Sarkar, P. K. (2008). Antioxidant activities of soybean as affected by Bacillus-fermentation to kinema. *Food Research International*, 41, 986-593.

[46] Shahidi, F., Janitha, P. K., & Wanasundara, P. D. (1992). Phenolic antioxidants. *Critical Reviews in Food Science and Nutrition*, 32, 67-103.

[47] Wojdylo, A., & Oszmainski, J. (2007). Comparison of the content phenolic acid, alpha-tocopherol and the antioxidant activity in oat naked and weeded. *Electronic Journal of Environmental, Agricultural and Food Chemistry*, 6(4), 1980-1988.

[48] Yuan, X., Wang, J., & Yao, H. (2006). Production of feruloyl oligosaccharides from wheat bran insoluble dietary fibre by xylanases from Bacillus subtilis. *Food Chemistry*, 95, 484-492.

[49] Mc Cue, P., & Shetty, K. (2003). Role of carbohydrate-cleaving enzymes in phenolic antioxidant mobilization from whole soybean fermented with Rhizopusoligosporus. *Food Biotechnology*, 17(1), 27-37.

[50] Amadou, I., Le Shi, G., et, Y., & al, . (2011). Reducing, radical scavenging, and chelation properties of fermented soy protein meal hydrolysate by Lactobacillus plantarum LP6. *International Journal of Food Properties*, 14, 654-665.

[51] Chen, H., Muramoto, K., & Yamauchi, F. (1995). Structural analysis of antioxidative peptides from soybean β-conglycinin. *Journal ofAgricutural and Food Chemistry*, 43, 574-578.

[52] Saito, K., Jin, D., Ogawa, T., & Muramoto, K. (2003). Antioxidative properties of tripeptide libraries prepared by the combinatorial chemistry. *Journal of Agricultural and Food Chemistry*, 51, 3668-3674.

[53] Oatway, L., Vasanthan, T., & Helm, J. H. (2001). Phytic acid. *Food Reviews International*, 17(4), 419-431.

[54] Wang, H. L., Swain, E. W., & Hesseltine, C. W. (1980). Phytase of molds used in oriental food fermentation. *Journal of Food Science*, 45, 1262-1266.

[55] Sudarmadji, S., & Markakis, P. (1977). The phytate and phytase of soybean tempeh. *Journal of the Science of Food and Agriculture*, 28, 381-383.

[56] Egounlety, M., & Aworth, O. C. (2003). Effect of soaking, dehulling, cooking and fermentation with Rhizopus oligosporus on the oligosaccharides, trypsin inhibitor, phytic acid and tannins of soybean (Glycine max Merr.), cowpea (Vignaunguiculata L. Walp) and groundbean (Macrotyloma geocarpa) Harm . *Journal of food engineering*, 56, 249-254.

[57] Kim, S. S., Galaz, G. B., Pham, et., & al, . (2009). Effects of dietary supplementation of a meju, fermented soybean meal, and Aspergillus oryzae for Juvenile Parrot Fish (Oplegnathusfasciatus). *Asian-Australasian Journal of Animal Science*, 22(6), 849-856.

[58] Matsui, T., Hirabayashi, M., Iwama, Y., et al. (1996). Fermentation of soya-bean meal with Aspergillus usami improves phosphorus availability in chicks. *Animal Feed Science and Technology*, 60(1-2), 131-136.

[59] Hirabayashi, M., Matsui, T., & Yano, H. (1998a). Fermentation of soybean meal with Aspergillus usami improves zinc availability in rats. *Biological Trace Element Research*, 61(2), 227-234.

[60] Hirabayashi, M., Matsui, T., Yano, H., et al. (1998b). Fermentation of soybean meal with Aspergillususamii reduces phosphorus excretion in chicks. *Poultry Science*, 77, 552-556.

[61] Grieshop, C. M., Kadzere, C. T., Clapper, G. M., et al. (2003). Chemical and nutritional characteristics of United States soybeans and soybean meals. *Journal of Agricultural and Food Chemistry*, 51, 7684-7691.

[62] Krause, D. O., Easter, R. A., & Mackie, R. I. (1994). Fermentation of stachyose and raffinose by hindgut bacteria of the weanling pig. *Letters in Applied Microbiology*, 18, 349-352.

[63] Zhang, L. Y., Li, D. F., Qiao, S. Y., et al. (2003). Effects of stachyose on performance, diarrhoea incidence and intestinal bacteria in weanling pigs. *Archives of Animal Nutrition*, 57, 1-10.

[64] Karr-Lilienthal, L. K., Kadzere, C. T., Grieshop, C. M., et al. (2005). Chemical and nutritional properties of soybean carbohydrates as related to nonruminants: a review. *Livestock Production Science*, 97, 1-12.

[65] Parsons, C. M., Yhang, Y., & Araba, M. (2000). Nutritional evaluation of soybean meals varying in oligosaccharides content. *Poultry Science*, 79, 1127-1131.

[66] Nout, M. J. R., & Rombouts, F. M. (1990). Recent developments in tempeh research. *Journal of Applied Bacteriology*, 69(5), 609-633.

[67] Yoon, M. Y., & Hwang, H. (2008). Reduction of soybean oligosaccharides and properties of alpha-D-galactosidase from Lactobacillus curvatus R08 and Leuconostoc mesenteriodes JK55. *Food Microbiology*, 25, 815-823.

[68] Wang, Y. C., Yu, R. C., Yang, H. Y., & Chou, C. C. (2002). Growth and survival of bifidobacteria and lactic acid bacteria during the fermentation and storage of cultured soymilk drinks. *Food Microbiology*, 19, 501-508.

[69] Garro, M. S., de Valdez, G. F., Oliver, G., & de Giori, G. S. (1996). Purification of alpha-galactosidase from Lactobacilus ferment. *Journal of Biotechnology*, 45, 103-109.

[70] Leblanc, J. G., & Garro de, Giori. G. S. (2004a). Effect of pH on Lactobacillus fermentum growth, raffinose removal, alpha-galactosidase activity and fermentation products. *Applied Microbiology and Biotechnology*, 65, 119-123.

[71] Le Blanc, J. G., Garro, Silvestroni. A., Connes, C., Piard, J. C., Sesma, F., & de Giori, G. S. (2004b). Reduction of alpha-galactooligosaccharides in soyamilk by Latobacillus-fermentum CRL 722: in vitro and in vivo evaluation of fermented soyamilk. *Journal of Applied Microbiology*, 97, 876-881.

[72] Mulimani, V. H., Thippeswamy, S., & Ramalingam, S. (1997). Enzymatic degradation of oligosaccharides in soybean flour. *Food Chemistry*, 59, 279-282.

[73] Friedman, M., & Brandon, D. L. (2001). Nutritional and health benefits of soy proteins. *Journal of Agricultural and Food Chemistry*, 49(3), 1069-1086.

[74] Hocine, L. L., & Boye, J. I. (2007). Allergenicity of soybean: new developments in identification of allergenic proteins, cross-reactivities and hypoallergenization technologies. *Critical Reviews in Food Science and Nutrition*, 47, 127-143.

[75] Rackis, J. J. (1981). Gumbmann MR. Protease inhibitors: physiological properties and nutritional significance. In: Ory RL. (ed.) Antinutrrents and natural toxicants in foods., *USA: Food & Nutrition Press*, Inc;, 203-237.

[76] Isanga, J., & Zhang, G. (2008). Soybean bioactive components and their implications to health- a review. *Food reviews International*, 24, 252-276.

[77] Kakade, M. L., Hoffa, D. E., & Liener, I. E. (1973). Contribution of trypsin inhibitors to the deleterious effects of unheated soybeans fed to rats. *Journal of Nutrition*, 103, 1172.

[78] Denter, J., Rehm, H., & Bisping, B. (1998). Changes in the contents of fat-soluble vitamins and provitamins during tempeh fermentation. *International Journal of Food Microbiology*, 45, 129-134.

[79] Sato, T., Yamada, Y., Ohtani, Y., et al., & 200, . (2001). Production of Menaquinone (Vitamine K2)- 7 by Bacillus subtilis. *Journal of Bioscience and Bioengineering*, 91(1), 16-20.

[80] Schurgers, I. J., Knapen, M. H. J., & Vermeer, C. (2007). Vitamin K2 improves bone strength in postmenopausal women. *International Congress Series*, 1297, 179-187.

[81] Tsukamoto, Y., Ichisse, H., Kakuda, H., et al. (2000). Intake of fermented soybean (natto) increases circulating vitamin K2 (menaquinone-7) and γ-carboxylatedosteocalcin concentration in normal individuals. *Journal of Bone and Mineral Metabolism*, 18, 216-222.

[82] Wu, W., & Ahn, B. (2011). Improved Menaquinone (Vitamin K2) production in Cheonggukjang by optimization of the fermentation conditions. *Food Science and Biotechnology*, 20(6), 1585-1591.

Chapter 11

Soybean and Prostate Cancer

Xiaomeng Li, Ying Xu and Ichiro Tsuji

Additional information is available at the end of the chapter

Part I: Consumption of Soybean and the incidence of Prostate Cancer in East and West

Prostate cancer is the most frequently diagnosed malignancy in men all over the world. The investigation from International Agency for Research on Cancer in World Healthy Organization of United Nations reported the data of 899, 000 new cases of prostate cancer in 2008, accounted for 13.6% of the total cancer cases. The investigation discovered the large difference of the incidence and the mortality of prostate cancer between in West and in East (Figure 1). In Western countries, the incidences and mortalities are significantly high. For example, the cases from Australia, Northern America or Europe, occupied nearly three-quarters of all the globe prostate cancer patients. In contrast, the incidences and mortalities in Asia are quite low, less than one tenth of that in Europe, the US or Australia of the West, especially in the South-Central Asia, showing the lowest incidence and mortality [1].

The significant difference of the incidences of prostate cancer addressed several important questions.

1. Why do Asian men have a much lower incidence of prostate cancer compared to men from the Western countries (the US and Europe)? Latent or clinically insignificant cancer of the prostate is found at autopsy at approximately the same rate in men from Asian countries as those from the USA (approximately 30% of men aged over 50 years), but there are large differences in the clinical incidence and mortality. Is there a strong possibility that diet and nutrition play a prominent role in accelerating or inhibiting the process by which clinically significant prostate cancer develops?

2. Is the fact that the hormone-dependent cancers of the prostate and breast show the same incidence and lifetime risk (the correlation r is 0.81 in 21 countries) related to diet?

3. East Asian countries, including Chinese and Japanese men, have the lowest incidence of prostate cancer in the world. But why, when Japanese men from those countries migrate to the North America, does their risk of developing prostate cancer increase 10-fold compared to their counterparts in Japan or China?

4. Although it is relatively rare in East, an increase in the incidence of prostate cancer has been reported in China, where the life style especially the diet structure is changing followed with the developed economy in recent years. What factors can account for the conspicuous increase?

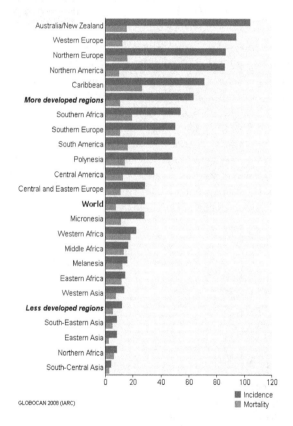

Figure 1. The incidence and mortality of Prostate cancer in the world, 2008(from International Agency for Research on Cancer)

Those questions highlight the critical roles of environmental dietary factors in the different risks of prostate cancer between East and West. Many epidemiological studies suggest that the different dietary, most probably one kind of traditional food in the East countries, soybean, may become a dominant reason for the protective effects against prostate cancer [2].

Soybean, recognized as a complete protein food, is a species of legume and an annual plant in Asia. As being a traditional food in Asia, the history of soybean in Asia is very long, more than 5000 years and even before written records. Soybean can be made to a large body of kinds of foods by fermented or non-fermented process. Typical soy foods include soy milk, bean curd and tofu skin which are non-fermented soy foods as well as miso sauce, natto and soy sauce which belong to fermented soy foods. Not only fermented soybean foods, but also non-fermented soybean products are very essential dietary ingredient for Asian people, especially Chinese, Japanese and Koreans. Some Asians often have and their fermented or non-fermented products almost everyday. Japanese nearly had the soy food daily by ingestion of miso sauce and Chinese approximately had them more than 100 grams per day.

The history of soybean used in western countries is quite short, compared to that in Asia. Soybean was introduced into America, Australia and New Zealand about in 17th century and into Canada in 1831 as a sauce named "A new dozen India Soy". Most westerners hardly consume soy foods in the diet, although soybean and its products are used largely in some other ways, for example soybean oil could be made into bio-diesel in the United States. However Soy food consumption in the western countries are quite low. Soy foods have been consumed for centuries in Asian countries Japanese nearly had the soy food daily by ingestion of miso sauce and Chinese approximately had them by Tofu, soy milk and tofu skin.. The mean daily intakes of soy protein are approximately 30 grams in Japan, 20 grams in Korea, 7 grams g in Hong Kong, and 8 grams but more than 100 grams per day in some area in China [3]. While the average daily intake in the United States is less than 1 gram [4].

Country	Prostate cancer mortality	Total energy	Fish energy	Soy energy	Animal energy
Australia	29.59	3055	22	0	1019
Austria	31.23	3575	15	0	1233
Canada	29.20	3340	29	0	1297
France	31.43	3529	34	0	1343
Italy	22.57	3688	25	0	913
USA	32.19	3641	23	1	1316
Hong Kong	5.44	2771	89	36	834
Japan	5.87	2852	195	93	590
Korea	0.90	3056	67	94	269
Singapore	7.47	3165	63	29	689
Thailand	0.53	2330	37	18	152

Table 1. Prostate cancer and amount of soy food consumption in East and West, 1998 (from JNCI 1998; 90: 1637-1647)

For clear comparison of soybean in the Asian and the western diet, a table published in JNCI (Table 1), clearly clarified Prostate cancer and amount of soy food consumption in East and West. In Asian regions, the soybean energy is 36 in Hong Kong, 29 in Singapore and 18 in Thailand; and for much more dramatically difference, the soybean energy is 93 in Japanese, and 94 in Korean diet [5]. In contrast, in western countries of Australia, Austria, Canada, France, or Italia, the energy from soybean in daily dietary was zero, even in USA, the soy energy is only 1. This indicates that soybean is the dominantly different diet factor between the Eastern countries and the Western countries.

Interestingly the dramatically increased prostate cancer mortality in western countries, combined together with almost no soybean energy consumption, compare with very low mortality in Asia (only 1/60 to 1/6 of that in West), combined with high soybean energy consumption. This provided the important evidence for the consumption of soybean related to the lower risk of prostate cancer mortality. Even another point of the higher animal energy consumption in western diet, it is not that dominant as the difference of soy food.

Soy food	Study site	Finding	Study type	OR/RR	P trend	Reference
Soy food	Japan	protection	cohort	0.52(0.29–0.90)	0.010	Kurahashi 2007
Miso soup				0.65(0.39–1.11)	0.220	
Soymilk	USA	protection	cohort	0.30(0.1 – 0.9)	0.020	Jacobsen 1998
Tofu	Hawaii	protection	cohort	0.35(0.08–1.43)	0.054	Severson 1989
Soy food	UK	protection	case-control	0.52(0.30–0.91)	0.340	Heald 2007
Tofu	Japan	protection	case-control	0.47(0.20–1.08)	0.160	
All soy products				0.53(0.24–1.14)	0.110	Sonoda 2004
Soyfoods	China	protection	case-control	0.51(0.28–0.95)	0.061	
Tofu				0.58(0.35–0.96)	0.032	Lee 2003
Soyfoods	Hawaii,San Francisco,	protection	case-control	0.62(0.44–0.89)	0.060	
All legumes	Los Angeles, British Columbia and Ontario			0.62(0.49–0.80)	0.0002	Kolonel 2000
Beans/lentils/ nuts	Canada	protection	case-control	0.69(0.53–0.91)	0.030	Jain 1999
Tofu, Soybean	Canada	protection	case-control	0.80(0.60–1.10)	0.290	Villeneuve 1999
Soybean foods	China	protection	case-control	0.29(0.11-0.79)	0.02	Li 2008
Baked beans	UK	protection	case-control	0.57(0.34–0.95)	N/A	Key 1997
Garden peas	UK			0.35(0.13–0.91)	N/A	

Table 2. N/A:no adequate dataEpidemiological studies on food intake of soy products and prostate caner risk(most data summrized from Mol Nutr Food Res. 2009; 53: 217-226)

Most epidemiological studies have suggested that the consumption of soy food is associated with a reduction in prostate cancer risk in humans. Eight case-control studies and three cohort studies have reported the protective effect of soy food, with odds ratios or relative risks ranging from 0.3 to 0.80, including in China and Japan, where people consume more soybean food, tofu, soymilk and natto. Here the epidemiological data are summarized in Table 2 [6].

Some cohort studies provided the convincing data on this issue. A population based prospective study recruited 43 509 Japanese men aged 45–74 years and followed them up for 10 years (1995 through 2004). For men aged 60 years, in whom soy food were associated with a dose-dependent decrease in the risk of localized cancer, with RRs for men in the highest quartile of soy food consumption compared with the lowest obtained a protective OR of 0.52 (95% confidence interval (CI) 0.29–0.90, P trend = 0.01) [7]. A cohort study in the USA with 225 incident cases of prostate cancer in 12395 California seventh-day adventist men showed frequent consumption (more than once a day) of soymilk was associated with 70% reduction of the risk of prostate cancer (RR = 0.3, 95%CI 0.1–0.9, P trend = 0.02) [8]. In an early cohort study among 7999 men of Japanese ancestry who were first examined between 1965 and 1968 and then followed through to 1986, 174 incident cases of prostate cancer were recorded. Increased consumption of tofu did not show statistical significant association with the risk of prostate cancer (RR = 0.35, 95% CI 0.08–1.43) [9].

Much more case-control studies provided more evidence for the reduced risk of prostate cancer associated with consumption of soy foods. First we concentrate some case-control studies conducted in men living in China. Our group carried out a population-based case-control study in China to investigate the possible correlation factors for prostate cancer, 28 cases from 3940 men over 50 years old with prostate-specific antigen screening in Changchun city in China, matched them with controls of low prostate-specific antigen value (< 4.1 ng/mL) by 1: 10 according to age and place of employment. In all ten food items, the consumption of soybeans was demonstrated the only factor to decrease the risk of prostate cancer. Men who consumed the soybean product of Tofu and soymilk more than once per day had a multivariate OR of 0.29 (95% CI, 0.11–0.79) compared with men who consumed soybean products less than once per week. The P for trend was 0.02. There was no significant difference for any other dairy food (2). Another case-control study in China of 133 cases and 265 age- and residential community-matched controls from 12 cities were recruited. Results showed that the age- and total calorie-adjusted OR of prostate cancer risk was 0.58 (95% CI 0.35-0.96, P= 0.032) in the highest tertile of Tofu intake comparing to the lowest tertile. There were also statistically significant associations of intake of soy foods (OR 0.51; 95% CI 0.28–0.95, P = 0.061) [10].

The case-control studies also provide the similar evidence for that the soy intake protect against prostate cancer. A case-control study of diet and prostate cancer in Japan demonstrated the possible protective effect of traditional Japanese soybean die plays a preventive role against prostate cancer in four geographical areas (Ibaraki, Fukuoka, Nara, and Hokkaido) of Japan. All 140 cases and 140 age -matched hospital controls were analyzed to confirm the consumption of fish and natto showed significantly decreasing linear trends for risk, with RR of 0.53 (95% CI, 0.24-1.14) (P < 0.05) for all soybean products, 0.47 (95% CI, 0.20-1.08) for tofu, and 0.25 (95% CI, 0.05-1.24) for natto [11].

Not only has the case-control research conducted in Asian men supported the hypothesis that the rich soybean products may protect against prostate cancer but many case-control studies conducted in western countries also provid the epidemiological evidence for the protective effects of soy food against prostate cancer. A multiethnic case-control study carried out in Hawaii, San Francisco, and Los Angeles in the USA, and British Columbia and Ontario in Canada, 1619 cases were diagnosed during 1987–1991 and were compared to 1618 controls of African-American, white, Japanese, and Chinese men. Controls were frequency- matched to cases on ethnicity, age, and region of residence of the case, in a ratio of approximately 1:1. Intake of soy foods was inversely related to prostate cancer with OR of 0.62 (95% CI, 0.44–0.89). Results were similar when restricted to prostate-specific antigen normal controls [12]. A further four case-control studies from USA, Canada and UK showed similar results with a protective effect of consuming legumes (beans, lentils, garden peas, etc.) against prostate cancer. A case-control study in Canada for a total study population consisted of 617 incident cases of prostate cancer and 636 population controls from Ontario, Quebec, and British Columbia. To obtain a decreasing, statistically significant association was found with increasing intakes of beans/lentils/nuts (OR = 0.69, 95% CI, 0.53-0.91) [13]. Another population-based case-control study conducted in eight Canadian provinces. Risk estimates were generated by applying multivariate logistic regression methods to 1623 histologically confirmed prostate cancer cases and 1623 male controls aged 50-74 to obtained a OR 0.80 (95% CI, 0.60-1.10) of Tofu and Soybeans [14]. Not only soybean but also other legumes had the inverse relationship against prostate cancer. Oxford group in UK found the baked beans had an OR of 0.57 (95% CI, 0.34-0.95); and garden peas had an OR of 0.35 (95% CI, 0.13-0.91) [15]. And a population-based case-control study of diet, inherited susceptibility and prostate cancer was undertaken in Scotland investigated a total of 433 cases of Scottish men and 483 controls aged 50-74 years, indicates the theoretical scope for reducing the risk from prostate cancer, with the consumption of soy foods (adjusted OR 0.52, 95 % CI 0.30-0.91) [16].

All these data from both East and West, summarized in Table 2, provide the convincing epidemiological evidence that the higher intake of soy products is associated with a reduced risk of prostate cancer in human. Some experimental data in animal model also demonstrated the possible protective effects for Soybean food against prostate cancer. For example, Zhou group demonstrated that dietary soy products inhibit the experimental prostate tumor growth through a combination of direct effects on tumor cells and indirect effects on tumor neovasculature in mice, confirming that soy foods could act as a preventive factor against prostate cancer in animal models [17].

Prostate cancer has marked geographic variations between countries. The data from both western and eastern countries support the critical roles of soybean food in the protection of prostate cancer.

Part II: Correlation between prostate cancer and soybean isoflavones from diet, serum epidemiological and laboratorial data

In recent decades, much evidence from epidemiological studies support the notion that frequent consumption of soybean foods, the most different diet between East and West, is beneficial for the protection against prostate cancer.

Soybeans are high in protein, for about 35% to 40% of the dry weight. They also contain 18% polyunsaturated fats, 30% carbohydrates, and some vitamins, minerals. They are the only legume that provides ample amounts of the essential omega-3 fatty acid alpha-linolenic acid.2 Soybeans are a rich source of isoflavones (or phytoestrogens), a subclass of flavonoids that bind to estrogen receptors (though not as strongly as estrogen). Isoflavones are also discovered in peanut, alfalfa; however, soybean contains the largest amounts of isoflavones in the nature. Soybeans and soy foods such as tofu, soymilk, and miso are the only significant dietary sources of these phytoestrogens. In soybeans, isoflavones bound to a sugar molecule as glycoside, and when soybean is fermented or digested, isoflavones could be released from the bounded sugar [18].

Table 3 summarized the relationship between isoflavones levels and soy protein content in a variety of soy-containing foods. The isoflavones in soymilk are lowest, 2.5mg per 100g soymilk, may because in the fluid soymilk, the nutrition of soybeans is dissolved in water and lower concentration of nutrition in soymilk, whereas isoflavones in soy flour are highest, 131.2 to 198.9mg per 100g. In tofu, a traditional soy food in Asia, the isoflavones are 27.9mg/100g. In the traditional and very important Japanese soy food, the isoflavones in miso are 42.6mg/100g. The isoflavones in natto are 20mg/100g. The information from Table 3 clearly provides the amounted isoflavones in some other soybean foods as well. In the most frequently diet, each gram of soy protein is associated with approximately 3.6 mg of isoflavones in Tofu, 3.7 mg of isoflavones in whole bean based soymilk [19].

As the special nutrition in soybean compared to other plants, the isoflavones have many health benefits, including protection against chronic diseases, menopausal symptoms, osteoporosis and breast cancer or prostate cancer [20]. The most effective soy isoflavones are well known as genistein and daidzein. Here, the relationship between each soybean isoflavones and prostate cancer would be further illustrated.

The epidemiological evidence for the association of soy isoflavones and the prostate cancer risk is still limited. Some epidemiological data, summarized in Table 4, evaluated the effects of soybean isoflavones on prostate cancer. To date, one study suggests a causal relationship between isoflavones and prostate cancer, but two cohorts and two case-control studies suggest that soy isoflavones, genistein and daidzein, have prophylactic effects on prostate cancer. In addition,, an accumulating body of evidence from laboratory studies in recent decades has suggested that diets rich in high concentration of isoflavones are associated with anti-tumor effects in prostate cancer.

	estimated protein (g/100 g)	total isoflavones (mg/100 g)	isoflavone (mg/g protein)
soy products			
soybean chips	54.2	35	0.6
soy links frozen raw	3.9	15	3.9
natto boiled fermented	46.4	20	0.4
tofu, silken, firm	7.8	27.9	3.6
tempeh	14.9	43.5	2.9
miso	12.5	42.6	3.4
soy cheese, Cheddar	7.2	28	3.9
soy cheese, mozarella	7.7	32	4.2
soy cheese, Parmesan	6.4	36	5.6
soy milks			
soy milk, isolate based, low isoflavone	3.3	2.5	0.7
soy milk, isolate based, high isoflavone	3.3	4.4	1.3
soy milk, whole bean based, low isoflavone	3.1	2.4	0.8
soy milk, whole bean based, high isoflavone	3.1	11.6	3.7
soy milk, entire bean	3.1	3.6	1.2
soy milk	3.5	9.7	2.8
soy protein materials			
soy protein isolate, aqueous extract, type 1	85	114	1.3
soy protein isolate, aqueous extract, type 2	85	78.7	0.9
soy protein isolate, aqueous extract, type 3	85	103.4	1.2
soy protein isolate, alcohol extract	85	81.9	1.0
soy protein concentrate, alcohol extract	65	12.5	0.2
soy flour, defatted	50	131.2	2.6
soy flour, full fat, raw	35	177.9	5.1
soy flour, full fat, roasted	39	198.9	5.1

Table 3. Relationship between isoflavone levels and soy protein content in a variety of soy-containing foods from J. Agric. Food Chem. 2003; 51: 4146-4155

2.1. Genistein

Genistein, firstly isolated from Genista tinctoria in 1899, was abundant in soybean. Even though a serum case-control study suggested the promotion roles of genistein for the prostate cancer risk; most studies demonstrated its protective effects against prostate cancer. Two cohort studies and two case control studies demonstrated the significant protective roles of genistein against prostate cancer, with multivariate RRs (ORs) of 0.71, 0.52, 0.58 or 0.53, summarized in Table 4.

Phynoestrogen	Study site	Findings	Study type	OR/RR	P_{trend}	Reference
genistein						
	Europe	protection	cohort	0.71(0.53-0.96)	0.030	
	Japan	protection	cohort	0.52(0.30-0.90)	0.030	
		protection	case-control	0.58(0.34-0.97)	0.040	Nagata 2007
		protection	case-control	0.53(0.29-0.97)	0.058	Lee 2003
	Japan	promotion	serum case-control			Akaza 2002
daidzein						
	Japan	protection	cohort	0.50(0.28-0.88)	0.040	Kurahashi 2007
	Japan	protection	case-control	0.55(0.32-0.93)	0.020	Nagata 2007
	China	protection	case-control	0.56(0.31-1.04)	0.116	Lee 2003
	Japan	protection	serum case-control			Akaza 2002

Table 4. Effects of photoestrogens on cacinogenesis of prostate cancer in epidemiological or animals experimental data.

The population based prospective study recruited 43 509 Japanese men aged 45–74 years and followed them up for 10 years (1995 through 2004). All 147 food items in 220 cases with organ localized cancers was investigated the relationship of isoflavones intake and the risk of prostate cancer. The increased consumption of genistein was found associated with the decreased risk of localized prostate cancer. These results were strengthened when analysis was restricted to men aged more than 60 years, in whom isoflavones and soy food were associated with a dose-dependent decrease in the risk of localized cancer, with RRs for men in the highest quartile of genistein consumption compared with the lowest of 0.52 (95% CI 0.30–0.90, P trend = 0.03) [7].

Another Japanese group examined associations between nutritional and the prevalence of prostate cancer in a case-control study of Japanese men. Two hundred patients and 200 age-matched controls were selected from 3 geographic areas of Japan. Isoflavones and their agly-cones (genistein and daidzein) were significantly associated with decreased risk, with the OR for genistein 0.58 (95% CI 0.34–0.97, P trend = 0.04), indicating that isoflavones might be an effective dietary protective factor against prostate cancer in Japanese men [21].

A case-control study in China showed an overall reduced risk of prostate cancer associated with consumption of soy foods and genistein. In this study, 133 cases and 265 age- and residential community-matched controls from 12 cities were recruited. Results showed that the age- and total calorie-adjusted OR of prostate cancer risk was 0.58 (95% CI 0.35–0.96, P trend = 0.032) comparing the highest tertile of tofu intake to the lowest tertile. There were also statistically significant associations of intake genistein (OR, 0.53; 95% CI, 0.29–0.97, P trend = 0.058) [10].

Notably, a recent prospective investigation of plasma phytoestrogens and prostate cancer in the European also found that higher plasma concentrations of genistein were associated with lower risk of prostate cancer. They examined plasma concentrations of phyto-oestrogens in relation to risk for subsequent prostate cancer in a case-control study nested in the European Prospective Investigation into Cancer and Nutrition. Concentrations of isoflavones genistein, daidzein and equol, and that of lignans enterolactone and enterodiol, were measured in plasma samples for 950 prostate cancer cases and 1042 matched control participants. Relative risks (RRs) for prostate cancer in relation to plasma concentrations of these phyto-oestrogens were estimated by conditional logistic regression. Higher plasma concentrations of genistein were associated with lower risk of prostate cancer, the RR among men in the highest vs. the lowest fifth was 0.71 (95%CI 0.53-0.96, P trend=0.03). After adjustment for potential confounders, this RR was 0.74 (95% CI 0.54-1.00, P trend=0.05). No statistically significant associations were observed for circulating concentrations of daidzein, equol, enterolactone or enterodiol in relation to overall risk for prostate cancer [22].

Laboratory studies revealed that Genistein has multiple functions in the antitumor effects against prostate cancer, with the concentration as the prominent associated factor. In 2002, an Italy group declared that at low concentration, about 1~10μM, genistein would stimulate androgen-dependent prostate cancer cell line LNCaP growth, but when genistein was at high concentration, more than 100μM, it would result in cell apoptosis in prostate caner [23]. The conclusions above were supported by the research of another Gao's group, published in the journal of the Prostate in 2004 [24].

In vitro studies in several prostate cancer cell lines have revealed that genistein directly inhibit the growth of prostate cancer cells or through inducing apoptosis, affecting the expression of a large number of genes that are related to the control of cell survival and physiologic behaviors [25,26]. As tyrosine kinase inhibitor and topoisomerase inhibitor, genistein induces cancer cell apoptosis by upregulating the expression of cyclin-dependent kinase inhibitor p21 or inhibiting the activity of nuclear factor kappa B (NF-kB) signaling pathways [27, 28]. Genistein is a potent inhibitor of protein–tyrosine kinase, which may attenuate the growth of cancer cells and decrease the level of oxidative DNA damage [29-31]. Genistein also reported to enhance the ability of endoglin, a component of the transforming growth factor beta receptor complex, in suppression of the motility of prostate cancer cell [32]. Moreover, genistein is also a potent inhibitor of angiogenesis and metastasis. It can effectively inhibit cell invasion by inhibiting transforming growth factor β-mediated phosphorylation of the p38 mitogen-activated protein kinase-activated protein kinase 2 and the 27 kDa heat shock protein [33].

The anti-tumor effects of genistein are demonstrated in the animal models. Animal experiments have also shown that dietary concentrations of genistein can inhibit metastasis of prostate cancer [34]. Lifetime consumption of isolate/isoflavones has prevented spontaneous development of metastasizing adenocarcinoma in Lobund–Wistar rat [35, 36]. Dietary genistein can also suppress the development of advanced prostate cancer in castrated transgenic adenocarcinoma of mouse prostate (TRAMP) mice [37]. Some other result also indicated biphasic role for genistein in the regulation of prostate cancer growth and metastasis. A low concentration of 500 nmol/L of genistein to 12-week-old TRAMP-FVB mice as evidenced by increased proliferation, invasion. But a pharmacologic dose (50 nmol/L) decreased proliferation, invasion, and MMP-9 activity (>2.0-fold) concomitant with osteopotin reduction [38]. With 250 mg genistein/kg diet in treatments (TRAMP) mice model, the most significant effect was seen in the TRAMP mice exposed to genistein throughout life (1-28 weeks) with a 50% decrease in poorly-differentiated cancerous lesions. In a separate experiment in castrated TRAMP mice, dietary genistein suppressed the development of advanced prostate cancer by 35% compared with controls. The data obtained in intact and castrated transgenic mice suggest that genistein may be a promising chemopreventive agent against androgen-dependent and independent prostate cancers [39]. This group further identified the associated signaling, and revealed that Genistein in the diet significantly inhibited the cell proliferation by down-regulating tyrosine kinase regulated proteins, EGFR, IGF-1R, and down-regulating the downstream mitogen-activated protein kinases, ERK-1 and ERK-2 in prostates in TRAMP mice [37].

Genistein is demonstrated to be synergy with other phyto-chemicals together to inhibit the growth of prostate cancer. For example, genistein and curcumin is reported could inhibit the growth of prostate cancer cells in a synergistic way. Genistein, together with DIM, was proved to repress the proliferation of androgen-dependent prostate cancer cell line LNCaP and androgen-independent prostate cancer cell line PC-3 [40]. Genistein, together with biochanin A, could also inhibit the growth of human prostate cancer cells [41]. Genistein is also reported acts as a radiosensitizer for prostate cancer both in vitro and in vivo inhibit metastasis of prostate cancer [34].

Based on the studies above, a conclusion is deduced that genistein showed protective effects against prostate cancer, in vivo and in vitro, at its high concentration.

2.2. Daidzein

Daidzein is another important soy isoflavone. Most of epidemiological studies prove that daidzein contributes to the reduction of prostate cancer risk and the prevention of prostate cancer. A cohort study and two case control studies demonstrated its significant preventive roles of daidzein for the prostate cancer prevention, with multivariate RRs (ORs) of 0.50, 0.55 or 0.56, summarized in Table 4.

Several studies indicated that daidzein might be an effective dietary protective factor against prostate cancer in Japanese men. One cohort study of the population based prospective study recruited 43 509 Japanese men aged 45–74 years mentioned above and followed them up for 10 years. From 147 food items in 220 cases with organ localized cancers was investigated and isoflavones intake was founded to be associated with the risk of prostate cancer. In men aged more than 60 years, RRs for men in the highest quartile of daidzein consumption compared with the lowest of 0.50 (95%CI 0.28-0.88, P trend = 0.04) [7]. Another Japanese group examined associations between nutritional and the prevalence of prostate cancer in a case-control study in Japanese men. Two hundred patients and 200 age-matched controls were selected from 3 geographic areas of Japan, demonstrated that daidzein were significantly associated with decreased risk, with the OR 0.55 (95% CI 0.32–0.93, P trend = 0.02) for daidzein [21].

A case-control study in China showed an overall reduced risk of prostate cancer associated with consumption of soy daidzein. In this study, 133 cases and 265 age- and residential community-matched controls from 12 cities were recruited. Results showed that the age- and total calorie-adjusted OR of prostate cancer risk was 0.58 (95% CI 0.35–0.96, P trend = 0.032) comparing the highest tertile of tofu intake to the lowest tertile. There were even not statistically significant associations, but the protective trend for the intake of daidzein (OR, 0.56; 95% CI, 0.31–1.04, P trend = 0.116) [10].

Daidzein displays the modest protective effect against prostate cancer from the experimental data. Daidzein could act as a radiosensitizer against prostate cancer and an inhibitor of cell growth. Daidzein inhibited cell growth and synergized with radiation, affecting APE1/Ref-1, NF-kappaB and HIF-1alpha, but at lower levels than genistein or soy [42].

In addition, the protective effect on prostate cancer was strengthened after daidzein was combined with genistein, compared to individual one. The study from Oregon State University demonstrated daidzein and genistein could also induce cell apoptosis in benign prostate hyperplasia (BPH) cells at concentration of 25 μM. Soy extractions were indicated more effective as chemopreventive agents than genistein or daidzein. A combination of active soy-derived compounds is demonstrated more efficacious and safer as chemopreventive agents than individual compound. Soy extracts also increased Bax expression in PC3 cells [43]. Isoflavones exert their chemopreventive properties by affecting apoptosis signaling TRAIL (tumor necrosis factor-related apoptosis-inducing ligand) pathways in prostate cancer LNCaP cells. The chemopreventive effects of soy foods on prostate cancer are associated with isoflavone-induced support of TRAIL-mediated apoptotic death [44].

Soy isoflavones, including daidzein and genistein, exert anticarcinogenic effects against prostate cancer, proposed that soy extracts, containing a mixture of soy isoflavones and other bioactive components, would be a more potent chemo-preventive agent than individual soy isoflavones.

Part III: The mechanisms of prevention effects of soybean phyto-estrogens against prostate cancer.

Isoflavones are the best-known phyto-estrogens, are a diverse group of naturally soy compounds, play important roles in prostate cancer inhibition. The consumptions of soy isoflavone are investigated to be an effective protection factor against certain diseases. And the products rich in isoflavones might protect against enlargement of the male prostate gland, slow prostate cancer growth and lead to prostate cancer cell death.

The chemical structure-based ligand and receptor binding is the most important primary step in generating downstream signal transduction. To reveal their mechanisms of prevention effects against prostate cancer, their specific structures for the receptor binding are illustrated. Isoflavones are classified to phyto-estrogens, because their structures are similar to estradiol (17β-oestradiol), also could mimic estrogenic effect to bind to estrogen receptors. However, their structures are similar not only to estradiol, but also to dihydrotestosterone (DHT), the most important steroid hormone in male. DHT and 17β-oestradiol, endogenous steroid hormones, have a four-ringed carbon backbone, where as non-steroid hormone thyroxin has a quite different structure from them (Figure 2). The isoflavones of genistein and daidzein are two phyto-oestrogens found at very high levels in soy formula. Flavone, another group of phyto-estrogens, is found abundant in many plants. The structures of genistein, daidzein and flavone showed similar properties: a hydrophobic core and one or two terminal polar groups, similar to 17β-oestradiol or DHT (Figure 2), have the ability to cause estrogenic or/and anti-testosterone effects through the modulation of androgen receptor (AR) transactivation.

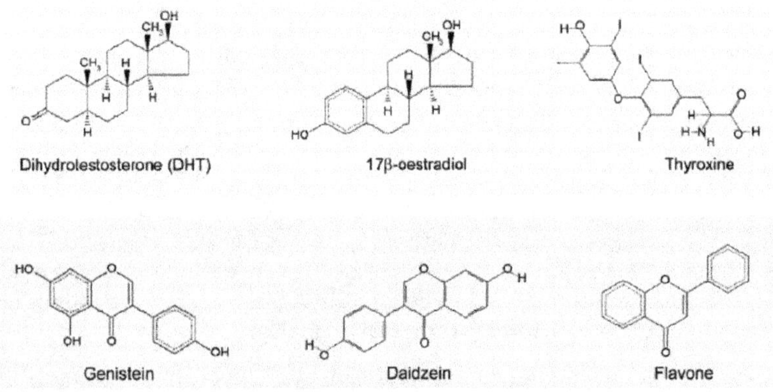

Figure 2. Molecular formulas of ligand compounds, two endogenous steroid hormones of dihydrotestosterone (DHT) and 17β-oestradiol, one endogenous non-steroidal hormone, thyroxin, and three phyto-estrogens of genistein, daidzein and flavone. from Asian J Androl. 2010; 12: 535-547

The normal development and maintenance of the prostate is dependent on androgen acting through the AR, which is driven by DHT plays a critical role in prostate cancer development and progression. AR expression is maintained throughout prostate cancer progression, and the majority of androgen-independent or hormone refractory prostate cancers with the overexpression or the mutation of AR. This progress may be affected by the soybean phyto-oestrogens, as mimic oestrogens, compete the same binding sites of AR that binds to DHT, further affect androgen-controlled AR mediated prostate cancer growth and development [45].

To reveal the mechanisms of these isoflavones for prostate cancer, our group adopted a computerized approach to examine the interaction of the human AR and isoflavone (genistein or daidzein), whereas the interaction of AR and flavone was set up as a positive control. Auto Dock method was adopted to summarize the roles of genistein, daidzein in AR activity regulation, further to evaluate the importance of isoflavones for the tumor repression against prostate cancer. Auto Dock applies a half-flexible docking method, which permits small molecular conformation changes. Based on a complex 'lock-and-key model', it is an excellent method to reveal ligand–receptor binding. The result of computer stimulation from Auto Dock contains two parts, one is the binding site of a ligand docked in the receptor and the other is the binding affinity when a ligand is docked in the receptor.

The 3D spatial structure of AR-LBD was obtained from RCSB Protein Data Bank and its PDB ID is 2ama (676-919AA). The positive-control docking result showed that 17β-oestradiol fit the ligand-binding site of AR, at the same position in AR as its natural ligand, DHT. The negative control, thyroxine, showed a quite different binding position with external docking site (Figure 3A). Comparing the three endogenous ligands, thyroxine was expected to have the weakest binding to AR and the highest affinity energy, which we measured at -5.4 kcal mol^{-1}. Very strong binding to AR with lower affinity energies was expected in the two steroid hormones, with affinity energies of -11.2 kcal mol^{-1} for DHT and -10.7 kcal mol^{-1} for 17β-oestradiol (Figure 3B) [46].

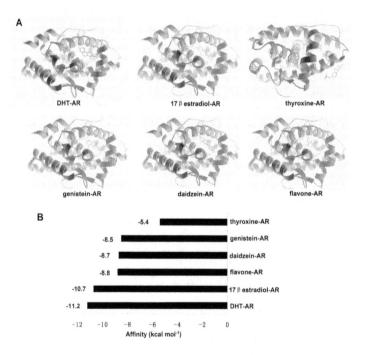

Figure 3. Auto docking analysis of endogenous hormones or phyto-estrogens binding to androgen receptor.(A) Positions of the steroid hormones of DHT or 17β-oestradiol, endogenous non-steroid hormone, thyroxin, phyto-estrogens of genistein, daidzein or flavone binding to androgen receptor.(B): The affinity energies of each ligand binding to androgen receptor. from Asian J Androl. 2010; 12: 535-547

Genistein and daidzein are abundant in soy formula and as healthy ingredient in soybean to protect against prostate cancer. To understand the role of the isoflavone compounds in prostate cancer, a computerized auto-dock system was adopted to examine the interactions between the human AR and phyto-oestrogens (genistein, daidzein, and flavone). As shown in Figure 3 genistein, daidzein and flavone fit in the middle region of the AR-LBD (Figure 3A), the same as DHT and 17β-estradiol. The affinities of them were expected to lie between the affinities of thyroxin and 17β-estradiol. Genistein and daidzein, soy isoflavones, showed affinity energies of -8.5 and -8.7 kcal mol^{-1}, respectively, which were very similar to the affinity energy of flavone of -8.8 kcal mol^{-1}. From that result, we concluded that the three (iso) flavones exhibit similar binding affinities to AR (Figure 3B). Considering their sharing of a binding site with estradiol, their affinities for AR and quantities potentially consumed in the diet, these phyto-estrogens could have significant effects on AR and AR-related cancers [46].

The phyto-estrogens (genistein, daidzein and flavone) can bind to AR in an Auto Dock model and can be regarded as androgenic effectors, suggesting important roles for them in AR-mediated cancers. Interestingly, all these phyto-estrogens are reported to be associated with prostate cancer, so we consider them AR-related phyto-estrogens. We summarized some recent data about the effects of them on AR-mediated transcriptional activity and on prostate tumorigenesis in Table 5. Genistein, daidzein and flavone, implicated as androgenic effectors in our research, indeed regulate AR-mediated PSA transcriptional activity. They have been demonstrated previously to either enhance AR-mediated transcriptional activity or inhibit DHT induced AR-mediated PSA activity. Moreover, isoflavones or soy beverage has already been shown in Phase II trials to decrease PSA levels in prostate cancer patients. It is noteworthy that two recent phase II trials showed that isoflavones or soy beverage can decrease PSA levels in prostate cancer patients, suggesting that androgen receptor target genes can be regulated by isoflavones or flavones [46].

Phynoestrogen	Findings	Study type	Reference
genistein	inhibition of R1881-induced AR mediated pPSA-luc activity	reporter assay	Davis 2002
	decrease AR binding to ARE	EMSA	
	inhibition of R1881-induced AR mediated pPSA-luc activity	reporter assay	Gao 2004
	enhancement of AR mediated pPSA/ARE/Probasin/ MMTV-luc		
daidzein	enhancement of AR (with ARA) mediated MMTV-luc	reporter assay	Chen 2007
flavone	inhibition of DHT-induced AR mediated pPSA-luc activity	reporter assay	Rosenberg 2002
soy food	decline of serum PSA	PhaseII trail	Kwan 2010 Pendleton 2008

Table 5. Effects of photoestrogens on AR mediated transcriptional activityfrom Asian J Androl. 2010; 12: 535-547

The mechanism of these phyto-estrogens against prostate cancer has been studied, some laboratory studies demonstrated that phyto-estrogens might not only mimic estrogenic activities but also interfere with other steroid hormones, for example DHT, the natural androgen in men. In human body, the activity of DHT binding with androgen receptor can be weaken in the presence of a large amount of these phyto-estrogens, genistein, daidzein and flavone.

The findings in Auto Dock interestingly demonstrate that phyto-estrogens are displayed the great binding abilities to AR, demonstrating their disrupt effects against DHT/AR binding as anti-androgenic effectors, supporting the epidemiological studies that soybean were the potential inhibitor for prostate cancer cell growth related to AR. Figure 4 illustrated clearly the mechanism of phyto-estrogens against prostate cancer. The different consumption of soy foods and the different concentrations of isoflavones possibly disrupt the endogenous DHT binding to AR in the prostate and thus inhibited DHT induced AR translocation and AR-mediated PSA transactivation, thus reduce prostate cancer risk. As the antagonists of androgen, phyto-estrogens are ale to inhibit the cell growth induced by androgen in prostate gland.

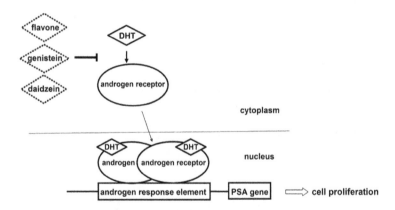

Figure 4. The mechanism of protective effect of genistein, daidzein or flavone against prostate cancer.

The reasons that Asian populations have lower rates of hormone-dependent cancers (breast, prostate) and lower incidences of menopausal symptoms and osteoporosis than Westerners are still need to be further revealed. However, the association of the large quantities of soy products consumption in Asian populations with the reduction in the risk of prostate cancer, provides a unique insight for the beneficial effects of soy foods. Soy food display completely different consumption between eastern and western populations. The geometric mean levels of plasma total isoflavonoids were demonstrated to be 7-10 times higher in Japanese men than in Finnish men. Asian immigrants living in Western nations also have increased risk of these maladies as they 'Westernize' their diets to include more protein and fat and reduce their soy intake. This provides a good explanation for much epidemiological data, indicating the significant protective effect of (iso) flavones for prostate cancer.

The much higher serum phyto-estrogen levels could hypothetically inhibit the growth of prostate cancer in Chinese and Japanese men, which may give a good explanation for the quite low incidence and mortality from prostate cancer in Japan or China. The novel important insights for soy food against prostate cancer need to be further illustrated. Soy-based food products are expected to introduce more to the western markets and to have more consumption in western daily diet.

Acknowledgements

This study was supported by Ministry of Science and Technology (No. 2010DFA31430), NCET-10-0316; the Fundamental Research Funds for the Central University (10JCXK004), and Changchun Science & Technology Department (No.2011114).

Author details

Xiaomeng Li[1*], Ying Xu[1] and Ichiro Tsuji[2]

*Address all correspondence to: lixm441@nenu.edu.cn

1 The Key Laboratory of Molecular Epigenetics of MOE, Institute of Genetics and Cytology, School of Life Sciences, Northeast Normal University, China

2 Department of Public Health, Tohoku University, Japan

References

[1] International Agency for Research on Cancer (IARC). (2008). WHO Mortality Database. *Lyon, France: IARC*.

[2] Li, X., Li, J., Tsuji, I., Nakaya, N., Nishino, Y., & Zhao, X. (2008). Mass screening-based case-control study of diet and prostate cancer in Changchun, China. *Asian J Androl*, 10(4), 551-60.

[3] Xiao, C. W. (2008). Health effects of soy protein and isoflavones in humans. *J Nutr*, 138(6), 1244S-9S.

[4] Messina, M., Nagata, C., & Wu, A. H. (2006). Estimated Asian adult soy protein and isoflavone intakes. *Nutr Cancer*, 55(1), 1-12.

[5] Hebert, J. R., Hurley, T. G., Olendzki, B. C., Teas, J., Hampl, Y., & Ma, J. S. (1998). Nutritional and socioeconomic factors in relation to prostate cancer mortality: a cross-national study. *J Natl Cancer Inst*, 90(21), 1637-47.

[6] Jian, L. (2009). Soy, isoflavones, and prostate cancer. *Mol Nutr Food Res*, 53(2), 217-26.

[7] Kurahashi, N., Iwasaki, M., Sasazuki, S., Otani, T., Inoue, M., & Tsugane, S. (2007). Japan Public Health Center-Based Prospective Study Group. Soy product and isoflavone consumption in relation to prostate cancer in Japanese men, Cancer Epidemiol. *Biomarkers Prev*, 16, 538-45.

[8] Jacobsen, B. K., Knutsen, S. F., & Fraser, G. E. (1998). Does high soy milk intake reduce prostate cancer incidence? The Adventist Health Study (USA). *Cancer Causes Control*, 9, 553-7.

[9] Severson, R. K., Nomura, A. M., Grove, J. S., & Stemmermann, G. N. (1989). A prospective study of demographics, diet, and prostate cancer among men of Japanese ancestry in Hawaii. *Cancer Res*, 49, 1857-60.

[10] Lee, M. M., Gomez, S. L., Chang, J. S., Wey, M., Wang, R. T., & Hsing, A. W. (2003). Soy and isoflavone consumption in relation to prostate cancer risk in China, Cancer Epidemiol. *Biomarkers Prev*, 12, 665-8.

[11] Sonoda, T., Nagata, Y., Mori, M., Miyanaga, N., Takashima, N., Okumura, K., Goto, K., Naito, S., Fujimoto, K., Hirao, Y., Takahashi, A., Tsukamoto, T., Fujioka, T., & Akaza, H. (2004). A case-control study of diet and prostate cancer in Japan: Possible protective effect of traditional Japanese diet. *Cancer Sci*, 95, 238-42.

[12] Kolonel, L. N., Hankin, J. H., Whittemore, A. S., Wu, A. H., Gallagher, R. P., Wilkens, L. R., John, E. M., Howe, G. R., Dreon, D. M., West, D. W., & Paffenbarger, R. S. Jr. (2000). Vegetables, fruits, legumes and prostate cancer: A multiethnic case-control study,. *Cancer Epidemiol. Biomarkers Prev*, 9, 795-804.

[13] Jain, M. G., Hislop, G. T., Howe, G. R., & Ghadirian, P. (1999). Plant foods, antioxidants, and prostate cancer risk: Findings from case-control studies in Canada. *Nutr Cancer*, 34, 173-84.

[14] Villeneuve, P. J., Johnson, K. C., Kreiger, N., & Mao, Y. (1999). Risk factors for prostate cancer: Results from the Canadian National Enhanced Cancer Surveillance System The Canadian Cancer Registries Epidemiology Research Group. *Cancer Causes Control*, 10, 355-67.

[15] Key, T. J., Silcocks, P. B., Davey, G. K., Appleby, P. N., & Bishop, D. T. (1997). A case-control study of diet and prostate cancer,. *Br J Cancer*, 76, 678-87.

[16] Heald, C. L., Ritchie, M. R., Bolton-Smith, C., Morton, M. S., & Alexander, F. E. (2007). Phyto-oestrogens and risk of prostate cancer in Scottish men,. *Br J Nutr*, 98, 388-96.

[17] Zhou, J. R., Li, L., & Pan, W. (2007). Dietary soy and tea combinations for prevention of breast and prostate cancers by targeting metabolic syndrome elements in mice. *Am J Clin Nutr*, 86(3), 882-8.

[18] Wu, Z., Rodgers, R. P., & Marshall, A. G. (2004). Characterization of vegetable oils: Detailed compositional fingerprints derived from electrospray ionization fourier

transform ion cyclotron resonance mass spectrometry. *J Agric Food Chem*, 52(17), 5322-8.

[19] Setchell, K. D., & Cole, S. J. (2003). Variations in isoflavone levels in soy foods and soy protein isolates and issues related to isoflavone databases and food labeling. *J Agric Food Chem*, 51(14), 4146-55.

[20] Food and Drug Administration. (1999). Food labeling: Health claims; soy protein and coronary heart disease. *Fed Regist.*, 64(206), 57700-33.

[21] Nagata, Y., Sonoda, T., Mori, M., Miyanaga, N., Okumura, K., Goto, K., Naito, S., Fujimoto, K., Hirao, Y., Takahashi, A., Tsukamoto, T., & Akaza, H. (2007). Dietary isoflavones may protect against prostate cancer in Japanese men. *J Nutr*, 137(8), 1974-9.

[22] Travis, R. C., Spencer, E. A., Allen, N. E., Appleby, P. N., Roddam, A. W., Overvad, K., Johnsen, N. F., Olsen, A., Kaaks, R., Linseisen, J., Boeing, H., Nöthlings, U., Bueno-de-Mesquita, H. B., Ros, M. M., Sacerdote, C., Palli, D., Tumino, R., Berrino, F., Trichopoulou, A., Dilis, V., Trichopoulos, D., Chirlaque, M. D., Ardanaz, E., Larranaga, N., Gonzalez, C., Suárez, L. R., Sánchez, M. J., Bingham, S., Khaw, K. T., Hallmans, G., Stattin, P., Rinaldi, S., Slimani, N., Jenab, M., Riboli, E., & Key, T. J. (2009). Plasma phyto-oestrogens and prostate cancer in the European Prospective Investigation into Cancer and Nutrition. *Br J Cancer*, 100, 1817-23.

[23] Maggiolini, M., Vivacqua, A., Carpino, A., Bonofiglio, D., Fasanella, G., Salerno, M., Picard, D., & Andó, S. (2002). The mutant androgen receptor T877A mediates the proliferative but not the cytotoxic dose-dependent effects of genistein and quercetin on human LNCaP prostate cancer cells. *Mol Pharmacol*, 62(5), 1027-35.

[24] Gao, S., Liu, G. Z., & Wang, Z. (2004). Modulation of androgen receptor-dependent transcription by resveratrol and genistein in prostate cancer cells. Prostate. May 1; , 59(2), 214-25.

[25] Li, Y., & Sarkar, F. H. (2002). Gene expression profiles of genisteintreated PC3 prostate cancer cells. *J. Nutr*, 132, 3623-3631.

[26] Li, Y., & Sarkar, F. H. (2002). Down-regulation of invasion and angiogenesis-related genes identified by cDNA microarray analysis of PC3 prostate cancer cells treated with genistein,. *Cancer Lett.*, 186, 157-64.

[27] Davis, J. N., Kucuk, O., & Sarkar, F. H. (1999). Genistein inhibits Nfkappa B activation in prostate cancer cells. *Nutr. Cancer.*, 35, 167-74.

[28] Raffoul, J. J., Wang, Y., Kucuk, O., Forman, J. D., Sarkar, F. H., & Hillman, G. G. (2006). Genistein inhibits radiation-induced activation of NF-kappaB in prostate cancer cells promoting apoptosis and G2/M cell cycle arrest,. *BMC Cancer*, 6, 107.

[29] Akiyama, T., Ishida, J., Nakagawa, S., Ogawara, H., Watanabe, S., Itoh, N., Shibuya, M., & Fukami, Y. (1987). Genistein, a specific inhibitor of tyrosine-specific protein kinases. *J Biol Chem.*, 262, 5592-5.

[30] Barnes, S., Peterson, T. G., & Coward, L. (1995). Rationale for the use of genistein-containing soy matrices in chemoprevention trials for breast and prostate cancer,. J. Cell Biochem. Suppl. , 22, 181-7.

[31] Djuric, Z., Chen, G., Doerge, D. R., Heilbrun, L. K., & Kucuk, O. (2001). Effect of soy isoflavone supplementation on markers of oxidative stress in men and women,. *Cancer Lett.*, 172, 1-6.

[32] Craft, C. S., Xu, L., Romero, D., Vary, C. P., & Bergan, R. C. (2008). Genistein induces phenotypic reversion of endoglin deficiency in human prostate cancer cells. *Mol. Pharmacol.*, 73, 235-42.

[33] Xu, L., & Bergan, R. C. (2006). Genistein inhibits matrix metalloproteinase type 2 activation and prostate cancer cell invasion by blocking the transforming growth factor beta-mediated activation of mitogen-activated protein kinase-activated protein kinase 2-27-kDa heat shock protein pathway,. Mol. Pharmacol. , 70, 869-77.

[34] Lakshman, M., Xu, L., Ananthanarayanan, V., Cooper, J., Takimoto, C. H., Helenowski, I., Pelling, J. C., & Bergan, R. C. (2008). Dietary genistein inhibits metastasis of human prostate cancer in mice. *Cancer Res.*, 68, 2024-32.

[35] Pollard, M., & Wolter, W. (2000). Prevention of spontaneous prostaterelated cancer in Lobund-Wistar rats by a soy protein isolate/isoflavone diet. *Prostate.*, 45, 101-5.

[36] Wang, J., Eltoum, I. E., & Lamartiniere, C. A. (2002). Dietary genistein suppresses chemically induced prostate cancer in Lobund-Wistar rats,. *Cancer Lett.*, 186, 11-8.

[37] Wang, J., Eltoum, I. E., & Lamartiniere, C. A. (2007). Genistein chemoprevention of prostate cancer in TRAMP mice. *J Carcinog.*, 6(3).

[38] El Touny, L. H., & Banerjee, P. P. (2009). Identification of a biphasic role for genistein in the regulation of prostate cancer growth and metastasis. Cancer Res. , 69, 3695-703.

[39] Wang, J., Eltoum, I. E., & Lamartiniere, C. A. (2004). Genistein alters growth factor signaling in transgenic prostate model (TRAMP). *Mol Cell Endocrinol.*, 219(1-2), 171-80.

[40] Smith, S., Sepkovic, D., Bradlow, H. L., & Auborn, K. J. (2008). Diindolylmethane and genistein decrease the adverse effects of estrogen in LNCaP and PC-3 prostate cancer cells. *J Nutr*, 138(12), 2379-85.

[41] Seo, Y. J., Kim, B. S., Chun, S. Y., Park, Y. K., Kang, K. S., & Kwon, T. G. (2011). Apoptotic effects of genistein, biochanin-A and apigenin on LNCaP and PC-3 cells by 21 through transcriptional inhibition of polo-like kinase-1. *J Korean Med Sci.*, 26(11), 1489-94.

[42] Singh-Gupta, V., Zhang, H., Yunker, C. K., Ahmad, Z., Zwier, D., Sarkar, F. H., & Hillman, G. G. (2010). Daidzein effect on hormone refractory prostate cancer in vitro and in vivo compared to genistein and soy extract: potentiation of radiotherapy. *Pharm Res.*, 27(6), 1115-27.

[43] Hsu, A., Bray, T. M., Helferich, W. G., Doerge, D. R., & Ho, E. (2010). Differential effects of whole soy extract and soy isoflavones on apoptosis in prostate cancer cells. *Exp Biol Med (Maywood)*, 235(1), 90-7.

[44] Szliszka, E., & Krol, W. (2011). Soy isoflavones augment the effect of TRAIL-mediated apoptotic death in prostate cancer cells. *Oncol Rep.*, 26(3), 533-41.

[45] Rahman, M., Miyamoto, H., & Chang, C. (2004). Androgen receptor coregulators in prostate cancer: mechanisms and clinical implications. *Clin Cancer Res.*, 10, 2208-19.

[46] Wang, H., Li, J., Gao, Y., Xu, Y., Pan, Y., Tsuji, I., Sun, Z., & Li, X. (2010). Xeno-oestrogens and Phyto-oestrogens are Alternative Ligands for the Androgen Receptor. *Asian J Androl.*, 12(4), 535-47.

Chapter 12

Evaluation of Soybean Straw as Litter Material in Poultry Production and Substrate in Composting of Broiler Carcasses

Valeria Maria Nascimento Abreu,

Paulo Giovanni de Abreu,

Doralice Pedroso de Paiva, Arlei Coldebella,

Fátima Regina Ferreira Jaenisch,

Taiana Cestonaro and Virginia Santiago Silva

Additional information is available at the end of the chapter

1. Introduction

The litter has never been a subject of extensive studies or considered a priority in large poultry company. However, due to the increasing lack of good litter material, more attention has been given to proper litter management, litter reuse, and to the search for new litter materials. In this context, the use of crop residues as poultry litter material seems to be promising. There is a current trend in poultry production to use alternative litter materials, that is, other than wood shavings. Despite being demonstrated by several authors that the use of these alternative materials do not interfere with flock live performance, most agree that they are more difficult to manage and may result in a higher incidence of carcass lesions. The effects of several many materials used as poultry litter substrate on poultry performance have been evaluated. It was shown that litter made of rice husks not only does not impair performance, but also reduced foot-pad and breast lesions [1, 2 – 9]. A study on the use of soybean crop residues as poultry litter did not show any influence of this material on broiler performance or on its agronomic value [10]. However, these bedding materials have not been evalatuated as to the evolution of darkling beetle and enteric parasite populations. *Alphitobius diaperinus* (darkling beetle) adults and larvae are considered a problem in intensive broiler and turkey production.

These beetles replicate in the litter, becoming potential vectors of pathogens and parasites both on site and to the neighboring farms. Those insects have been associated to many pathogenic agents, and there are reports they carry *Escherichia coli*, *Salmonella spp*, avian leucosis virus, as well as internal parasites, such as coccidian, avian tapeworms, and helminths [11,12], enterobacteria [13] and *Clostridium perfringens* [14]. Attempts to control darkling beetles have been made by changing the litter pH using hydrated lime [15]) or applying insecticides in the entire poultry house during downtime [16] or as a management complement during rearing, with application in specific spots, as suggested by [17]. The continuous contact of birds with the excreta in the litter poses a risk of infection with parasites, and coccidiosis caused by *Eimeria spp* is one of the most significant diseases in this production system. The evolution of oocyst population may determine the need to change the litter, and an evaluation may aid decision-making regarding this need. Litter reutilization for more than one flock is practiced in several countries. However, aspects relative to the potential health risk posed by these alternative litter materials have been discussed, and their use may limit the international chicken meat trade due to the requirements of showing equivalence of production processes practiced among exporting and importing countries [18]. The study was divided into three parts, where the use of soybean straw as litter in the poultry production was compared with rice husk under two ventilation conditions, as follows.

2. The study

The experiment was carried out at the experimental field of Suruvi, belonging to Embrapa Swine &Poultry, Concórdia, Santa Catarina, Brazil. Four 12m×10m broiler houses were internally divided in four pens each (total of 16 pens), at a density of 200 birds/pen (28kg meat/m^2), totaling 3,200 birds/flock. Four consecutive flocks were followed up. Each flock was reared to 42 days of age, and an interval between flocks (downtime) of 15 days was applied. Two ventilation systems (stationary or oscillating), reaching a distance of 10m, and two litter materials (soybean straw or rice husks) were tested. Rice husks and ventilation system using stationary fans are considered as standards as they are commonly used in broiler production. Fans were activated by a thermostat when the environmental temperature reached 25°C, and were equipped with a potentiometer and speed regulator matching the broiler house size. Treatments were distributed as follows (Figure 1): house 1 – stationary ventilation system, pens 2 and 3 with soybean straw; pens 1 and 4 with rice husks; house 2 – oscillating ventilation system, pens 2 and 3 with soybean straw; pens 1 and 4 with rice husks; house 3 - oscillating ventilation system, pens 1 and 4 with soybean straw; pens 2 and 3 with rice husks; house 4 – stationary ventilation system, pens 1 and 4 with soybean straw; pens 2 and 3 with rice husks. Birds and feeds were weekly weighed and the following parameters were evaluated: body weight, weight gain, feed intake and feed conversion ratio when birds were 21, 35 and 42 days of age. Performance data were analyzed according to the theory of mixed models for repeated measures, considering the effects of flock, ventilation system, litter material, bird age, and the interaction of these parameters up to third order, and 16 types of variance and covariance matrices, using PROC MIXED procedure of SAS statistical package [19], according to [20].

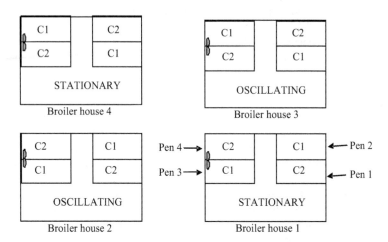

Figure 1. Fan (stationary and oscillating) and litter material (C1 – soybean straw and C2 – rice husks) distribution in the experimental facilities

The variance and covariance structure used for analysis was chosen based on the lowest value of the Akaike Information Criterion (AIC). The estimation method was that of restricted maximum likelihood. Mortality was daily recorded and assigned to ascitis, sudden death, or other causes. Total mortality was also evaluated. Breast and foot-pad lesions were evaluated by gross examination the last time birds were weighed, and scored as present or absent. Because mortality and presence of foot-pad lesion data have binomial distribution, these data were analyzed by logistic regression, using the LOGISTIC procedure of SAS statistical package [19], and considering the effects of flock, litter material, ventilation system, and their interactions. The overdispersion of the presence of foot-pad lesion was adjusted by the dispersion parameter estimated by Pearson's χ^2 statistics divided by degrees of freedom. Litter samples were collected at each flock, when chicks were housed (day 0) and when they were removed (day 42), and submitted to enterobacteria quantitative exam. Litter samples weighing 10g were diluted at 1:10 in PBS (phosphate buffer saline solution) and serially diluted to 10^{-7}. Aliquots of 100 μL of 10^{-3} to 10^{-7} dilutions were seeded in Mac Conkey agar and incubated at 37ºC for 48h, and submitted to colony-forming unit (CFU) counting in plates containing 30 to 300 colony-forming units. CFU data analysis considered the effects of flock, litter material, ventilation system, and their interactions and used the theory of mixed models for repeated measures and nine variance and covariance matrix structures applied to the PROC MIXED procedure of SAS statistical package [19]. The following psychrometric parameters were recorded at the center of each pen and in the external environment: dry- and wet-bulb temperature, black globe temperature, and air velocity. Temperatures were collected using copper-constantan thermocouples connected to a potentiometer and a 20-channel selecting key. The wet bulb thermometer was characterized by securing a cotton wick attached to the thermocouple terminal to the mercury bulb and immersing it in a flask with distilled water. Black globe temperature was collected by placing thermocouples inside a hollow 15cm-diameter PVC sphere painted with black mat spray paint. air velocity was recorded using a digital anemometer (Pacer® Model DA40V), with a resolution of 0.01m/s. Data were collected every three hours, from 08:00 to 18:00h, when broilers were 4, 5, and 6 weeks. Based on the data collected at each time, wet bulb globe temperature (WBGT) and radiant heat load and a radiant heat load (RHL). Litter temperature was also measured using an infrared thermometer Raytec® in five different spots in each pen (two near the lateral pen wall, two near the central aisle, and at the geometrical center of the pen) every three hours, from 08:00 to 18:00 when broilers were 6 weeks of age. Based on the average data in each spot, isothermal maps of litter temperature using the kriging method of the SURFER software program were built. Internal thermal environmental parameters were evaluated as to the effects of flock, ventilation, litter, week, hour, and the interactions among the last four factors using mixed models for repeated measures and 19 variance and covariance matrix structures, applying PROC MIXED procedure of SAS statistical package [19], according to [20]. The structure used for analysis was chosen based on the lowest value of the Akaike Information Criterion (AIC). The estimation method was that of restricted maximum likelihood. The effect of hour was detailed using orthogonal polynomial analysis up to the polynomial of the third degree. For the external environment, parameter means were calculated as a function of hour and week in order to compare the internal with the external thermal environment. The soybean straw was chopped into approximately 3cm long particles. Litter was initially 10cm high and was reused for four consecutive flocks. Litter quality was evaluated in terms of physical-chemical composition, moisture, and compactness. The development of the population of darkling beetles and the evolution of the number of parasite eggs/oocysts was also observed. In order to evaluate darkling beetles population, three traps per pen made with 20 × 5cm PVC tubes filled with 20 × 50cm rolled-up corrugated cardboard, totaling 48 traps [21]. Traps were placed under the litter in three sites in each pen: at the center, between the external wall and the first line of feeders, and the other two between the feeders, equally distant from the drinkers. Traps remained in the pens for seven days and were removed after catching of each flock. Traps were then placed in 1L thick plastic bags, closed with thin plastic-coated wire, and submitted to the laboratory. *Alphitobius diaperinus* adults and larvae were identified and counted, and the total number of individuals was recorded. Litter samples were collected to count number of endoparasite eggs/oocysts per gram of litter (epg), and also submitted to physical-chemical analysis. Two samples were collected per flock: on the day chicks were housed and after catching. In each pen, 15 litter samples, weighing 50g each, were collected on the surface and under the surface of the litter centrally to the feeders, drinkers, and external and internal limits of the pens. In the laboratory, after homogenization, 50g aliquots were used for endoparasite counting. The remaining sample was submitted to the physical-chemical analyses lab to determined dry matter, ashes, and phosphorus contents. Phosphorus was colorimetrically determined by the molybdovanadate method [22]. Pre-dry matter or moisture was determined at 65ºC. Copper, zinc, calcium, manganese, and iron contents were determined by flame atomic absorption spectrometry after nitro-perchloric digestion [23]. Nitrogen was determined by the Kjeldahl method. Litter pH was measured according to the method described by [24]. Organic carbon was titered after chemical oxidation with sulfochromic solution [25]. Potassium was determined by flame photometry [24]. Litter temperature was recorded using an infrared thermometer (Raytec®) in five different spots in each pen (two near the lateral pen wall, two near the central aisle, and at the geometrical center of the pen). Darkling beetle counts were transformed into log (y+1) and submitted to analysis of variance using a model that considered the effects of litter material, ventilation, flock, and the interaction among these factors. The GLM procedures of SAS statistical package were used [19]. Parasite egg and oocyst counts were used to characterize the presence of absence of parasite in the pens. Parasite presence was analyzed by logistic regression considering the effects of litter material, ventilation and flock, using LOGISTIC procedures of SAS statistical package [19]. Litter quality data were analyzed by mixed models for repeated measures, considering the effects of litter material, ventilation, and the interaction between litter material and ventilation (plot), sample sampling and respective interactions (subplot, and period (subplot), and three matrix structures of variance and covariance, using the PROC MIXED procedure of SAS statistical package [19], according to [20]. The structure used for analysis was chosen based on the lowest value of the Akaike Information Criterion (AIC). The estimation method was that of restricted maximum likelihood.

2.1. Evaluation poultry production using different ventilation systems and litter material: I – general performance

According to statistical analysis results, flock, litter, and age significantly ($p<0.05$) influenced all evaluated parameters, whereas ventilation system had no effect on any variable. Two interactions significantly affected almost all parameters: litter x age and flock x ventilation x age. The details of the litter x age interactions indicated better results for rice husks litter as compared to soybean straw litter in all studied parameters at all ages (Table 1). These results are similar to those described by Mizubuti et al. (1994), who evaluated rice husks, guinea grass, and napier grass as litter material. On the other hand, opposite results were obtained by [27, 5, 27, 28, 9, 8, 10], who tested different broiler litter materials and did not find any differences in body weight or feed intake. Moreover, [29] studied five broiler litter materials and observed significant reduction of body weight, feed intake, and antibody titers in broilers reared on rice husks litter. [30] compared the use of rice husks, coconut hulls, and wood shavings as broiler litter material and low- and high-density diets, and concluded that weight gain and feed intake during the total experimental period were higher in broilers reared on coconut hulls litter. The evaluation of the details of the triple interaction among flock, ventilation and age indicated that there was no consistent effect of ventilation system on performance parameters. The analysis of the presence of foot-pad lesion showed a significant influence ($p<0.0001$) of the interaction between flock and litter and of the main effects litter and flock. Independently from the interaction, soybean straw litter caused higher incidence of foot-pad lesions as compared to rice husks, in all flocks (Figure 2).

Litter material	Bird age (days)		
	21	35	42
	Feed intake (g)		
Rice husks	1196 ± 3.90a	3373 ± 8.88a	4764 ± 13.15a
Soybean straw	1179 ± 3.90b	3310 ± 8.88b	4692 ± 13.15b
	Feed conversion ratio		
Rice husks	1.278 ± 0.004a	1.529 ± 0.003a	1.666 ± 0.004a
Soybean straw	1.306 ± 0.004b	1.543 ± 0.003b	1.683 ± 0.004b
	Weight gain (g)		
Rice husks	894 ± 3.75a	2164 ± 6.05a	2819 ± 9.44a
Soybean straw	861 ± 3.75b	2104 ± 6.05b	2749 ± 9.44b
	Body weight (g)		
Rice husks	937 ± 3.76a	2207 ± 6.05a	2862 ± 9.45a
Soybean straw	903 ± 3.76b	2146 ± 6.05b	2791 ± 9.45b

Means followed by different letters in the same columns are different ($p < 0.05$) by the F test.

Table 1. Feed intake, feed conversion ratio, weight gain and body weight of broiler reared on two different litter materials and two different ventilation systems

These results are opposite to those found in other studies evaluating litter materials that concluded that litter material has not effect on carcass lesions [31]. The obtained results indicate that litter material causes foot-pad lesions, resulting in carcass condemnation in the processing plant and consequent economic losses. [1] also found higher incidence of foot-pad lesions when napier grass and coast-cross grass hays were used as litter material. However, [7] used sunflower crop residue and Brachiaria hay as litter, but did not find any effects on breast, hock, and foot-pad lesions. In the present study, it was also observed that the percentage of foot-pad lesions markedly increased from the first to the second flock, decreased in the third flock, and that this reduction was maintained in the fourth flock. We believe that, as the number of flocks increased, caked litter was removed and the remaining litter was turned, making litter softer, and thereby, reducing leg lesions. Mortality was influenced by flock ($p < 0.0001$) and by the interaction between flock and litter ($p < 0.05$). However, mortality was only significantly affected by litter material in the third flock in favor of soybean straw. Evaluating the use of wood shavings, rice husks, sugarcane residue and carnauba palm residue as broiler litter material, [32] did not find any differences in mortality, although in absolute numbers, values were higher in broilers reared on wood shavings and sugarcane residue. [5] and [29] reported similar results, that is, no mortality differences in studies on alternative broiler litter materials. According to [33], bacteria present in the litter may have different effects. Many Gram-positive bacteria, such as lactobacilli and bifidobacteria, are present in broiler excreta and in the litter, but are not necessarily related to problems. On the other hand, the frequent presence of pathogens in the litter, particularly of enterobacteria and zoonotic bacteria in general, is a reason of concern due to possible diseases transmitted both to the broiler flock itself and to consumers.

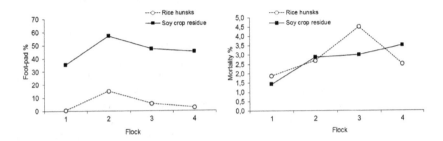

Figure 2. Presence of foot-pad lesions and mortality rate of broilers reared on two different litter materials for four consecutive flocks

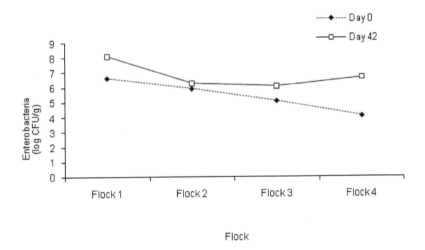

Figure 3. Enterobacteria (log CFU/g) in rice husks litter reused for four broiler flocks (flock housing and removal – Days 0 and 42)

Moreover, according to [33], the composition of the bacterial population in the litter is usually very similar to the composition of the physiological microbiota of the ileum of broilers, and consists approximately of 70% lactobacilli, 11% *Clostridium* spp., 6.5% *Streptococcus* spp., and 6.5% *Enterococcus* spp. The litter presents, in average, 10 times less bacteria than the digesta, but this is still a high concentration of microorganisms. The digest contains between 10^8 and 10^{10} of Gram-positive bacteria and 10^6 to 10^7 Gram-negative bacteria per gram. As the concentration of bacteria in the litter can increase 10 times per reared flock, it may achieve the same levels as the digesta. From the practical perspective, it may be assumed that bacterial concentration in the litter of broilers is similar to that in feces. In the present study, total enterobacteria counts (expressed as colony-forming units, CFU) per gram of litter for each treatment were carried out when each of the four flocks was removed (day 42) and when the following flocks were housed (day 0), aiming at evaluating the effect of downtime on litter bacterial load. Significant effects ($p<0.05$) of flock, evaluation day (0 or 42), and their interaction was observed, whereas litter material (rice husks or soybean straw) and ventilation system had no influence on litter enterobacterial load (Figures 3 and 4). The evaluation carried out on day 42 always presented better results as compared to that performed on day 0, as expected. However, the difference increased as the number of flocks increased (Figure 5), as shown by the lowest UFC count on day 42 (immediately after flock removal) in the third flock, whereas on day 0 (day of housing, after downtime), bacterial load the lowest only in the fourth flock. This suggests that the reduction of the load of enterobacteria after downtime (day 0), may continue to occur in subsequente flocks, as the reduction was linear and did not achieve a stabilization point.

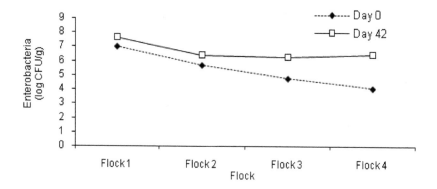

Figure 4. Enterobacteria (log CFU/g) in soybean straw litter reused for four broiler flocks (flock housing and removal – Days 0 and 42)

It must be mentioned that, during downtime, flame-gun was used twice on the litter: immediately after bird removal and before the following flock was housed. However, during the 15-d downtime, litter was not submitted to any management practice for the reduction or control of undesirable bacteria, including potentially pathogenic bacteria. Despite the reduction in the bacterial load along the four flocks, enterobacteria counts were high in all flocks, and therefore some method of litter treatment during downtime is recommended to reduce pathogen load. The initial enterobacteria counts in the litters used in the experiment before the first flocks were housed were considered high. When evaluating litter treatment methods for bacterial load reduction, [18] observed that average loads of enterobacteria and total mesophilles were high in new litters. According to those authors, these results call the attention to the quality of new litter, as high bacterial loads in that material are associated to their origin, possibly due to production, preservation, storage, and transport conditions to the broiler house. In addition, these high bacterial loads are a hazard to birds that will be housed in that environment, particularly considering that these will be day-old chicks. [34] determined bacterial levels in pine shavings and sand used as broiler litter, and found a marked increase in bacterial counts after birds were housed. Pine shavings reached a level of 10_8 CFU/g of aerobic bacteria in the second week, and this level remained stable up to six weeks, after which ir increased approximately 1 log the next week, and remained at this level until birds were removed at the seventh week. Enteric bacteria in pine shavings reach a plateau at 10_7 to 10_8 CFU/g in the second week, and these values showed little variation until the seventh week. In the present study, enterobacteria counts were higher, in general, than those obtained in untreated wood shavings litter by [18], which were about 10^5 CFU/g. Therefore, the need to treat the litter during downtime is stressed, independently of the material used.

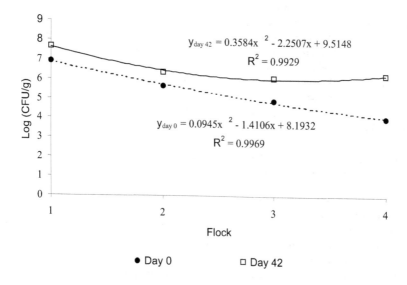

Figure 5. Enterobacteria load on days 0 and 42 of the evaluated flocks, considering all treatments

2.2. Poultry production using different ventilation systems and litter materials: II – thermal comfort

The main effects of flock, week and hour were significant (p<0.05) for all evaluated parameters. Litter material significantly influenced (p<0.05) only air relative humidity, whereas ventilation system did not affect any of the parameters. The hour × week interaction significantly affected all parameters, whereas the effect of the ventilation × hour interaction was significant only for air temperature and WBGT. Heat load inside the broiler house did not show much variation, and the results were more favorable to the birds as compared to the external environment. When rice husks litter was used, higher air relative humidity values were detected (Figure 6). According to [35] optimal air temperature values are between 23°C and 26°C; 20°C and 23°C; and 20°C for 4-, 5-, and 5-week-old broilers, respectively. On the other hand, the recommended air relative humidity is 60 to 70%, regardless bird age. In the present study, air relative humidity remained higher than the recommendations for broiler production in all studied weeks, times, and litter materials (Figure 6). Controlling air and litter humidity is important to reduce pathogens, ammonia, and parasite, such as coccidia, in the poultry house environment [36]. [37] did not find significant differences in breast blisters as a function of bird density (10 birds/m² or 14 birds/m²), litter material (wood shavings or saw dust) or by their interaction, as this part of the broiler body is not in permanent contact with the litter, and therefore is no influenced by litter humidity. Litter temperature presented the same behavior as air temperature, as expected, with a linear correlation of 0.95. At all evaluated times, litter temperature was, in average, 2°C to 3°C higher than air temperature. It must be noted that litter temperature was not affected by the used material, that is, independently of using soybean straw or rice husks, litter temperature remained similar (Figure 6). According to [38], temperature measured 5cm below litter surface was 23.5°C and 31.3°C at housing densities of 19 and 40 kg/m², respectively, whereas air temperature 1m above litter surface was 22°C, indicating that litter temperatures were 1.5°C and 9.3°C higher than the environmental temperature. The higher litter temperature at higher bird densities may be explained by different effects. As bird density increases, nitrogen and humidity levels increase in the litter, allowing higher microbial activity and the heat transference from the litter surface to the environmental air is prevented when the litter surface area is covered by birds, particularly as birds reach market weight.

2.3. Poultry production using different ventilation systems and litter materials: III – effect of litter reuse on the populations of *Alphitobius diaperinus* and intestinal parasites

Considering the composition of rice husks and soybean straws before its utilization and the values established in [39] for simple organic fertilizers (Table 2), both materials are very different, except for organic carbon content, and do not comply with the recommendations of IN-23 for organic fertilizers. Rice husks humidity and pH values are different from those described by [40], who evaluated the reutilization of rice husks as litter material by 18 broiler flocks and reported 9.4% humidity, pH 7.0, 0.47% nitrogen, 0.03% phosphorus, and 0.27% potassium in the beginning of the experiment, that is, composition before utilization. According to [10], considering that plant nutrient requirements vary as a function of cultivar, soil, expected yield, etc., and exceeding supplied quantities remain in the soil and are susceptible to leaching and percolation, it is essential to balance soil and litter compositions. Based on these considerations, the knowledge on the quality (physical-chemical composition, compactness, and reutilization) of litter materials used as alternative to wood shavings is essential as the disposal of such material is part of good production practices. Litter quality was evaluated according to chemical element levels and physical parameters. Litter chemical composition parameters were significantly affected (p<0.05) by litter material, except for humidity and zinc. Sampling affected only ashes and potassium content, whereas period significantly influenced (p<0.05) all parameters. The interaction litter material x period x sampling affected organic carbon, copper, iron, potassium, pH and zinc. In order to study litter chemical composition, samplings were performed in three different periods, as follows: period 1 – 15-day interval (downtime) between removal of flock 1 and housing of flock 2; period 2 - 15-day interval (downtime) between removal of flock 2 and housing of flock 3; and period 3 - 15-day interval (downtime) between removal of flock 3 and housing of flock 4.

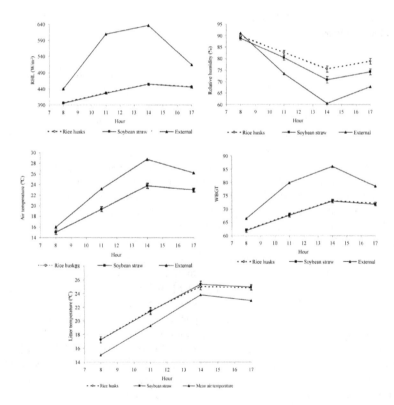

Figure 6. Air temperature, wet bulb globe temperature (WBGT), radiant heat load (RHL), air relative humidity and litter temperature inside broiler houses with two different litter materials

	Rice husks	Soybean straw	IN-23
Ashes (%)	15.18	4.33	
Organic carbon (%)	35.13	33.33	≥ 20
Copper (%)	traces	0.00065	
Iron (mg/kg)	433.25	607.63	
Potassium (%)	0.075	0.108	
Manganese (mg/kg)	169.85	32.75	
Humidity (%)	10.35	14.33	≤ 30
Nitrogen (%)	0.344	0.662	≥ 1
Phosphorus (%)	0.056	0.092	
pH	6.56	7.60	a.d.*
Zinc (mg/kg)	11.43	14.70	

* c.d. – as declared.

Table 2. Chemical composition of rice husks and soybean straw before their utilization as litter and values established in Normative Instruction n. 23

The chemical composition of rice husks and soybean straw after utilization (Table 3) indicates a differentiation pattern of chemical element values. In general, values were high after the removal of flock, and decreased after 15 days of downtime in all studied periods. An exception was pH, which was higher at the end of downtime. Humidity values at flock removal were high, exceeding the recommendations both for poultry rearing and for organic fertilizer; however, at the end of downtime, values returned to acceptable levels. [40] observed that the moisture of rice husks litter varied between 23.4 and 29.1%, averaging 26.4%. Those authors found that litter pH increased from 7.05 to 8.59 after the first flock, and significantly increased between the first and the second flock, after which only minor changes were observed, with an average litter pH of 8.80 after 18 flocks. Similar pH values were obtained by [41] for rice husks litter reused by four flocks (pH 8.75); by [42], with pH values of 8.4-8.5; and by [10], with an average pH of 8.79. For soybean straw, [10] obtained an average pH of 8.97. Ammonia and nitrates are the most common chemical forms of nitrogen found in poultry waste. Nitrates can significantly contaminate underground waters when excessive levels of broiler litter as used as crop fertilizer [43]. Moreover, according to that author, phosphorus is found in large amounts in poultry excreta, and its excessive application for crop fertilization may exceed soil and plant capacity to use that nutrient, resulting in leaching and subsequent contamination of underground waters. [10] studied six different litter materials reused for six consecutive flocks and found nitrogen values of 2.46 and 2.63 and 0.84 and 1.00 of phosphorus in rice husks and soybean straw, respectively. [40] showed that nitrogen, phosphorus, and potassium contents significantly increased in the first seven to eight flocks. Nitrogen, phosphorus, and potassium values obtained in the litter of the fourth flock of the study of [40] were 3.56, 1.59, and 3.12%, respectively, and are used for the comparison with those obtained in the present study.

Period	Rice husks		Soybean straw	
	Removal	Housing	Removal	Housing
	Ashes (%DM)			
1	20.86 ± 0.78	-	15.71 ± 0.35	-
2	20.84 ± 0.17 aA	21.13 ± 0.23 aA	17.29 ± 0.32 bA	17.04 ± 0.35 bA
3	21.40 ± 0.20 aA	21.77 ± 0.12 aA	18.95 ± 0.29 bA	18.39 ± 0.30 bA
	Organic carbon (%DM)			
1	32.93 ± 0.32 bB	36.43 ± 0.24 bA	36.08 ± 0.43 aB	38.69 ± 0.33 aA
2	32.72 ± 0.45 bA	34.95 ± 1.53 aA	35.92 ± 0.28 aA	32.02 ± 0.64 bB
3	32.14 ± 1.41	-	28.45 ± 0.66	-
	Copper (%DM)			
1	45.19 ± 0.84 bA	39.33 ± 1.33 bB	63.31 ± 1.26 aA	49.44 ± 1.64 aB
2	56.05 ± 1.09 bA	52.99 ± 1.65 bB	67.55 ± 0.91 aA	62.72 ± 1.20 aB
3	63.27 ± 1.34 bA	59.13 ± 0.98 bB	74.75 ± 1.22 aA	73.03 ± 1.43 aA
	Iron (mg/kg – DM)			
1	911 ± 26 bA	608 ± 29 bB	1492 ± 57 aA	1469±117 aA
2	1014 ± 28 bA	738 ± 16 bB	1452 ± 44 aA	1361± 59 aA
3	1035 ± 21 bA	507 ± 18 bB	1364 ± 49 aA	736± 44 aB
	Potassium (%DM)			
1	2.04 ± 0.05 bA	1.97 ± 0.05 bA	2.78 ± 0.05 aA	2.73 ± 0.05 aA
2	2.35 ± 0.04 bA	2.00 ± 0.06 bB	2.81 ± 0.04 aA	2.78 ± 0.04 aA
3	2.16 ± 0.05 bB	2.57 ± 0.08 aA	2.69 ± 0.07 aA	2.62 ± 0.03 aA
	Manganese (mg/kg – DM)			
1	351 ± 8 aA	356 ± 7 aA	311 ± 6 bA	252 ± 12 bB
2	427 ± 9 aB	448 ± 9 aA	386 ± 6 bA	375 ± 12 bA
3	420 ± 8 aA	421 ± 7 aA	404 ± 7 aA	383 ± 8 bB
	Nitrogen (%DM)			
1	2.31 ±0.09 bA	1.77 ± 0.05 bB	2.92 ± 0.02 aA	2.30 ± 0.06 aB
2	2.67 ± 0.04 bA	2.32 ± 0.02 bB	3.26 ± 0.04 aA	2.83 ± 0.03 aB
3	2.90 ± 0.07 bA	2.44 ± 0.03 bB	3.32 ± 0.04 aA	2.88 ± 0.05 aB
	pH			
1	8.59 ± 0.08 aB	9.20 ± 0.01 bA	8.39 ± 0.09 aB	9.35± 0.02 aA
2	8.39 ± 0.04 bA	8.98 ± 0.02 bA	8.68 ± 0.05 aB	9.11± 0.03 aA
3	8.54 ± 0.06 aB	8.86 ± 0.01 bA	8.71± 0.04 aB	8.97± 0.02 aA
	Phosphorus (%DM)			
1	1.38 ± 0.07 aA	0.88 ± 0.05 aB	1.54 ± 0.07 aA	1.06 ± 0.07 aB
2	1.65 ± 0.07 bA	1.43 ± 0.07 bB	1.91 ± 0.10 aA	1.66 ± 0.04 aB
3	1.43 ± 0.05 aA	1.35 ± 0.06 bA	1.46 ± 0.09 aA	1.54 ± 0.08 aA
	Humidity (%)			
1	42.71 ± 3.22 aA	25.56 ± 1.31 a B	44.09 ±1.01 aA	27.38 ± 0.94 aB
2	31.12 ± 1.01 aA	16.51 ± 0.50 b B	33.85 ± 0.98 aA	19.94 ± 0.49 aB
3	33.44 ± 0.79 aA	16.55 ± 0.21 b B	35.35 ± 0.93 aA	18.62 ± 0.22 aB
	Zinc (mg/kg DM)			
1	227 ± 19 bA	145 ± 15 aB	286 ± 12 aA	80.57 ±9.87 bB
2	266 ± 5 aA	254 ± 21 aA	285 ± 6 aA	254 ± 14 aA
3	266 ± 5 bA	141 ± 2 bB	299 ± 7 aA	158 ± 5 aB

Means followed by different small letters in the same row are different, within sampling, (p<0.05) by the F test.

Table 3. Chemical composition of broiler litter made of rice husks or soybean straw

There is little information on copper (Cu), iron (Fe), zinc (Zn), and manganese (Mn) levels in different poultry litter materials. According to [43], poultry feeds are rich in iron, and high levels of this mineral are commonly found in broiler litter. Excessive levels of those mineral in the soil affect plant root growth. That author presented a table with findings of several authors relative to the levels of those trace minerals in poultry litter, and those values (in mg/kg, dry weight) are used to compare with the results of the present study. That table presents average concentrations and ranges of 77 and 58-100 for copper, 1,625 and 1,026-2,288 for iron, 348 and 125-667 for manganese, and finally 315 and 106-669 for zinc. It must also be mentioned that the chemical characteristics of the litter materials after three flocks comply, in terms of their nutritional aspects, with the legislation relative to simple or-ganic fertilizer. However, it is recommended that the litter removed from the poultry house is distributed in rows for an additional composting period in order to eliminate or reduce health risks. Despite the effect of litter material on the evaluated parameters, with higher averages promoted by soybean straw, litter made of these crop residues were similar and presented similar characteristics to those described in literature for wood shavings [44], showing that after three flocks, both materials presented excellent fertilizing characteristics. Relative to the physical aspects, when the first flock was removed, soybean straw litter was more compacted and caked as compared to rice husks litter, and this condition remained for the following three flocks, requiring labor interference to break the caked parts, including during the rearing period. At flock removal, soybean straw required more labor to allow its reuse due to the formation of a caked layer on the top of the litter. When the fourth and last flock was removed, the litter made with soybean straw had reached its maximum limit of reutilization, with decomposition of the lower layer, presenting fiber breakdown and humic matter formation. Rice husks litter, on the other hand, presented reuse conditions after the removal of the fourth flock. [45] recommended monitoring insect populations in poultry housed as a routine procedure of the management program, independently from the strat-egy used for insect control. However, according to [46], it is difficult to carry out darkling beetle population studies because their population is usually very high in poultry houses, and they have cryptic behavior. In the present study, one of the objectives was to know which litter material was more favorable to the dissemination of darkling beetles, as well as the evolution of this population as the litter was reused by several consecutive flocks. Dar-kling beetle count was significantly influenced ($p<0.05$) by litter material, flock, and the in-teractions between litter material and ventilation and between litter material and flock. Rice husks presented lower darkling beetle count ($p<0.05$) as compared to soybean straw, with oscillating ventilation in all flocks, and after the second flock, with stationary ventilation (Figure 7). Darkling beetle count increased from the first to the third flock, and decreased in the fourth flock, independently of litter material or ventilation. [47], studying darkling beetle distribution and population dynamics, observed that in broiler houses with cement floor covered with wood shavings litter reused for four flocks and equipped with automatic feed-ers the average numbers of insects trapped in the first flock were 385.4 larvae and 24.5 adults, and these figures increased to 615.3 larvae and 208.7 adults in the next flock. Those authors found that the population tended to become stable in the third flock (651.3 larvae and 248 adults), and found an apparent reduction to 422 larvae and 160.2 adults per trap in the fourth flock. They also found that average litter temperature in flock 1 was 30.7°C, which favors the multiplication of saprophyte microorganisms. Air and litter temperatures directly influence the population of darkling beetles [48]: temperatures of 22 and 31°C determine incubation periods of 8.9 and 3 days, respectively. Below 17°C, eggs do not hatch. The larval stage can take 70.1 days when environmental temperature is 22°C or 33.2 dias at 28°C, whereas the pupa stage takes 4 days at 31°C and 9.7 days at 22°C. Therefore, according those authors, the complete life cycle of the darkling beetle at a constant temperature of 28°C is 42.5 days, and considering that a new flock is housed every 50 days, a new generation of darkling beetles may occur at each flock housed in the farm.

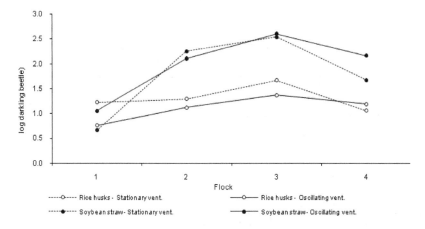

Figure 7. Average darkling beetle counts, transformed into log(y+1), in the litter of broilers reared on two different litter materials and under two ventilation systems.

In the present study, average air and litter temperatures were mild (Table 5), but nevertheless allowed the multiplication of darkling beetles. When the presence of parasites was investigated, only *Eimeria sp.* occysts were identified, and significant effects (p<0.05) of litter material and of the interaction between litter material and ventilation on the presence of oocystes were determined. Rice husks submitted to oscillating ventilation presented the higher percentage of contaminated pens and 18.78 more chances of being contaminated as compared to soybean straw submitted to the same ventilation (Table 4). When ventilation was stationary, no differences between litter materials were observed. When comparing ventilation types, it was found that rice husks litter contamination was significantly higher when ventilation was oscillating (odds ratio = 7.22), whereas there was no significant influence of (p>0.05) ventilation type on the contamination of soybean straw by *Eimeiria spp*. Other factors may have influenced litter contamination, such as the higher nutrient levels – particularly nitrogen levels – in soybean straw, explaining the lower oocyst count due to the negative effects of the release of ammonia levels that are lethal to oocysts [49]. As optimal *Eimeria* spp oocyst sporulation, which makes it infective, occurs at temperatures of 28 to

30°C [49], data were tested to verify if oocyst counts could have changed with litter temperature variations (Table 5), but the results showed that litter temperature did not have any significant influence on oocyst counts.

Litter material	Ventilation		Odds ratio	P^x/>χ2
	Oscillating	Stationary		
Rice husks	81.3%	37.5%	7.22	0.0163
Soybean straw	18.8%	31.3%	0.50	0.4182
Odds ratio	18.78	1.32		
P''/>χ²	0.0012	0.7100		

Table 4. Percentage of pens contaminated with *Eimeiria spp* and odds ratio (probability) of being contaminated

Hour	Week			
	4	5	6	Média
Air temperature (°C)				
8:00	16.15 ± 0.45	13.03 ± 0.47	15.93 ± 0.59	15.04 ± 0.30
11:00	19.86 ± 0.48	17.08 ± 0.51	21.10 ± 0.63	19.34 ± 0.33
14:00	24.30 ± 0.49	21.14 ± 0.52	25.93 ± 0.65	23.79 ± 0.33
17:00	23.23 ± 0.39	20.64 ± 0.41	24.97 ± 0.52	22.95 ± 0.27
Litter temperature (°C)				
8:00	17.12 ± 0.36	14.86 ± 0.50	19.77 ± 0.51	17.25 ± 0.27
11:00	20.87 ± 0.38	19.39 ± 0.53	24.12 ± 0.54	21.46 ± 0.28
14:00	24.35 ± 0.37	23.48 ± 0.51	27.71 ± 0.52	25.18 ± 0.27
17:00	24.20 ± 0.27	22.84 ± 0.38	27.53 ± 0.38	24.86 ± 0.20

Means followed by different letters in the same column are different ($p \leq 0.05$) by the F test.

Table 5. Means and standard error of the parameters air temperature (°C) and litter temperature (°C) as a function of
measurement week and time

According to [43], poultry production generates nutrient-rich residues that can be utilized to
generate energy or to fertilize crops; however, their application to the soil must follow nutrient management plans in order to prevent environmental impacts. Considering the composition of rice husks and soybean straws before its utilization and the values established in
Normative Instruction n. 23 for simple organic fertilizers, both materials are very different,
except for organic carbon content, and do not comply with the recommendations of IN-23
for organic fertilizers. According to [10], considering that plant nutrient requirements vary

as a function of cultivar, soil, expected yield, etc., and exceeding supplied quantities remain in the soil and are susceptible to leaching and percolation, it is essential to balance soil and litter compositions. Based on these considerations, the knowledge on the quality (physical-chemical composition, compactness, and reutilization) of litter materials used as alternative to wood shavings is essential as the disposal of such material is part of good production practices. It must also be mentioned that the chemical characteristics of the litter materials after three flocks comply, in terms of their nutritional aspects, with the legislation relative to simple organic fertilizer. However, it is recommended that the litter removed from the poultry house is distributed in rows for an additional composting period in order to eliminate or reduce health risks. Despite the effect of litter material on the evaluated parameters, with higher averages promoted by soybean straw, litter made of these crop residues were similar and presented similar characteristics to those described in literature for wood shavings [44], showing that after three flocks, both materials presented excellent fertilizing characteristics. Relative to the physical aspects, when the first flock was removed, soybean straw litter was more compacted and caked as compared to rice husks litter, and this condition remained for the following three flocks, requiring labor interference to break the caked parts, including during the rearing period. At flock removal, soybean straw required more labor to allow its reuse due to the formation of a caked layer on the top of the litter. When the fourth and last flock was removed, the litter made with soybean straw had reached its maximum limit of reutilization, with decomposition of the lower layer, presenting fiber breakdown and humic matter formation. Rice husks litter, on the other hand, presented reuse conditions after the removal of the fourth flock.

At the same time also the same comparison was made using the two materials as substrate for the composting of broiler carcasses.

2.4. Rice husks and soy straw as substrate for composting of broiler carcasses.

Considering the possibility of using rice husk and soybean straw as litter, the present study aimed to evaluate these products as substrates for broiler carcasses composting. The appropriate destination of waste from poultry production is a challenge for producers. Carcasses of broilers dead during the rearing period need to be managed so as not to cause problems like unpleasant odors and attraction of flies. One alternative for destination of carcasses considered economically and environmentally acceptable is the composting [50], a natural process of organic matter decay performed by bacteria and fungi that turn the carcasses into a useful product, the compost.

The experiment was conducted in the Experimental Field of Suruvi, of Embrapa Swine and Poultry, in Concórdia (Santa Catarina State) using a composter with six cameras, with internal measures: 0.80 m wide, 1.20 m depth and 1.50 m height of the wall. Cameras were constructed with concrete floor, wooden walls, and asbestos tiles. Two types of substrate were tested for composting: soybean straw (T1) and rice husk (T2). The experiment started with new substrates, which were reused in the composting, accompanying four flocks of broilers. Three repetitions of each treatment for each composting period were randomly selected. By the end of each flocks, each camera received 10 newly slaughtered broilers adding up a total of 60 animals per flocks. The total of 10 broilers was weighed, calculating the amount of water to be added, equal to 30% of broilers weight. The composting pile was arranged on a 30 cm-layer of new substrate, placing at the beginning 5 carcasses in a same layer, and the other 5 in a second layer, covered by a new 20 cm-layer of the same substrate. After a composting period of 15 days, pile was tumbled carcasses and substrates were weighed separately, piles were rearranged and water was added in amount corresponding to 30% of carcasses weight. All the process was performed using equipment for individual protection (rubber gloves, dust mask, boots, hat and overalls). An electronic scale with capacity of 100 kg (Toledo 2124-C5) has weighed substrates and carcasses. Carcasses were placed into thick plastic bags (20 kg) and substrates into raffia bags (60 kg). For the substrate removal, it was used a rounded tip cupped shovel and a watering can to add water. The removal of carcasses waste was made with a garden spade and a polyethylene broom. Thermocouples (S. E. Test tools A20) were installed to monitor the temperature of the composting pile, with readers inserted into each camera, in the central portion of the pile, with reading at three points (top, middle and bottom) and record of data at 7, 15, 22 and 30 days after mounting the pile. On the 30th day after the start of composting, the second weighing of carcasses and substrate was performed separately, being mounted a new pile with the same substrate and the remaining waste was divided into two layers, allowing the composting for more 15 days. The procedure was repeated for four periods, reusing the same substrate, forming from the second period, three layers of carcasses, being the bottom made up by the remaining waste of the previous batch, and the other two with five newly slaughtered broilers each. By the end of 30 days of composting, samples were collected from each camera, at nine points per layers (subsamples) and taken a pool of these points at 5 layers per camera (totaling 30 samples per batch) for analyses. It was analyzed the content of dry matter, ash, phosphorus, Cu; Zn; Mn; Fe, Nitrogen, pH; Calcium and Magnesium. For the analysis of the organic carbon, with the homogenization of part of five samples of layers of each camera, it was formed a new sample for each camera, totaling six samples per period, being the analysis performed by the titration method after chemical oxidation with sulfochromic solution [25]. The variable "Carcass decomposition" was obtained by the ratio between the difference of the initial and final weight, divided by the initial weight multiplied by 100. The variable "Substrate decomposition" was obtained by the ratio between the difference of the initial and final weight of the substrate. These data were examined by an analysis of variance for the model considering the effect of the composting cycle, substrate and the interaction, using the procedure GLM of SAS [19]. Data of composting temperature were analyzed using the mixed model theory for repeated measures, considering the effects of the composting cycle, substrate, composting week and interaction of these latter two variables and 16 types of structures of matrices of variance and covariance, using the PROC MIXED of SAS [19]. The structure employed in the analysis was chosen based on the Akaike's Information Criterion (AIC). The estimation method used was the restricted maximum likelihood.

The analysis of variance for the percentage of decomposition of carcasses and substrates, a significant effect ($p<0.05$) was detected for the interaction between period and substrate for the variable "carcass decomposition", whereas for the "substrate decomposition", a significant effect was observed only for the period. The Table 6 presents the mean values and standard errors of the % of decomposition of carcasses and substrates according to the type of substrate and period. A significant effect (p<0.05) was observed for the carcass decomposition on the 4th period in relation to the substrate type, being the soybean straw the substrate with the highest value. The variable "substrate decomposition" had values above 100 because it was added to the substrate part of the decomposed carcasses besides the addition of water and the natural process of formation of humic matter, characterized by being a complex of several elements [51] which works in the supply of nutrients to the plants, on the structure and compatibility of the soil and water retention capacity [52]. Nevertheless, the change in the shape and color of substrate particles indicate the decomposition of the substrate. The Figure 8 shows the mean profile of composting temperature according to substrates. No significant effect was observed for the interaction between week and treatment, but a significant effect (p<0.05) of the week and treatment separately. The rice husk maintained higher temperatures than the soybean straw, and in both substrates there was an increase of temperature with the tumbling of the pile from the 2nd to the 3rd week.

Substrate	Period			
	1	2	3	4
Carcass Decomposition (%)				
Rice husk	64.29 ± 1.90 a	65.25 ± 1.90a	60.18 ± 1.90 a	55.76 ± 1.90 b
Soybean straw	63.74 ± 1.90a	59.79 ± 1.90a	64.47 ± 1.90 a	62.70 ± 1.90a
Substrate Decomposition (%)				
Rice husk		106.79 ± 3.73	96.17 ± 3.73	102.79 ± 3.73
Soybean straw		104.97 ± 3.73	91.54 ± 3.73	100.90 ± 3.73

Means followed by different letters in the columns are significantly different by the F-test (p≤0.05)

Table 6. Mean values and standard errors of the % of decomposition of carcasses and substrates according to the type of substrate and period.

Nevertheless, the absolute maximum temperature registered inside the piles was higher in the camera with soybean straw, reaching 73.3°C, while the absolute maximum of the camera with rice husk reached 65.9°C. Both were close to the temperatures reported by [53] for composted mass on the fifth day after mounting the piles (57 - 71°C). [54] in the composting experiment with agroindustrial waste have obtained temperatures within the range of 40 - 60°C, with some peaks, with the highest temperature at 71°C. Biologically, the operating limits for the temperature can be classified as: > 55°C to maximize the sanitization; 45 - 55°C to maximize the biodegradation rate; and between 35 - 45°C to maximize the microbial diversity [55]. In this study, temperatures remained between 35 – 45ºC, but as abovementioned, there were temperature peaks above 55ºC promoting the sanitization of the composting mass. In the first week it was also found values indicative of the maximization of the degradation rate. [56] in a composting study using different stations obtained temperatures higher than 55ºC for over 3 consecutive days, achieving the maximum reduction of pathogenic microorganisms and indicating the biosafety of the composting. [57] recommended that the conditions of time-temperature for the compost to meet the biosafety standards should be any of the following procedures: 53ºC for 5 days, 55ºC for 2.6 days or 70ºC for 30 min. [58] with poultry carcasses composting in the climate of the United Kingdom, during the autumn and winter, for 8 weeks, have obtained positive results of carcass decomposition, as well as appropriate temperatures (60ºC - 70ºC) for the control of pathogens. [59] in studies on the biosafety of composting quoted that little attention has been given to strategies to evaluate the microbiological safety of composting systems, once there are different zones of the compost (e.g. the external edge of the pile) that usually have less organic matter and lower temperature. Thus the challenges for this type of benefit are greater than conventional. The analysis of variance for physical and chemical variables of the compost, are presented a significant effect (p<0.05) of the interaction between period and substrate for all variables, except for pH and K. The major effect of substrate was not significant only for the dry matter and the main effect of the period was significant for all variables. The significance effect of the interaction demonstrates that the effect of the substrate type depends on the period. The levels of physical and chemical variables measured in the substrate at the time zero, before its use as a composting substrate. In this way, it can be calculated the C/N ratio of the two substrates used, of 50.37 for the soybean straw and of 101.86 for the rice husk. Meanwhile, studies were performed using different sources of waste and residue from livestock and crop production, presenting a large variation in the initial C/N ratio, from 5/1 to 513/1 [60]. The mean values and standard errors of physical and chemical variables of composts according to periods and substrate type can be found in the Table 7. There was a significant difference between the substrate types for all variables in all periods, except for the dry matter, in the periods 2 and 4. Also there was an increase in the concentrations of the different parameters, expected given the addition of carcasses at each cycle. This increase was higher and statistically significant for the soybean straw, in the variables organic C, Cu, Fe, K, N, P, Zn and pH, besides the dry matter, although this difference had not been significant. The rice husk only presented levels of ash and Mn significantly higher than the soybean straw. The products obtained with carcass composting using two substrates tested can be classified as class D "compost organic fertilizer", according to the Normative Instruction 23 [39] meeting the requirements set concerning the minimum levels of N, organic C and moisture. But in relation to the C/N ration considered an indicative of the process maturity level [61], only the compost with soybean straw presented desired levels on the 3rd (17.75) and 4th (13.29) periods. The compost with rice husk would need to be subjected to a secondary composting to reduce this ratio and meet the requirements of the IN (maximum C/N of 18) or be used for composting new carcasses until reaching the suitable C/N ratio.

Substrate	Period			
	1	2	3	4
Ash (%)				
Rice husk	14.45 ± 0.30a	15.55 ± 0.30a	15.05 ± 0.30 a	15.46 ± 0.30a
Soybean straw	6.19 ± 0.30 b	6.61 ± 0.30b	8.52 ± 0.30b	9.67 ± 0.30b
Organic C (g/kg)				
Rice husk	306 ± 3.19b	326 ± 3.19b	301 ± 3.19 b	296 ± 3.19b
Soybean straw	345 ± 3.19a	384 ± 3.19a	355 ± 3.19a	336 ± 3.19 a
Cu (mg/kg)				
Rice husk	1.78 ± 0.33b	2.95 ± 0.33b	2.29 ± 0.33b	3.62 ± 0.33 b
Soybean straw	7.04 ± 0.33a	6.82 ± 0.33a	6.88 ± 0.33a	9.92 ± 0.33a
Fe (mg/kg)				
Rice husk	224 ± 71.37 b	256±71.37b	263±71.37b	463±71.37b
Soybean straw	1384 ± 71.37a	852±71.37a	980±71.37a	1544±71.37 a
K (mg/kg)				
Rice husk	1828 ± 373.0b	2706±373.0b	3398±373.0 b	3688±373.0 b
Soybean straw	11983 ± 373.0a	13574±373.0a	13717±373.0 a	15226±373.0 a
Dry matter (%)				
Rice husk	81.74 ± 0.59 a	86.83 ± 0.59a	80.65 ± 0.59a	82.82 ± 0.59 a
Soybean straw	78.96 ± 0.59 b	86.70 ± 0.59a	83.02 ± 0.59 b	83.71 ± 0.59a
Mn (mg/kg)				
Rice husk	211 ± 7.48a	282 ± 7.48a	196 ± 7.48a	236 ± 7.48 a
Soybean straw	56.45 ± 7.48 b	48.21 ± 7.48b	61.25 ± 7.48 b	64.04 ± 7.48b
N (mg/kg)				
Rice husk	5058 ± 711.7b	7416±711.7b	11962±711.7b	14020±711.7b
Soybean straw	10233 ± 711.7a	13491±711.7a	19993±711.7a	25281±711.7a
P (mg/kg)				
Rice husk	714 ± 166.4b	1017±166.4b	1870±166.4a	1387±166.4b
Soybean straw	1295 ± 166.4a	1757±166.4 a	1247±166.4b	2096±166.4a
pH				
Rice husk	8.88 ± 0.06b	8.99 ± 0.06b	8.64 ± 0.06b	8.33 ± 0.06b
Soybean straw	9.20 ± 0.06a	9.30 ± 0.06a	8.95 ± 0.06a	8.64 ± 0.06a
Zn (mg/kg)				
Rice husk	13.87 ± 1.16 b	14.61 ± 1.16b	20.03 ± 1.16b	24.63 ± 1.16b
Soybean straw	19.52 ± 1.16a	20.21 ± 1.16a	29.36 ± 1.16a	41.89 ± 1.16a

Means followed by different letters in the columns are significantly different by the F-test (p≤0.05)

Table 7. Mean values and standard errors of physical and chemical variables of the composts according to the periods and type of substrate.

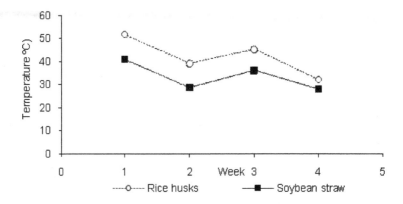

Figure 8. Mean profile of composting temperature according to substrates

Regarding the pH, both substrates presented varied values during the experimental period and reached the levels required by the IN-23 (pH 6.0) by the end of the 4th period. [62] found in the final compost in similar experiments, values of water pH in the range between 8.20 and 9.34. The levels of nitrogen increased at the end of composting periods. [54] achieved a compost with values of N, P, K between 17,710 -26,700 mg/Kg, 4,810-6,600 mg/Kg and 5,000-13,000 mg/Kg, respectively.

3. Conclusion

As compared to soybean straw, the use of rice husks as broiler litter material promotes better live performance of broilers up to 42 days of age. The use of soybean straw litter increases the incidence of footpad lesions relative to rice husks litter. Enterobacteria counts in broiler litter are reduced after downtime (day 0) when the litter is consecutively used by four flocks. Air relative humidity was higher when rice husks were used as litter material. Broiler litter used for three flocks, in average, complies with the minimal legal requirements to be traded as simple organic fertilizer, independently from the material. Soybean straw can be used as litter for rearing up to four flocks of broilers. At this same number of flocks, rice husks remain usable, whereas soybean straw is deteriorated and under humification. The number of darkling beetles was higher in the soybean straw litter and the rice husks litter presented 18.78 more chances of being contaminated with oocysts when ventilation was oscillating as compared to soybean straw litter. The soybean straw can represent an alternative of substrate for poultry carcasses composting, reaching values of C/N required by legislation with three reuses. Likewise, the rice husk can be used in composting broilers carcasses and can be reused for a greater number of times. The soybean straw presented a higher percentage of carcasses decomposition by the end of the 4th composting period (p<0.05).

Acknowledgements

The authors thank Fundação de Apoio à Pesquisa de Santa Catarina - FAPESC, for funding this study, Unifrango Agroindustrial de Alimentos Ltda., in the person of the farmer Mr. Arsênio for supplying the litter material, and Roster Ind. e Com. Ltda for lending the fans.

Author details

Valeria Maria Nascimento Abreu*, Paulo Giovanni de Abreu, Doralice Pedroso de Paiva, Arlei Coldebella, Fátima Regina Ferreira Jaenisch, Taiana Cestonaro and Virginia Santiago Silva

*Address all correspondence to: Valeria.abreu@embrapa.br

Brazilian Agricultural research Corporation, Embrapa Swine & Poultry, Concórdia, SC, Researh DT CNPq, Brazil

References

[1] Angelo, J.C.; Gonzales, E.; Kondo, N. et al. Material de cama: qualidade, quantidade e efeito sobre o desempenho de frangos de corte. Revista Brasileira de Zootecnia, v. 26, n.1, p.121-130, 1997.

[2] Benabdeljelil, K.; Ayachi, A. Evaluation of alternative litter materials for poultry. Journal Applied Poultry Science, v.5, p.203-209, 1996.

[3] Mizubuti, I.Y.; Fonseca, N.A.N.; Pinheiro, J.W. Desempenho de duas linhagens comerciais de frangos de corte, criadas sob diferentes densidades populacionais and diferentes tipos de camas. Revista da Sociedade Brasileira de Zootecnia, v.23, n.3, p. 476-484, 1994.

[4] Mouchrek, E.; Linhares, F.; Moulin, C.H.S. et al. Identificação de materiais de cama de frango de corte criados em diferentes densidades na época fria. In: Reunião Anual Da Sociedade Brasileira De Zootecnia, 29, 1992, Lavras. Anais... Lavras: SBZ, 1992. p. 344.

[5] Santos, E.C.; Teixeira, C.J.T.B.; Muniz, J.A. et al. Avaliação de alguns materiais usados como cama sobre o desempenho de frangos de corte. Ciência e Agrotecnologia, v.14, n.4, p.1024-1030, 2000

[6] Willis, W.L.; Murray, C.; Talbott, C. Evaluation of leaves as a litter material. Poultry Science, v.76, p.1138-1140, 1997.

[7] Oliveira, M.C.; Carvalho, I.D. Rendimento e lesões em carcaça de frangos de corte criados em diferentes camas e densidades populacionais. Ciência e Agrotecnologia, v.26, n.5, p.1076-1081, 2002.

[8] Araújo, J.C.; Oliveira, V.; Braga, G.C. Desempenho de frangos de corte criados em diferentes tipos de cama e taxa de lotação. Ciência Animal Brasileira, v.8, n.1, p.59-64, 2007.

[9] Atapattu, N.S.B.M; Wickramasinghe, K.P. The use of refused tea as litter material for broiler chickens. Poultry Science, v.86, n.5, p.968-972, 2007.

[10] Avila, V.S.; Oliveira, U.; Figueiredo, E.A.P. et al. Avaliação de materiais alternativos em substituição à maravalha como cama de aviário. Revista Brasileira de Zootecnia / Brazilian Journal of Animal Science, v.37, p.273-277, 2008.

[11] Arends, J.J. Control, management of the litter beetle. Poultry Digest, v.46, n.542, p. 172-176, 1987.

[12] Arends, J.J. External parasites and poultry pests. In: CALNEK, B.W. Diseases of Poultry. 9. ed. Ames: Iowa State Univ. Press. 1991. p.703-730 (710-712)

[13] Chernaki-Leffer, A.M.; Biesdorf, S.M.; Almeida, L.M. et al. Isolamento de enterobactérias em Alphitobius diaperinus e na cama de aviários no oeste do Estado do Paraná, Brasil. Revista Brasileira de Ciência Avícola/Brazilian Journal of Poultry Science, v.4, n.3, p.243-247, 2002.

[14] Vittori, J.; Schocken-Iturrino; R.P.; Trovó, K.P. et al. Alphitobius diaperinus como veiculador de Clostridium perfringens em granjas avícolas do interior paulista-Brasil. Ciência Rural, v.37, n.3, p.894-896, 2007.

[15] Watson, D.W.; Denning, S.S.; Zurek, L. et al. Effects of lime hidrated on the growth and development of darkling beetle, Alphitobius diaperinus. International Journal of Poultry Science, v.2, n.2, p.91-96, 2003.

[16] Salin, C.; Delettre, Y.R.; Vernon, P. Controling the mealworm Alphitobius diaperinus (Coleoptera: Tenebrionidae) in broiler and turkey houses: field trials with a combined inseticide treatment: Insect Growth Regulator an Pyrethroid. Journal of Economic Entomology, v.96, n.1, p.126-130, 2003.

[17] Surgeoner, G.A.; Romel, K. Control of the lesser mealworm in poultry houses (Coleoptera: Tenebrionidae). University of Ghelph. Disponível em: <http://www.uoguelph.ca/pdc/ Factsheets/FactsheetList.html>. Acesso em: 10/11/05.

[18] Silva, V.S.; Voss, D.; Coldebella, A. et al. Efeito de tratamentos sobre a carga bacteriana de cama de aviário reutilizada em frangos de corte. Concórdia: Embrapa Suínos e Aves, 2007. 4p. (Embrapa Suínos e Aves. Comunicado Técnico, 467).

[19] Sas Institute Inc. System for Microsoft Windows: release 9.1. Cary, 2002-2003. 1 CD-ROM.

[20] Xavier, L.H. Modelos univariado e multivariado para análise de medidas repetidas e verificação da acurácia do modelo univariado por meio de simulação. 2000. Tese (Mestrado) - Escola Superior de Agricultura "Luiz de Queiroz", Piracicaba.

[21] Safrit, R.D.; Axtell, R.C. Evaluations of sampling methods for darkling beetles (*Alphitobius diaperinus*) in the litter of turkey and broiler houses. Poultry Science, v. 63, p. 2368-2375, 1984.

[22] Windham, W.R. (Ed.) Animal feed. In: CUNNIFF, P. (Ed.) Official methods of analysis of AOAC international. 16. ed. Arlington: AOAC International, 1995. v. 1, cap. 4, p.1, 4, 27 (Method 965.17).

[23] Sindicato Nacional Dos Fabricantes De Ração. Compêndio brasileiro de alimentação animal. São Paulo: SINDIRAÇÕES, 1998.

[24] BRASIL. Ministério da Agricultura, Pecuária and Abastecimento (MAPA) Manual de métodos analíticos oficiais para fertilizantes minerais, orgânicos, organominerais and corretivos. Disponível em: <http://www.agricultura.gov.br>. Acesso em: 23 out.2007

[25] Tedesco, M.J.; Gianelo, C.; Bissani, C.A. et al. Análise de solo, plantas and outros materiais. 2. ed. Porto Alegre: UFRGS/Dep de Solos, 1995.

[26] Mendes, A.A.; Patricio, I.S.; Garcia, E.A. Utilização de fenos e gramíneas como material de cama para frangos de corte. In: Congresso Brasileiro De Avicultura, 10, 1987, Natal. Anais... Campinas: UBA, 1987. p.135.

[27] Chamblee, T.N.; Yeatman, J.B. Evaluation of Rice hull ash as broiler litter. The Journal of Applied Poultry Research, v.12, n.4, p.424-427, 2003.

[28] Grimes, J.L.; Carter, T.A.; Godwin, J.L. Use of a litter material made from cotton waste, gypsum and old newsprint for rearing broiler chickens. Poultry Science, v.85, v.4, p.563-568, 2006.

[29] Toghyani, M.; Gheisari, A.; Modaresi, M. et al. Effect of different litter material on performance and behavior of broiler chickens. Applied Animal Behaviour Science, v. 122, p.48-52, 2010.

[30] Huang, Y.; Yoo, J.S.; Kim, H.J. et al. Effect of bedding tipes and different nutrient densities on growth performance, visceral organ weight, and blood characteristics in broiler chickens. The Journal of Applied Poultry Research, v.18, n.1, p.1-7, 2009.

[31] Paganini, F.J. Reutilização de cama na produção de frangos de corte: porquê, quando e como fazer. In: Conferência Apinco 2002 De Ciência E Tecnologia Avícolas, 2002, Campinas, SP. Anais... Campinas, SP: APINCO, 2002. p.194 – 206.

[32] Azevedo, A.R.; Costa, A.M.; Alves, A.A. et al. Desempenho produtivo de frango de corte de linhagem Hubbard, criados sobre diferentes tipos de cama. Revista Científica de Produção Animal, v.2, n.1, p.52-57, 2000.

[33]   Fiorentin, L. Reutilização da cama de frangos e as implicações de ordem bacteriológi-
       ca na saúde humana e animal. Concórdia: Embrapa suínos e Aves, 2005. 23p. (Em-
       brapa Suínos e Aves. Documentos, 94).

[34]   Macklin, K.S.; Hess, J.B.; Bilgili, S.F. et al. Bacterial levels of pine shavings na sand
       used as poultry litter. The Journal of Applied Poultry Research, v.14, n.2, p.238-245,
       2005.

[35]   Abreu P.G.; Abreu, V.M.N. Conforto térmico para aves. Concórdia: Embrapa Suínos
       e Aves, 2004, 5p. (Embrapa Suínos e Aves, Comunicado técnico, 365)

[36]   Soliman, E.S.; Taha, E.G.; Sobieh, M.A.A. et al. The influence of ambient environmen-
       tal conditions on the survival of salmonella enteric serovar typhimurium in poultry
       litter. International Journal of Poultry Science, v.8, n.9, p.848-852, 2009.

[37]   Oliveira, M.C.; Goulart, R.B.; Silva, J.C.N. Efeito de duas densidades e dois tipos de
       cama sobre a umidade da cama e a incidência de lesões na carcaça de frango de corte.
       Ciência Animal Brasileira, v.3, n.2, p.7-12, 2002

[38]   BESSEI, W. Welfare of broilers: a review. World's Poultry Science Journal, Vol. 62,
       September 2006, n3, pp 455-466

[39]   BRASIL. Ministério da agricultura, Pecuária e Abastecimento. Secretária de Defesa
       Agropecuária. Instrução Normativa nº 23 de 31 agosto, 2005. Diário Oficial da Re-
       pública Federativa do Brasil, Brasília, 8 set. 2005. Seção 1. p. 12-15.

[40]   Coufal,C.D.; Chavez, C.; Niemeyer, P.R. et al. Effect of top-dressing recycled broiler
       litter on litter production, litter characteristic, and nitrogen mass balance. Poultry Sci-
       ence, v. 85, p.392-397, 2006.

[41]   Moore Jr, P.A.; Daniel, T.C; Eduards, D.R. et al. Evaluation of chemical amendments
       to reduce ammonia volatilization from poultry litter. Poultry Science, v.75, p.315-320,
       1996.

[42]   Singh, A.; Bicudo, J.R.; Tinoco, A.L. et al. Characterization of nutrientes in built-up
       broiler litter using trench and random walk sampling methods. The Journal of Ap-
       plied Poultry Research, v.13, p.426-432, 2004.

[43]   Oviedo-Rondón, E.O. Tecnologias para mitigar o impacto ambiental da produção de
       frangos de corte. Revista Brasileira de Zootecnia, v.37, suplemento especial, p.
       239-252, 2008.

[44]   Turazi, C.M.V.; Junqueira, A.M.R.; Oliveira, A.S. et al. Horticultura Brasileira, v.24, n.
       1, p.65-70, 2006.

[45]   Pinto, D.M.; Ribeiro, P.B.; Bernardi, E. Flutuação populacional de *Alphitobius diaperi-
       nus* (Panzer, 1879) (Coleoptera: Tenebrionidae), capturados por armadilha do tipo
       sanduíche, em granja avícola, no Município de Pelotas, RS. Arquivos do Instituto Bi-
       ológico, v.72, n.2, p.199-203, 2005.

[46] Godinho, R.P.; Alves, L.F.A. Método de avaliação de população de cascudinho *(Alphitobius diaperinus)* panzer em aviários de frango de corte. Arquivos do Instituto Biológico, v.76, n.1, p.107-110, 2009

[47] Uemura, D.H.; Alves, L.F.A.; Opazo, M.U. et al. Distribuição and dinâmica populacional do cascudinho *Alphitobius diaperinus* (Coleoptera: Tenebrionidae) em aviários de frango de corte. Arquivos do Instituto Biológico, v.75, n.4, p.429-435, 2008.

[48] Chernaki, A.M; Almeida, L.M. Exigências térmicas, period de desenvolvimento and sobrevivência de imaturos de *Alphitobius diaperinus* (Panzer) (Coleoptera: Tenebrionidae). Neotropical Entomology, v.30, n.3, p.365-8, 2001

[49] Shirley, M.W. Epizootiologia. In: Simpósio Internacional Sobre Coccidiose, 1994, Santos. Anais... Santos: [s.n.], 1994, p.11-22.

[50] Macsafley, L. M.; DUPOLDT, C.; GETER, F. Agricultural waste management system component design. In: Krider, J. N.; Rickman, J. D. (Ed.). Agricultural waste management field handbook. Washington, D.C.: Department of Agriculture, Soil Conservation Service, 1992. p. 1-85.

[51] Diniz Filho, E. T.; Mesquita, L. X.; Oliveira, A. M.; Nunes, C. G. F.; Lira, J. F. B. A prática da compostagem no manejo sustentável de solos. Revista Verde de Agroecologia e Desenvolvimento Sustentável, v. 2, n. 2, p. 27-36, 2007.

[52] Budziak, C. R.; Maia, C. M. B. F.; Mangrich, A. S. Transformações químicas da matéria orgânica durante a compostagem de resíduos da indústria madeireira. Química Nova, v. 27, n. 3, p. 399-403, 2004.

[53] Rynk, R. ed. On farm composting handbook. Ithaca: Northeast Regional Agricultural Engineering Service, 1992. 186p. (Cooperative Extension. NRAES, 54).

[54] Fiori, M. G. S.; Schoenhals, M.; Follador, F.A. C. Análise da evolução tempo-eficiência de duas composições de resíduos agroindustriais no processo de compostagem aeróbia. Engenharia Ambiental, v. 5, n. 3, p. 178-191, 2008.

[55] Hassen, A.; Belguith, K.; Jedidi, N.; Cherif, A. Microbial characterization during composting of municipal solid waste. Bioresource Technology, v. 80, n. 3, p. 217-25, 2001.

[56] Sivakumar, K.; Kumar, V. R. S.; Jagatheesan, P. N. R.; Viswanathan, K.; Chandrasekaran, D. Seasonal variations in composting process of dead poultry birds. Bioresource Technology, v. 99, n. 2, p. 3708-3713, 2008.

[57] Haug, R. T. The practical handbook of compost engineering. Boca Raton: Lewis Publishers Press Inc., 1993.

[58] Lawson, M. J.; Keeling, A. A. Production and physical characteristics of composted poultry carcases. British Poultry Science, v. 40, n. 5, p. 706-708, 1999.

[59] Wilkinson, K. G. The biosecurity of on-farm mortality composting. Journal of Applied Microbiology, v. 102, n. 3, p. 609-618, 2007.

[60] Valente, B. S.; Xavier, E. G.; Morselli, T. B. G.A.; Jahnke, D. S.; Brum Jr., B. S.; Cabrera, B. R.; Moraes, P. O.; Lopes, D. C. N. Fatores que afetam o desenvolvimento da compostagem de resíduos orgânicos. Archivos de Zootecnia, v. 58, n. 1, p. 59-85, 2009.

[61] Reis, M. F. P.; Escosteguy, P. V.; Selbach, P. Teoria e prática da compostagem de resíduos sólidos urbanos. Passo Fundo: UPF, 2004.

[62] Kumar, V. R. S.; Sivakumar, K.; Purushothaman, M. R.; Natarajan, A.; Amanullah, M. M. Chemical changes during composting of dead birds with caged layer manure. Journal of Applied Sciences Research, v. 3, n. 10, p. 1100-1104, 2007.

Permissions

The contributors of this book come from diverse backgrounds, making this book a truly international effort. This book will bring forth new frontiers with its revolutionizing research information and detailed analysis of the nascent developments around the world.

We would like to thank Hany A. El-Shemy, for lending his expertise to make the book truly unique. He has played a crucial role in the development of this book. Without his invaluable contribution this book wouldn't have been possible. He has made vital efforts to compile up to date information on the varied aspects of this subject to make this book a valuable addition to the collection of many professionals and students.

This book was conceptualized with the vision of imparting up-to-date information and advanced data in this field. To ensure the same, a matchless editorial board was set up. Every individual on the board went through rigorous rounds of assessment to prove their worth. After which they invested a large part of their time researching and compiling the most relevant data for our readers. Conferences and sessions were held from time to time between the editorial board and the contributing authors to present the data in the most comprehensible form. The editorial team has worked tirelessly to provide valuable and valid information to help people across the globe.

Every chapter published in this book has been scrutinized by our experts. Their significance has been extensively debated. The topics covered herein carry significant findings which will fuel the growth of the discipline. They may even be implemented as practical applications or may be referred to as a beginning point for another development. Chapters in this book were first published by InTech; hereby published with permission under the Creative Commons Attribution License or equivalent.

The editorial board has been involved in producing this book since its inception. They have spent rigorous hours researching and exploring the diverse topics which have resulted in the successful publishing of this book. They have passed on their knowledge of decades through this book. To expedite this challenging task, the publisher supported the team at every step. A small team of assistant editors was also appointed to further simplify the editing procedure and attain best results for the readers.

Our editorial team has been hand-picked from every corner of the world. Their multi-ethnicity adds dynamic inputs to the discussions which result in innovative outcomes. These outcomes are then further discussed with the researchers and contributors who give their valuable feedback and opinion regarding the same. The feedback is then collaborated with the researches and they are edited in a comprehensive manner to aid the understanding of the subject.

Apart from the editorial board, the designing team has also invested a significant amount of their time in understanding the subject and creating the most relevant covers. They scrutinized every image to scout for the most suitable representation of the subject and create an appropriate cover for the book.

The publishing team has been involved in this book since its early stages. They were actively engaged in every process, be it collecting the data, connecting with the contributors or procuring relevant information. The team has been an ardent support to the editorial, designing and production team. Their endless efforts to recruit the best for this project, has resulted in the accomplishment of this book. They are a veteran in the field of academics and their pool of knowledge is as vast as their experience in printing. Their expertise and guidance has proved useful at every step. Their uncompromising quality standards have made this book an exceptional effort. Their encouragement from time to time has been an inspiration for everyone.

The publisher and the editorial board hope that this book will prove to be a valuable piece of knowledge for researchers, students, practitioners and scholars across the globe.

List of Contributors

Mauricio G. Fonseca
INMETRO – National Institute of Metrology, Quality and Technology, Metrological Chemistry Division Xerém, Duque de Caxias, Rio de Janeiro, Brazil

Luciano N. Batista
INMETRO – National Institute of Metrology, Quality and Technology, Metrological Chemistry Division Xerém, Duque de Caxias, Rio de Janeiro, Brazil

Viviane F. Silva
INMETRO – National Institute of Metrology, Quality and Technology, Metrological Chemistry Division Xerém, Duque de Caxias, Rio de Janeiro, Brazil

Erica C. G. Pissurno
INMETRO – National Institute of Metrology, Quality and Technology, Metrological Chemistry Division Xerém, Duque de Caxias, Rio de Janeiro, Brazil

Thais C. Soares
INMETRO – National Institute of Metrology, Quality and Technology, Metrological Chemistry Division Xerém, Duque de Caxias, Rio de Janeiro, Brazil

Monique R. Jesus
INMETRO – National Institute of Metrology, Quality and Technology, Metrological Chemistry Division Xerém, Duque de Caxias, Rio de Janeiro, Brazil

Georgiana F. Cruz
UENF – North Fluminense University, Engineering and exploitation of petroleum laboratory, Macaé, Rio de Janeiro, Brazil

Stefano Tavoletti
Dipartimento di Scienze Agrarie, Alimentari ed Ambientali – Università Politecnica delle Marche - Via Brecce Bianche, 60131 ANCONA, Italy

B. Pampaloni
Department of Internal Medicine, School of Medicine, University of Florence, Florence, Italy

C. Mavilia
Department of Internal Medicine, School of Medicine, University of Florence, Florence, Italy

E. Bartolini
Department of Internal Medicine, School of Medicine, University of Florence, Florence, Italy

F. Tonelli
Department of Clinical Physiopathology, School of Medicine, University of Florence, Florence, Italy

M.L. Brandi
Department of Internal Medicine, School of Medicine, University of Florence, Florence, Italy

Federica D'Asta
Department of Neurosciences, Psychology, Drug Area and Child Health, School of Medicine, University of Florence, Florence, Italy

Farinaz Safavi and Abdolmohamad Rostami
Department of Neurology, Thomas Jefferson University, Philadelphia, PA, USA

Ednilton Tavares de Andrade
Federal Fluminense University/TER/PGEB/PGMEC, Brazil

Luciana Pinto Teixeira
Federal Fluminense University/TER/PGMEC, Brazil

Ivênio Moreira da Silva
Federal Fluminense University/TER/PGMEC, Brazil

Roberto Guimarães Pereira
Federal Fluminense University/TEM/PGMEC/PGEB/MSG, Brazil

Oscar Edwin Piamba Tulcan
National University of Colombia, Bogota, Colombia

Danielle Oliveira de Andrade
Federal Fluminense University/TER/PGMEC, Brazil

Pilar Teresa Garcia
Area Bioquimica y Nutricion. Instituto Tecnologia de Alimentos. Centro de Investigacionen Agroindustria. Instituto Nacional Tecnologia Agropecuaria. INTA Castelar, CC77(B1708WAB) Moron, Pcia Buenos Aires, Argentina

Jorge J. Casal
Universidad de Moron. Facultad de Ciencias Agroalimentarias. Cabido 137. Moron, Pcia Buenos Aires, Argentina

Joanna McFarlane
Energy and Transportation Science Division, Oak Ridge National Laboratory, Oak Ridge, USA

Michel Mozeika Araújo
Nuclear and Energy Research Institute (IPEN), Brazil
Strasbourg University (UdS), France

Gustavo Bernardes Fanaro
Nuclear and Energy Research Institute (IPEN), Brazil

Anna Lucia Casañas Haasis Villavicencio
Nuclear and Energy Research Institute (IPEN), Brazil

Susana Hernández López and Enrique Vigueras Santiago
Research and Development of Advanced Materials Laboratory, Chemistry Faculty of the State of Mexico Autonomous University, Toluca, Mexico

Liyan Chen
Bioprocessing and Renewable Energy Laboratory, Department of Grain Science and Industry, Kansas State University, Kansas, USA

Ronald L. Madl
Bioprocessing and Renewable Energy Laboratory, Department of Grain Science and Industry, Kansas State University, Kansas, USA

Praveen V. Vadlani
Bioprocessing and Renewable Energy Laboratory, Department of Grain Science and Industry, Kansas State University, Kansas, USA

Li Li
Department of Food Science, South China University of Technology, Guangzhou, China

Weiqun Wang
Department of Human Nutrition, Kansas State University, Manhattan, Kansas, USA

Xiaomeng Li
The Key Laboratory of Molecular Epigenetics of MOE, Institute of Genetics and Cytology, School of Life Sciences, Northeast Normal University, China

Ying Xu
The Key Laboratory of Molecular Epigenetics of MOE, Institute of Genetics and Cytology, School of Life Sciences, Northeast Normal University, China

Ichiro Tsuji
Department of Public Health, Tohoku University, Japan

Valeria Maria Nascimento Abreu, Paulo Giovanni de Abreu, Doralice Pedroso de Paiva
Arlei Coldebella, Fátima Regina Ferreira Jaenisch, Taiana Cestonaro and Virginia Santiago Silva
Brazilian Agricultural research Corporation, Embrapa Swine & Poultry, Concórdia, SC, Researh DT CNPq, Brazil